Cell Death in Biology and Diseases

Series Editors

Xiao-Ming Yin
Zheng Dong

More information about this series at http://www.springer.com/series/8908

David M. Hockenbery
Editor

Mitochondria and Cell Death

Humana Press

Editor
David M. Hockenbery
Clinical Research Divison
Fred Hutchinson Cancer Research Center
Seattle, WA, USA

Cell Death in Biology and Diseases
ISBN 978-1-4939-3610-6 ISBN 978-1-4939-3612-0 (eBook)
DOI 10.1007/978-1-4939-3612-0

Library of Congress Control Number: 2016934848

Printed on acid-free paper

This Humana Press imprint is published by Springer Nature
The registered company is Springer Science+Business Media LLC New York

Series Preface

Cell death, or conversely cell survival, is a major biological phenomenon. Just as with cell proliferation and cell differentiation, cell death is a choice that a cell has to make, sometimes voluntarily, other times accidentally. As such, cell death serves a purpose in the biology of a multicellular organism. The machinery of cell death and that of cell protection are evolutionarily conserved and their elements can even be found in single-celled organisms. The disruption of cell death mechanisms can often cause developmental abnormalities. Factors that can trigger cell death are diverse, and the cell death process is intricately connected with other biological processes. Cell death directly contributes to the pathogenesis of many diseases, including cancer, neurodegenerative diseases, and tissue injury in organ failure.

The study of cell death and cell survival has become a multidisciplinary subject, which requires expertise from all the fields of modern biology. Exploring the role of cell death in disease development and the modulation of cell death for the prevention and treatment of devastating disease demands constant updating of our knowledge through the broadest interactions among all investigators, basic and clinical. The rapid expansion of our knowledge in this field has gone beyond what could be summarized in a single book. Thus, this timely series *Cell Death in Biology and Diseases* summarizes new developments in different areas of cell death research in an elaborate and systemic way. Each volume of this series addresses a particular topic of cell death that either has a broad impact on the field or has an in-depth development in a unique direction. As a whole, this series provides a current and encyclopedic view of cell death.

We would like to sincerely thank the editors of each volume in the series and the authors of each chapter in these volumes for their strong commitment and great effort towards making this mission possible. We are also grateful to our team of professional Springer editors. They have worked with us diligently and creatively from the initiation and are continuing this on the development and production of each volume of the series. Finally we hope that the readers will enjoy the reading, find the content helpful to their work, and consider this series an invaluable resource.

Indianapolis, IN, USA Xiao-Ming Yin, M.D., Ph.D.
Augusta, GA, USA Zheng Dong, Ph.D.

Preface

This volume provides an in-depth, up-to-date collection of papers on mitochondrial biology in the context of the cellular environment, and the roles of mitochondria in cell death and the various cellular programs that can intersect with cell death pathways (intracellular signaling, homeostasis, and pathogen resistance). There has been a surge in interest in mitochondrial functions in the last 5 years, with substantial new understanding in the areas of mitochondrial organelle and protein homeostasis, mitochondrial dynamics, interorganelle associations, and integration within diverse cellular programs. This progress has led to novel insights into the important place of mitochondria in orchestrating these programs. The importance of mitochondrial dysfunction in pathogenesis of clinical diseases, including neurodegenerative diseases, cancer, metabolic syndrome, and aging, illustrates the strong relevance of these topics in "bench to bedside" investigations. The chapter authors are renowned experts in their fields, actively contributing to the expansion of modern mitochondrial research.

The earliest and best known function of mitochondria, generation of ATP from proton motive force, led to the recognition that the same electrochemical gradient could be used to move other ions and metabolites. The study of mitochondrial Ca^{2+} transport revealed the permeability transition, initially in isolated mitochondria, but later confirmed in situ and its role in the clinical setting of ischemia-reperfusion injury and necrotic cell death. Fontaine and Bernardi discuss current conceptions of the permeability transition pore and the evidence for physiological functions in Ca^{2+} handling. Among the recent advances in mitochondrial biology is the functional identification of new mitochondrial proteins. Boyman, Williams, and Lederer provide a detailed review of mitochondrial Ca^{2+}, with special emphasis on the recently identified Ca^{2+} uniporter, MCU, and Na^+/Ca^{2+} exchanger, NCLX, and the pathophysiology of ischemia/reperfusion injury.

The central role of mitochondria in apoptotic cell death was first suggested by the novel localization of Bcl-2 oncoproteins to mitochondria. Bcl-2 proteins regulate mitochondrial outer membrane permeabilization (MOMP) with release of cytochrome c and other proteins with second functions as proapoptotic factors from the intermembrane space. Flanagan, Lucantoni, and Prehn describe the unique features of this mitochondrial compartment and the intersecting lethal roles of resident proteins once relocated to the nucleocytoplasm. Current understanding of the mechanism of BAX/BAK-dependent MOMP and important remaining questions are reviewed by Edlich and Martinou. Microscopic techniques have provided key insights into remodeling of the mitochondrial network and cristae topology during apoptosis, and Perkins and Ellisman discuss the emerging roles of mitochondrial fission and fusion GTPases in the mechanism(s) of cytochrome c release.

The majority of reactive oxygen species (ROS) produced in a cell originate in mitochondria. Although ROS can damage proteins, lipids, and DNA in cells and have been implicated in aging, chronic inflammatory states, and cancer, healthy cells utilize ROS in signaling pathways. The detailed understanding of ROS signaling and its role in cell growth, longevity, and stress resistance, and how signaling strength is regulated, are under active investigation. Allen and Spitz provide a global perspective on these functions and the key factors involved in maintaining physiological balancing of the oxidant/antioxidant arms of these pathways.

Mitochondria are the only organelles with their own genome (mtDNA), with less developed DNA repair pathways and the added issue of heteroplasmy compared to the nuclear chromosomes. Valente and Bielas discuss the universal existence of "microheteroplasmy" with ultrasensitive mutation assays, the need for a new functional definition of homoplasmy, and the relevance of mtDNA mutations to cancer. Quality control is also required for the mitochondrial proteome in an oxidizing environment, and Germain reviews the recent progress in mapping the mitochondrial unfolded protein response, UPR^{mt}, in mammalian cells. Raimundo, Fernandez-Mosquera, and Yambire present a conceptual framework for mitochondrial signaling, including communication with other organelles in addition to mitochondria-to-nucleus regulation of gene expression.

One of the newest, and unexpected, roles for mitochondria is in innate antiviral immunity. Thomas and Gale describe the generation of "innate immune synapses" by mitochondria and mitochondria-associated membranes (MAM), as well as the varied strategies of viruses to disable this signaling platform. Finally, Oberst, Ichim, and Tait present an intriguing hypothesis of the evolutionary history of cytochrome c as a pathogen-associated molecular pattern (PAMP) associated with ancestral bacterial endosymbionts, linking MOMP in apoptosis to innate immune sensors, with the additional possibility of graded responses to MOMP producing nonlethal effects.

The intended audience for this book is students, new and established researchers, and others interested in learning about the fascinating field of mitochondrial biology. Although impressive progress has been made recently, important questions in mitochondrial biology remain unanswered, and the recruitment of new investigators to this field has been a critical component of recent discoveries. It is hoped that this book will stimulate future scientists to join this effort.

I, personally, would like to acknowledge the support and guidance of my mentor, the late Stanley Korsmeyer, who laid much of the early groundwork for the explosion of interest in mitochondria and apoptotic cell death. I am grateful to the co-editors of the *Cell Death in Biology and Diseases* series from Springer, Xiao-Ming Yin and Zheng Dong, for their help in the conception of this book, and Joseph Quatela and Aleta Kalkstein from Springer for their editorial efforts.

Seattle, WA, USA David M. Hockenbery

Contents

1 **Lethal and Nonlethal Functions of the Permeability Transition Pore** 1
Eric Fontaine and Paolo Bernardi

2 **Mitochondrial Calcium and Ischemia: Reperfusion Injury in Heart** 17
Liron Boyman, George S.B. Williams, and W. Jonathan Lederer

3 **Mislocalization of Mitochondrial Intermembrane Space Proteins** 45
Lorna Flanagan, Federico Lucantoni, and J.H.M. Prehn

4 **Bcl-2 Protein Interplay on the Outer Mitochondrial Membrane** 69
Frank Edlich and Jean-Claude Martinou

5 **Remodeling of Mitochondria in Apoptosis** 85
Guy A. Perkins and Mark H. Ellisman

6 **Physiologic and Pathologic Functions of Mitochondrial ROS** 111
Bryan G. Allen and Douglas R. Spitz

7 **Mitochondrial Mutagenesis in Cancer** 131
William J. Valente and Jason H. Bielas

8 **Mitochondrial UPR in Cancer** 149
Doris Germain

9 **Mitochondrial Signaling** 169
Nuno Raimundo, Lorena Fernandez-Mosquera, and King Faisal Yambire

10 Mitochondria and Antiviral Immunity ... 187
Sunil Thomas and Michael Gale Jr.

11 Mitochondrial Permeabilization: From Lethality to Vitality ... 213
Andrew Oberst, Gabriel Ichim, and Stephen W.G. Tait

Index ... 227

Contributors

Bryan G. Allen, M.D., Ph.D. Division of Free Radical and Radiation Biology, Department of Radiation Oncology, University of Iowa Hospitals and Clinics, Iowa City, IA, USA

The University of Iowa Carver College of Medicine, Iowa City, IA, USA

Paolo Bernardi Department of Biomedical Sciences, Consiglio Nazionale delle Ricerche Neuroscience Institute, University of Padova, Padova, Italy

Jason H. Bielas Fred Hutchinson Cancer Research Center, Seattle, WA, USA

Liron Boyman, Ph.D. Center for Biomedical Engineering and Technology, University of Maryland School of Medicine, Baltimore, MD, USA

Department of Physiology, University of Maryland School of Medicine, Baltimore, MD, USA

Frank Edlich Institute for Biochemistry and Molecular Biology, University of Freiburg, Freiburg, Germany

BIOSS, Centre for Biological Signaling Studies, University of Freiburg, Freiburg, Germany

Mark H. Ellisman Center for Research in Biological Systems, National Center for Microscopy and Imaging Research, University of California at San Diego, La Jolla, CA, USA

Department of Neurosciences, University of California at San Diego, La Jolla, CA, USA

Department of Bioengineering, University of California at San Diego, La Jolla, CA, USA

Lorena Fernandez-Mosquera Institute for Cellular Biochemistry, Universitätsmedizin Göttingen, Göttingen, Germany

Lorna Flanagan Centre for Systems Medicine, Department of Physiology and Medical Physics, Royal College of Surgeons in Ireland, Dublin, Ireland

Eric Fontaine LBFA, University Grenoble Alpes, Grenoble, France

INSERM, Grenoble, France

Michael Gale Jr. Department of Immunology, Center for Innate Immunity and Immune Disease, University of Washington School of Medicine, Seattle, WA, USA

Doris Germain Division of Hematology/Oncology Tisch Cancer Institute, Icahn School of Medicine at Mount Sinai, New York, NY, USA

Gabriel Ichim Cancer Research-UK Beatson Institute, Institute of Cancer Sciences, University of Glasgow, Glasgow, UK

W. Jonathan Lederer, M.D., Ph.D. Center for Biomedical Engineering and Technology, University of Maryland School of Medicine, Baltimore, MD, USA

Department of Physiology, University of Maryland School of Medicine, Baltimore, MD, USA

Federico Lucantoni Centre for Systems Medicine, Department of Physiology and Medical Physics, Royal College of Surgeons in Ireland, Dublin, Ireland

Jean-Claude Martinou Department of Cell Biology, University of Geneva, Genève, Switzerland

Andrew Oberst Department of Immunology, University of Washington, Seattle, WA, USA

Guy A. Perkins Center for Research in Biological Systems, National Center for Microscopy and Imaging Research, University of California at San Diego, La Jolla, CA, USA

J.H.M. Prehn Centre for Systems Medicine, Department of Physiology and Medical Physics, Royal College of Surgeons in Ireland, Dublin, Ireland

Nuno Raimundo Institute for Cellular Biochemistry, Universitätsmedizin Göttingen, Göttingen, Germany

Douglas R. Spitz Division of Free Radical and Radiation Biology, Department of Radiation Oncology, University of Iowa Hospitals and Clinics, Iowa City, IA, USA

Stephen W.G. Tait Cancer Research-UK Beatson Institute, Institute of Cancer Sciences, University of Glasgow, Glasgow, UK

Sunil Thomas Department of Immunology, Center for Innate Immunity and Immune Disease, University of Washington School of Medicine, Seattle, WA, USA

William J. Valente Fred Hutchinson Cancer Research Center, Seattle, WA, USA

George S.B. Williams, Ph.D. Center for Biomedical Engineering and Technology, University of Maryland School of Medicine, Baltimore, MD, USA

Department of Physiology, University of Maryland School of Medicine, Baltimore, MD, USA

King Faisal Yambire Institute for Cellular Biochemistry, Universitätsmedizin Göttingen, Göttingen, Germany

Chapter 1
Lethal and Nonlethal Functions of the Permeability Transition Pore

Eric Fontaine and Paolo Bernardi

1.1 Introduction

The process of scientific discovery is not linear. Peter Mitchell was awarded the Nobel Prize for Chemistry in 1978 for "his contribution to the understanding of biological energy transfer through the formulation of the chemiosmotic theory." This prestigious award should not send into oblivion all the difficulties Mitchell had to face from 1961 (Mitchell 1961), when he first proposed his hypothesis. If his theory took over 15 years to be fully recognized, this is also because the very idea that energy could be stored as a proton gradient was a discontinuity with prevailing concepts of energy conservation, and the scientific community—at least a large part of it—simply could not understand the concept. But as the bioenergetics community fully accepted the chemiosmotic theory, the proton gradient became a new paradigm; and such a fundamental one that dissipation of the proton gradient other than by uncoupling proteins in brown fat was ruled out from mitochondrial physiology. Indeed, an integral part of chemiosmotic energy conservation is a low permeability of the inner membrane to charged species and solutes.

An inner membrane permeability increase caused by Ca^{2+} and Pi, and counteracted by Mg^{2+} and adenine nucleotides, had been described since 1953 (Raaflaub 1953a, b), studied throughout the 1960s and 1970s and proposed to play a role in steroidogenesis (Pfeiffer et al. 1976), but research on this topic lost momentum in parallel with the success of chemiosmotic principles. In this scientific and historical context, the work

E. Fontaine (✉)
University Grenoble Alpes, LBFA, Grenoble 38000, France

INSERM, U1055, Grenoble 38000, France
e-mail: eric.fontaine@ujf-grenoble.fr

P. Bernardi
Department of Biomedical Sciences, Consiglio Nazionale delle Ricerche Neuroscience Institute, University of Padova, Padova 35121, Italy

D.M. Hockenbery (ed.), *Mitochondria and Cell Death*, Cell Death in Biology and Diseases, DOI 10.1007/978-1-4939-3612-0_1

X.-M. Yin and Z. Dong (Series eds.), *Cell Death in Biology and Diseases*

of Hunter and Haworth (Haworth and Hunter 1979; Hunter and Haworth 1979a, b;Hunter et al. 1976) concluding that the Ca^{2+}-dependent "permeability transition" was due to the opening of a regulated mitochondrial channel did not obtain much consensus. How could a large mitochondrial channel, whose opening dissipates the proton gradient, prevents ATP synthesis, depletes substrates and pyridine nucleotides, and causes mitochondrial swelling (Zoratti and Szabo 1995), be compatible with life?

Thirty-five years after these pioneering contributions, the existence of a permeability transition pore (PTP) is no longer questioned, and the recent demonstration that the F-ATP synthase forms channels with the same properties of the PTP has opened new perspectives (Giorgio et al. 2013; Carraro et al. 2014; von Stockum et al. 2015; Alavian et al. 2014). Yet the permeability transition has not revealed all its secrets. While acknowledging the existence of the permeability transition in isolated mitochondria, many researchers considered the phenomenon as a side effect of the isolation procedure, i.e., an in vitro artifact of little, if any, relevance. For the few who made the assumption that the permeability transition was not an artifact, several issues needed to be resolved. Is the permeability transition due to the opening of a channel? If so, are there one or several channels? How are they regulated? How can the occurrence of a permeability transition be detected in intact cells? What is the molecular nature and the function of the channel(s)? The purpose of this chapter is to provide an answer to this last question.

1.2 What Are the Consequences of the Opening of a Specific But Nonselective High-Conductance Inner Membrane Channel?

The permeability transition is an inner membrane permeability increase that can be traced to the opening of a specific but nonselective high-conductance inner membrane channel. "Specific" means that the channel is devoted to this function, while "nonselective" indicates that once the channel opens, permeability increases for many ions and solutes. The internal diameter of the open pore ($\approx$3 nm) has been estimated by swelling experiments (Haworth and Hunter 1979, 1980; Massari and Azzone 1972) (see below) and by patch clamp techniques (Kinnally et al. 1989; Petronilli et al. 1989; Szabo and Zoratti 1991). Both have concluded that solutes of molecular weight up to 1.5 kDa can pass through an open PTP, which would basically only prevent diffusion of proteins, RNA and glycogen. On the contrary, the outer membrane is naturally permeable to all solutes below 6.5 kDa, which would basically prevent diffusion of larger proteins.

Mitochondrial matrix volume reaches a steady state when tonicity is equal on both sides of the mitochondrial inner membrane. Importantly, tonicity is influenced only by solutes that cannot cross the membrane, as only these exert an effective osmotic pressure. Consequently, PTP opening may dramatically change the intramitochondrial tonicity without much affecting intramitochondrial osmolarity. This results in the classical large amplitude mitochondrial swelling observed after

PTP opening when isolated mitochondria are incubated in an isosmotic medium based on solutes of low molecular weight (sucrose or KCl, for example). Indeed, two types of mitochondrial swelling can occur (Azzone and Azzi 1965), i.e. active swelling, where the energy-dependent accumulation of an ion or solute is followed by water diffusion, and passive swelling. Passive swelling in turn can be due to incubation of mitochondria in hyposmotic medium, which is followed by water equilibration; to the addition of a permeant solute (like urea); and to increased membrane permeability to solutes due to membrane damage or to PTP opening.

It must be stressed that PTP opening is not necessarily always followed by mitochondrial swelling. Indeed, for brief open times swelling is negligible and can rapidly be compensated by activation of proton/cation antiporters; and in media made with dextrans larger than the diameter of the pore, PTP opening occurs without any mitochondrial swelling. Therefore, PTP opening is not expected to induce large mitochondrial swelling in intact cells unless an osmotic gradient exists between the matrix, the intermembrane space, and the cytosol.

Whether or not PTP opening affects mitochondrial volume, it dramatically changes the composition of the media inside and outside mitochondria. In normal conditions (before PTP opening), the matrix solute composition is different from that of the cytosol (or of the incubation medium in isolated mitochondria). This is because some ions and solutes do not cross the membrane due to the absence of selective channels/transport systems, while others are actively taken up or released through selective channels or transporters that are directly or indirectly driven by the proton gradient. Protons being the smallest solutes, PTP opening immediately collapses the proton gradient; however, because of its size an open PTP provides a nonselective channel that can be used by solutes that usually do not cross the membrane, while eliminating the driving force for ions and solutes that use the proton gradient for uptake or release. Thus, each solute of molecular weight less than 1.5 kDa will cross the inner membrane according to its concentration gradient before pore opening, and the balance between solute uptake and release will determine the net flux of water, hence mitochondrial volume.

Although it seems obvious, it is important to remind ourselves that the final composition of the intramitochondrial medium after PTP opening will dramatically depend on the composition and volume of the external medium. Functionally, this point is of major importance. For example, after PTP opening, all the enzymes of the TCA cycle remain entrapped inside mitochondria, but their substrates are diluted in a larger volume. With isolated mitochondria, the final distribution volume is much larger than the matrix volume, so that TCA cycle stops working because the concentrations of most substrates become too low for binding their enzymes. Consequently, mitochondrial respiration after PTP opening is inhibited even in the presence of millimolar concentrations of complex I substrates (like pyruvate and malate) because of pyridine nucleotide depletion ($NADH+H^+$ is the actual substrate of complex I) (Fontaine et al. 1998).

The situation is very different after PTP opening in intact cells. On one hand, substrate dilution is much lower (the total mitochondrial volume represents up to approximately 20 % of the total cellular volume). On the other, matrix and cytosolic concentrations of NAD^+ are similar, and therefore inhibition of respiration after PTP opening in intact cells may not be as prominent as it is in isolated mitochondria. However, PTP opening does partly inhibit mitochondrial respiration in intact cells as well, possibly as a consequence of NAD^+ degradation by glycohydrolase and PARP, which are often activated in parallel to PTP opening (Boyer et al. 1993; Di Lisa et al. 2001; Di Lisa and Ziegler 2001), and probably also because of a direct effect of PTP opening on complex I (see below) (Fontaine et al. 1998; Batandier et al. 2004).

Respiratory inhibition has been assessed by measuring oxygen consumption before and after PTP opening in intact cells (Dumas et al. 2009). This kinetic constraint on the respiratory chain accounts for the observation of depolarized mitochondria with high concentration of NAD(P)H after PTP opening (Dumas et al. 2009; Lablanche et al. 2011; Lamarche et al. 2013), which is usually observed after the addition of either uncouplers or inhibitors of ATP hydrolysis in the presence of a respiratory chain inhibitor. In isolated mitochondria, the PTP opening-induced respiratory chain inhibition has been localized at complex I because the maximal rate of oxygen consumption after PTP opening remains inhibited despite the presence of saturating amount of NADH in the incubation medium (Batandier et al. 2004).

Another unexpected consequence of PTP opening is a large increase in the production of mitochondrial reactive oxygen species (ROS). Initially observed in intact cells (Zorov et al. 2000), this effect has been reproduced in isolated mitochondria, where it has been shown to occur at complex I (Batandier et al. 2004). Both transient and permanent PTP openings induce ROS production. It is believed that transient PTP opening participates in the physiological ROS production, whereas the long-lasting ROS production during permanent PTP opening may play a key role in cell death. Interestingly, it is well-acknowledged that PTP opening can occur due to oxidative stress. In other words, PTP is both a target and a trigger of ROS production, and thus participates in the phenomenon named ROS-induced ROS release (Zorov et al. 2014). Both in intact cells and in isolated mitochondria, PTP opening causes the collapse of proton gradient (Zoratti and Szabo 1995). All the exchanges driven by the protonmotive force are thus abolished and ATP synthesis is inhibited. In addition, because ATP synthase in a reversible enzyme, ATP hydrolysis occurs (Zoratti and Szabo 1995), worsening energy depletion.

1.3 Transient But Not Necessarily Reversible

One of the main reasons that allowed Hunter and Haworth to postulate the existence of a channel was that they clearly established that the permeability transition was reversible (Haworth and Hunter1979). Indeed, how can a mere damage of the inner mitochondrial membrane have always the same size and be reversible? Not all the consequences of PTP opening are readily reversible, however. On the one hand, the proton gradient is rebuilt after PTP closure, allowing restoration of the active transport of ions and solutes that have specific transport systems (such as Ca^{2+}). On the other hand, closure of the pore confines inside or outside mitochondria solutes that can only cross the membrane through an open PTP. This accounts for the observations that PTP opening in isolated mitochondria incubated in either sucrose- or salt-based medium leads to mitochondrial swelling and depolarization; and that PTP closure allows repolarization in both cases, while matrix shrinkage is only possible in salt-based media (Petronilli et al. 1994). The widespread use of sucrose media in studies with isolated mitochondria substantially contributed to the erroneous idea that pore opening leads to irreversible damage of mitochondria.

Reversible and irreversible consequences of PTP opening can also be observed in intact cells, which can also lead to misunderstandings. PTP opening in intact cells is specifically visualized using compounds that do not enter or leave mitochondria unless the PTP is open. The most popular technique consists of loading intact cells with calcein acetoxymethyl ester and cobalt (Petronilli et al. 1999), the former diffusing in cytosol and mitochondria, the latter diffusing in cytosol only, due to a lack of mitochondrial channel for cobalt. Because cobalt quenches the fluorescence of calcein, fluorescence remains compartmentalized within mitochondria until PTP opening permits the redistribution of cobalt and/or calcein, which results in the quenching of calcein fluorescence in mitochondria as well. Obviously, this phenomenon is irreversible and the "calcein/cobalt" technique does not discriminate between transient or permanent PTP opening. The observation that lethal stresses cause a decompartmentalization of calcein fluorescence in a number of cells much higher than the number of cells entering into apoptosis has sometimes been interpreted as an argument against the involvement of PTP in cell death. The alternative interpretation is that permanent but not transient PTP opening leads to cell death, suggesting that PTP opening has lethal and nonlethal functions.

Mitochondrial membrane potential is another relevant parameter to study the PTP status in intact cells. Although not specific to PTP opening, mitochondrial depolarization occurs during PTP opening while PTP closure restores the mitochondrial membrane potential. In order to discriminate between transient and permanent PTP opening in intact cells, the two techniques (i.e., calcein/cobalt and membrane potential) can be coupled. Observations where calcein decompartmentalization occurred without mitochondrial depolarization indicate that transient PTP openings occur in intact cells Petronilli et al. (2001). Instead of measuring cobalt entry, it is possible to measure NAD(P)H release through an open pore. Coupled with the measure of membrane potential, this technique has confirmed that both transient and permanent PTP opening can occur in intact cells (Dumas et al. 2009; Lablanche et al. 2011; Lamarche et al. 2013).

1.4 PTP Opening and Cell Death

1.4.1 Evidence Supporting the Involvement of PTP in Cell Death

Since the discovery of pro-apoptotic proteins in the mitochondrial intermembrane space, there is no longer any doubt that mitochondria play a key role in the programmed cell death process. Because the mitochondrial pro-apoptotic proteins must redistribute from the intermembrane space into the cytosol, the first hypothesis was the existence of a channel in the outer membrane involving proteins from the BCL2 family (such BAX/BAK and BID) that was formed or opened by lethal stimuli. A large amount of work has been done and research is still ongoing in this field, partly slowed by the lack of pharmacological inhibitor of the pathway. Meanwhile, numerous, consistent, and reproducible data supporting the involvement of PTP in cell demise were produced. As discussed (Bernardi et al. 2001), the outer membrane channel(s) and the PTP can cooperate in mediating release of proapoptotic proteins and cell death (see below).

Abundant literature from different laboratories using cellular or animal models shows that PTP inhibition prevents some types of cell death, especially those involving oxidative stress. The results are particularly convincing for cell death induced by ischemia-reperfusion injury of the heart and brain (Baines et al. 2005; Nakagawa et al. 2005; Schinzel et al. 2005; Piot et al. 2008) and for fiber cell death in muscular dystrophy linked to collagen VI mutations (Irwin et al. 2003; Angelin et al. 2007; Merlini et al. 2008; Palma et al. 2009; Tiepolo et al. 2009; Zulian et al. 2014). While the majority of studies concluded that PTP opening is involved in cell death due to the protective effects of different PTP inhibitors, some have directly observed the entire sequence of events (Chauvin et al. 2001; Detaille et al. 2005; Guigas et al. 2004), from PTP opening to cell death via the release of pro-apoptotic proteins, all these events being prevented by PTP inhibition. However, the observation that transient PTP opening does not induce cell death suggests that this terminal event only follows persistent PTP opening. To the best of our knowledge, there is no evidence that cells can survive permanent PTP opening in a large population of mitochondria.

1.4.2 Mechanism by Which PTP Opening Induces Cell Death

Although long-lasting PTP opening allows release of pro-apoptotic proteins (cytochrome c for example) (Petronilli et al. 2001), the mechanism for release remains debated. In isolated mitochondria, cytochrome c release follows mechanical rupture of the outer membrane following mitochondrial swelling. In such case, the release of cytochrome c occurs within minutes of PTP opening (Kantrow and Piantadosi 1997). The process may be different in intact cells. As discussed above, PTP opening is not expected to induce a large mitochondrial swelling unless cytosolic solutes of high molecular weight are either catabolized or released through a leaky plasma membrane. In line with this argument, no large-amplitude mitochondrial swelling has been reported within minutes following PTP opening in intact cells (Dumas et al. 2009; Lablanche et al. 2011; Lamarche et al. 2013; De Giorgi et al. 2002). In addition, a period of up to 6 h between PTP opening and cytochrome c release has been reported in intact cells (De Giorgi et al. 2002).

Such a period of time is, however, consistent with a signaling cascade between the two events (PTP opening and cytochrome c release). Because PTP opening dramatically increases ROS production, while oxidative stress has been shown to participate in the formation or opening of an outer membrane channel, it has been proposed that PTP opening might be one of the events leading to the permeabilization of the outer membrane without real outer membrane rupture. In this scenario, PTP opening represents an event responsible for, not an alternative to, the permeabilization of the outer membrane.

It has also been proposed that PTP opening leads to cristae remodeling, leading to recruitment of pro-apoptotic proteins (Scorrano et al. 2002). Indeed, it has been shown that 10–15 % of cytochrome c is released after the removal of the outer membrane, while the remainder is mobilized only after matrix swelling (Bernardi and Azzone1981). In other words, a large part of cytochrome c (and perhaps of other pro-apoptotic proteins) may remain localized within the cristae after outer membrane permeabilization. In this scenario, PTP opening represents an event responsible for amplification of pro-apoptotic protein release after outer membrane permeabilization.

1.5 Nonlethal Functions of the Permeability Transition Pore

1.5.1 Calcium Homeostasis

In energized mitochondria, the Ca^{2+} electrochemical gradient (Scarpa and Azzone1970) drives the uptake of Ca^{2+} via the mitochondrial Ca^{2+} uniporter, MCU (Baughman et al. 2011; De Stefani et al. 2011; Kirichok et al. 2004), largely because of the inside-negative membrane potential. Ca^{2+} uptake is charge-compensated by an increase in H^+ pumping by the respiratory chain (Scarpa and Azzone 1970). This results first in an increase in ΔpH, which is transient because of the buffering of matrix pH by uptake of Pi, acetate, and/or bicarbonate, which allow regeneration of the initial membrane potential (Bernardi and von Stockum 2012). In such conditions, thermodynamic equilibrium predicts an enormous accumulation of Ca^{2+} (Azzone et al. 1977) which is never observed, indicating that Ca^{2+} release must occur at a rate comparable to that of Ca^{2+} uptake.

Importantly, if Ca^{2+} uptake requires energy owing to the electrophoretic nature of transport on MCU, Ca^{2+} release also requires energy. Ca^{2+} release via the Na^+/Ca^{2+} exchanger occurs with a stoichiometry of 3 Na^+ for 1 Ca^{2+} (Palty et al. 2010), which dissipates the electrical membrane potential, Na^+ being eventually released by the Na^+/H^+ exchanger with a stoichiometry of 1 Na^+ for 1H^+. Ca^{2+} release may also occur via a putative H^+/Ca^{2+} exchanger with a stoichiometry probably higher than 2H^+ per Ca^{2+}, whose operation is favored by the membrane potential (Bernardi and Azzone 1982, 1983). In other words, mitochondrial Ca^{2+} cycling (that is, the process of Ca^{2+} uptake and release through the above carriers) consumes the proton gradient and must be kept under kinetic control in order to prevent energy dissipation.

In order to limit Ca^{2+} cycling, the combined V_{max} of the efflux pathways is about 70-fold lower than the V_{max} of the MCU (Bernardi and von Stockum 2012). In turn, this arrangement exposes mitochondria to the hazards of Ca^{2+} overload when the rate of Ca^{2+} uptake exceeds that of the combined efflux pathways, for example, during sharp increases of cytosolic [Ca^{2+}] (Bragadin et al. 1979). Because mitochondrial Ca^{2+} overload did not occur despite the kinetic limitation in Ca^{2+} efflux through the well-recognized Ca^{2+} exchangers, it has been proposed that mitochondrial Ca^{2+} overload may be prevented by transient PTP opening, which does mediate mitochondrial depolarization and fast Ca^{2+} release in vitro, and possibly in vivo (Bernardi and Petronilli1996).

It has been argued that lack of selectivity is a problem for a mitochondrial "Ca^{2+} release" channel. We think on the contrary that it is an essential feature for a fast release of matrix Ca^{2+} (Bernardi and von Stockum2012). Indeed, Ca^{2+} efflux down its concentration gradient via a Ca^{2+}-selective channel would be opposed by the buildup of a Ca^{2+} diffusion potential (Akerman1978). According to the Nernst equation, the magnitude of the Ca^{2+} diffusion potential when no charge-compensating species are present is −30 mV per decade of the matrix/cytosol [Ca^{2+}] ratio. The Ca^{2+} efflux rate can be increased by any charge-compensating current (influx of positive charges, efflux of negative charges, or both). Since the inner membrane has a very low permeability to charged species, the rate of Ca^{2+} efflux would be extremely slow and essentially limited by the H^+ permeability. In other words, to obtain a significant rate of Ca^{2+} efflux via a Ca^{2+}-selective channel, the inner membrane permeability should be increased as well. An unselective pore of large size like the PTP confers the advantage of providing charge compensation within the channel itself, thus allowing maximal Ca^{2+} flux. Moreover, this would make Ca^{2+} release possible even for small [Ca^{2+}] gradients, and regulation could be achieved through modulation of the pore open time. Any increase in H^+ permeability will obviously lead to an increase in H^+ pumping by the respiratory chain, leading to energy dissipation. Therefore, the partial inhibition of Complex I concomitant with PTP opening might be viewed as an adaptive response to limit energy dissipation.

Is there any direct or indirect evidence that PTP opening participates to Ca^{2+} homeostasis? From 1992, it has been reported that the classic PTP inhibitor cyclosporin A (CsA) significantly increased net Ca^{2+} uptake and decreases Ca^{2+} efflux in intact cells, as measured by radiolabeled $^{45}Ca^{2+}$ (Altschuld et al. 1992). The effect of CsA was concentration-dependent and specific to mitochondria, as ATP-dependent Ca^{2+} uptake by the sarcoplasmatic reticulum was not affected. The probability of PTP opening can be prevented by genetic ablation of the *Ppif* gene coding for cyclophilin D, the molecular target of CsA (Baines et al. 2005; Nakagawa et al. 2005; Schinzel et al. 2005; Basso et al. 2005). In adult cortical neurons from wild-type and *Ppif*$^{-/-}$ mice, experimental conditions have been found where stimuli responsible for a robust increase of cytosolic [Ca^{2+}] resulted in much higher levels of mitochondrial [Ca^{2+}] in the *Ppif*$^{-/-}$ neurons, suggesting that the threshold for PTP activation had been reached in the wild-type but not in the cyclophilin D-null mitochondria in intact cells (Barsukova et al. 2011). It should be noted that this robust but transient increase in cytosolic [Ca^{2+}] did not induce cell death either in wild-type or in *Ppif*$^{-/-}$ neurons, suggesting that the mitochondrial response inhibited by cyclophilin D ablation (i.e., transient PTP opening) was part of a physiological process. Interestingly, less robust increases in cytosolic [Ca^{2+}] did not lead to different levels of mitochondrial [Ca^{2+}] in wild-type and *Ppif*$^{-/-}$ neurons, suggesting that the PTP could be silent unless high Ca^{2+} loads saturate the first line Ca^{2+} efflux pathways, allowing matrix [Ca^{2+}] to rise enough to trigger pore opening.

In adult mice, ablation of cyclophilin D increases resistance to acute ischemia-reperfusion injury both in the heart and brain (Baines et al. 2005; Nakagawa et al.2005; Schinzel et al. 2005; Wang et al. 2009). However, if transient PTP opening plays a physiological role, one may expect that *Ppif*$^{-/-}$ mice should have some health problems. Interestingly, an age-related phenotype has been discovered in the hearts of *Ppif*$^{-/-}$ mice, which displayed a decrease in maximum contractile reserve matched by increased shortening and relaxation times with longer decay of cytosolic Ca^{2+} transients (Elrod et al. 2010). *Ppif*$^{-/-}$ mice were also unable to compensate for the increase in afterload caused by transaortic constriction, displaying a larger reduction in fractional shortening, decompensation, ventricular dilation, fibrosis, and congestive heart failure. Indicating that the maladaptive phenotype of *Ppif*$^{-/-}$ mice depends on a primary disturbance of myocyte mitochondria rather than on an underlying systemic response, all consequences of cyclophilin D ablation were prevented by heart-selective re-expression of cyclophilin D (Elrod et al. 2010). Finally, direct measurement of total mitochondrial Ca^{2+} content of *Ppif*$^{-/-}$ hearts showed a 2.6-fold increase compared with wild-type heart (Elrod et al. 2010).

CsA does not always affect Ca^{2+} fluxes or increase mitochondrial [Ca^{2+}] in intact cells or in tissues (Eriksson et al. 1999). It must however be recalled that PTP opening requires that matrix [Ca^{2+}] reaches a threshold value, which is increased or decreased by PTP inducers and PTP inhibitors, respectively. In other words, PTP inhibition does not mean that PTP opening never occurs, but rather that more Ca^{2+} is required to open the PTP. Note also that depending on the cell line or tissue studied, the effect of CsA at PTP inhibition dramatically changes from no effect to a large increase in Ca^{2+} loads required for PTP opening (Li et al. 2012). Therefore, the lack of effect of CsA does not necessarily indicate that PTP opening in not involved in the process under study. It remains possible that the Ca^{2+} load was either too low to induce PTP opening, or too large to be effectively inhibited by CsA.

1.5.2 Transient Uptake-Release of Compounds Without Specific Channels

Solutes for which no transport system is known are present in mitochondria. Some are metabolites of other solutes that possess specific transporters. This is the case, for example, of oxaloacetate, which is indirectly exchanged via the malate/aspartate shuttle. The presence of magnesium inside mitochondria at a concentration similar to that observed in cytosol is generally explained by the existence of the ATP-Mg/Pi carrier (Aprille 1988; Fiermonte et al. 2004), which catalyzes the reversible electroneutral transport of ATP^{4-}-Mg^{2+} and Pi^{2-}. Because ATP and Pi recycle through the inner membrane via the adenine nucleotide translocator (ANT) and the phosphate carrier, respectively, this carrier is responsible for the net transport of adenine nucleotides across the inner mitochondrial membrane. In other words, the matrix Mg^{2+} and adenine nucleotide concentration must vary in a similar manner, and changes in matrix [Mg^{2+}] that are independent of modifications of adenine nucleotide concentration must involve another channel, which might be the PTP.

More generally, this rationale applies to all non-metabolized compounds without a specific channel in the inner membrane and present both in matrix and cytosol at a similar concentration.

1.5.3 Last Forces into Battle: Non-glycolytic Anaerobic ATP Production

Glycolysis plays a key role when ATP requirement exceeds ATP production by oxidative phosphorylation. If need be, ATP can also be produced by creatine kinase until the pool of phosphocreatine is consumed, and by adenylate kinase until the reaction reaches thermodynamic equilibrium. Depending on the AMP concentration, AMP is further catabolized to xanthine and hypoxanthine, thus relieving inhibition of adenylate kinase. From a theoretical point of view, apart from the ATP synthase, ATP (or its energy equivalent GTP) can be produced inside mitochondria by succinate-CoA ligase. However, its substrate (succinyl-CoA) is an intermediary of the TCA cycle that produces NADH, which is then oxidized by complex I in normal conditions. In the absence of oxygen, NADH accumulates inside mitochondria, inhibiting the TCA cycle. In other words, the production of ATP by succinate-CoA ligase is inhibited by the accumulation of NADH inside mitochondria. Assuming that transient PTP opening allows the release of NADH, it also allows ATP production inside mitochondria in the absence of oxygen. Note that this pathway has the same efficiency as anaerobic glycolysis (which produces 1 ATP per pyruvate), because 1 pyruvate gives 1 succinyl-CoA, and thus 1 ATP. Moreover, contrary to glycolysis, which strictly depends on carbohydrate availability, this pathway can run with other substrates because acetyl-CoA and alpha-ketoglutarate are provided by lipid and amino acid catabolism.

It has recently been shown (Chouchani et al. 2014) that succinate accumulates inside ischemic tissues. Interestingly, some TCA cycle metabolites located upstream of succinyl-CoA (such as citrate and/or alpha-ketoglutarate) decreased, supporting the proposal of a production of ATP by succinyl-CoA ligase. Surprisingly, isotopic enrichment suggested that succinate accumulated also due to the catabolism of aspartate into fumarate followed by a reverse succinate dehydrogenase activity. Note that the entry of aspartate into the matrix implies that the malate-aspartate shuttle runs in reverse, which generates a membrane potential. Therefore, whether succinate accumulation occurs via a forward or a reverse activity of the TCA cycle, it helps maintain energy homeostasis.

Production of ATP by succinyl-CoA ligase may not be so convenient if it runs in parallel with glycolysis. It is generally noticed that respiratory chain inhibitors do not dramatically decrease mitochondrial membrane potential in intact cells. This is because ATP produced by glycolysis is used by the reversible ATP synthase to maintain the proton gradient. In this particular case, PTP opening would depolarize mitochondria, further increasing ATP consumption by the ATP synthase. However, when glucose is exhausted, mitochondrial membrane potential cannot be sustained further and PTP opening would not depolarize already depolarized mitochondria. Therefore, this pathway must be viewed as the last opportunity for a cell to deliver ATP before terminal energy failure. After reperfusion, however, mitochondria will oxidize succinate, a condition in which ROS production occurs via reverse electron flux through complex I both in vitro (Korshunov et al. 1997; Kushnareva et al. 2002; Kwong and Sohal1998; Votyakova and Reynolds2001) and in vivo (Chouchani et al. 2014). Because such a production of ROS is inhibited by complex I inhibitors and strongly depends on a high membrane potential (Batandier et al. 2006), transient PTP opening may prevent such reperfusion-induced ROS production, both by inhibiting complex I and by a mild decrease in the membrane potential. The frontier between what is "good" and what becomes "bad" when PTP opens seems tenuous, and we hope that our considerations will be useful to further address the potential role of the pore in pathophysiology.

References

Akerman KE (1978) Charge transfer during valinomycin-induced Ca2+ uptake in rat liver mitochondria. FEBS Lett 93:293–296

Alavian KN, Beutner G, Lazrove E, Sacchetti S, Park HA et al (2014) An uncoupling channel within the c-subunit ring of the F1FO ATP synthase is the mitochondrial permeability transition pore. Proc Natl Acad Sci U S A 111:10580–10585

Altschuld RA, Hohl CM, Castillo LC, Garleb AA, Starling RC et al (1992) Cyclosporin inhibits mitochondrial calcium efflux in isolated adult rat ventricular cardiomyocytes. Am J Physiol 262:H1699–H1704

Angelin A, Tiepolo T, Sabatelli P, Grumati P, Bergamin N et al (2007) Mitochondrial dysfunction in the pathogenesis of Ullrich congenital muscular dystrophy and prospective therapy with cyclosporins. Proc Natl Acad Sci U S A 104:991–996

Aprille JR (1988) Regulation of the mitochondrial adenine nucleotide pool size in liver: mechanism and metabolic role. FASEB J 2:2547–2556

Azzone GF, Azzi A (1965) Volume changes in liver mitochondria. Proc Natl Acad Sci U S A 53:1084–1089

Azzone GF, Pozzan T, Massari S, Bragadin M, Dell'Antone P (1977) H+/site ratio and steady state distribution of divalent cations in mitochondria. FEBS Lett 78:21–24

Baines CP, Kaiser RA, Purcell NH, Blair NS, Osinska H et al (2005) Loss of cyclophilin D reveals a critical role for mitochondrial permeability transition in cell death. Nature 434:658–662

Barsukova A, Komarov A, Hajnoczky G, Bernardi P, Bourdette D et al (2011) Activation of the mitochondrial permeability transition pore modulates Ca2+ responses to physiological stimuli in adult neurons. Eur J Neurosci 33:831–842

Basso E, Fante L, Fowlkes J, Petronilli V, Forte MA et al (2005) Properties of the permeability transition pore in mitochondria devoid of Cyclophilin D. J Biol Chem 280:18558–18561

Batandier C, Leverve X, Fontaine E (2004) Opening of the mitochondrial permeability transition pore induces reactive oxygen species production at the level of the respiratory chain complex I. J Biol Chem 279:17197–17204

Batandier C, Guigas B, Detaille D, El-Mir MY, Fontaine E et al (2006) The ROS production induced by a reverse-electron flux at respiratory-chain complex 1 is hampered by metformin. J Bioenerg Biomembr 38:33–42

Baughman JM, Perocchi F, Girgis HS, Plovanich M, Belcher-Timme CA et al (2011) Integrative genomics identifies MCU as an essential component of the mitochondrial calcium uniporter. Nature 476:341–345

Bernardi P, Azzone GF (1981) Cytochrome c as an electron shuttle between the outer and inner mitochondrial membranes. J Biol Chem 256:7187–7192

Bernardi P, Azzone GF (1982) A membrane potential-modulated pathway for Ca2+ efflux in rat liver mitochondria. FEBS Lett 139:13–16

Bernardi P, Azzone GF (1983) Regulation of Ca2+ efflux in rat liver mitochondria. Role of membrane potential. Eur J Biochem 134:377–383

Bernardi P, Petronilli V (1996) The permeability transition pore as a mitochondrial calcium release channel: a critical appraisal. J Bioenerg Biomembr 28:131–138

Bernardi P, von Stockum S (2012) The permeability transition pore as a Ca(2+) release channel: new answers to an old question. Cell Calcium 52:22–27

Bernardi P, Petronilli V, Di Lisa F, Forte M (2001) A mitochondrial perspective on cell death. Trends Biochem Sci 26:112–117

Boyer CS, Moore GA, Moldeus P (1993) Submitochondrial localization of the NAD+ glycohydrolase. Implications for the role of pyridine nucleotide hydrolysis in mitochondrial calcium fluxes. J Biol Chem 268:4016–4020

Bragadin M, Pozzan T, Azzone GF (1979) Kinetics of Ca2+ carrier in rat liver mitochondria. Biochemistry 18:5972–5978

Carraro M, Giorgio V, Sileikyte J, Sartori G, Forte M et al (2014) Channel formation by yeast F-ATP synthase and the role of dimerization in the mitochondrial permeability transition. J Biol Chem 289:15980–15985

Chauvin C, De Oliveira F, Ronot X, Mousseau M, Leverve X et al (2001) Rotenone inhibits the mitochondrial permeability transition-induced cell death in U937 and KB cells. J Biol Chem 276:41394–41398

Chouchani ET, Pell VR, Gaude E, Aksentijević D, Sundier SY et al (2014) Ischaemic accumulation of succinate controls reperfusion injury through mitochondrial ROS. Nature 515(7527):431–435

De Giorgi F, Lartigue L, Bauer MK, Schubert A, Grimm S et al (2002) The permeability transition pore signals apoptosis by directing Bax translocation and multimerization. FASEB J 16:607–609

De Stefani D, Raffaello A, Teardo E, Szabo I, Rizzuto R (2011) A forty-kilodalton protein of the inner membrane is the mitochondrial calcium uniporter. Nature 476:336–340

Detaille D, Guigas B, Chauvin C, Batandier C, Fontaine E et al (2005) Metformin prevents high-glucose-induced endothelial cell death through a mitochondrial permeability transition-dependent process. Diabetes 54:2179–2187

Di Lisa F, Ziegler M (2001) Pathophysiological relevance of mitochondria in NAD(+) metabolism. FEBS Lett 492:4–8

Di Lisa F, Menabo R, Canton M, Barile M, Bernardi P (2001) Opening of the mitochondrial permeability transition pore causes depletion of mitochondrial and cytosolic NAD+ and is a causative event in the death of myocytes in postischemic reperfusion of the heart. J Biol Chem 276:2571–2575

Dumas JF, Argaud L, Cottet-Rousselle C, Vial G, Gonzalez C et al (2009) Effect of transient and permanent permeability transition pore opening on NAD(P)H localization in intact cells. J Biol Chem 284:15117–15125

Elrod JW, Wong R, Mishra S, Vagnozzi RJ, Sakthievel B et al (2010) Cyclophilin D controls mitochondrial pore-dependent Ca(2+) exchange, metabolic flexibility, and propensity for heart failure in mice. J Clin Invest 120:3680–3687

Eriksson O, Pollesello P, Geimonen E (1999) Regulation of total mitochondrial Ca2+ in perfused liver is independent of the permeability transition pore. Am J Physiol 276:C1297–C1302

Fiermonte G, De Leonardis F, Todisco S, Palmieri L, Lasorsa FM et al (2004) Identification of the mitochondrial ATP-Mg/Pi transporter. Bacterial expression, reconstitution, functional characterization, and tissue distribution. J Biol Chem 279:30722–30730

Fontaine E, Eriksson O, Ichas F, Bernardi P (1998) Regulation of the permeability transition pore in skeletal muscle mitochondria. Modulation by electron flow through the respiratory chain complex i. J Biol Chem 273:12662–12668

Giorgio V, von Stockum S, Antoniel M, Fabbro A, Fogolari F et al (2013) Dimers of mitochondrial ATP synthase form the permeability transition pore. Proc Natl Acad Sci U S A 110:5887–5892

Guigas B, Detaille D, Chauvin C, Batandier C, De Oliveira F et al (2004) Metformin inhibits mitochondrial permeability transition and cell death: a pharmacological in vitro study. Biochem J 382:877–884

Haworth RA, Hunter DR (1979) The Ca2+-induced membrane transition in mitochondria. II. Nature of the Ca2+ trigger site. Arch Biochem Biophys 195:460–467

Haworth RA, Hunter DR (1980) Allosteric inhibition of the Ca2+-activated hydrophilic channel of the mitochondrial inner membrane by nucleotides. J Membr Biol 54:231–236

Hunter DR, Haworth RA (1979a) The Ca2+-induced membrane transition in mitochondria. I. The protective mechanisms. Arch Biochem Biophys 195:453–459

Hunter DR, Haworth RA (1979b) The Ca2+-induced membrane transition in mitochondria. III. Transitional Ca2+ release. Arch Biochem Biophys 195:468–477

Hunter DR, Haworth RA, Southard JH (1976) Relationship between configuration, function, and permeability in calcium-treated mitochondria. J Biol Chem 251:5069–5077

Irwin WA, Bergamin N, Sabatelli P, Reggiani C, Megighian A et al (2003) Mitochondrial dysfunction and apoptosis in myopathic mice with collagen VI deficiency. Nat Genet 35:367–371

Kantrow SP, Piantadosi CA (1997) Release of cytochrome c from liver mitochondria during permeability transition. Biochem Biophys Res Commun 232:669–671

Kinnally KW, Campo ML, Tedeschi H (1989) Mitochondrial channel activity studied by patch-clamping mitoplasts. J Bioenerg Biomembr 21:497–506

Kirichok Y, Krapivinsky G, Clapham DE (2004) The mitochondrial calcium uniporter is a highly selective ion channel. Nature 427:360–364

Korshunov SS, Skulachev VP, Starkov AA (1997) High protonic potential actuates a mechanism of production of reactive oxygen species in mitochondria. FEBS Lett 416:15–18

Kushnareva Y, Murphy AN, Andreyev A (2002) Complex I-mediated reactive oxygen species generation: modulation by cytochrome c and NAD(P)+ oxidation-reduction state. Biochem J 368:545–553

Kwong LK, Sohal RS (1998) Substrate and site specificity of hydrogen peroxide generation in mouse mitochondria. Arch Biochem Biophys 350:118–126

Lablanche S, Cottet-Rousselle C, Lamarche F, Benhamou PY, Halimi S et al (2011) Protection of pancreatic INS-1 beta-cells from glucose- and fructose-induced cell death by inhibiting mitochondrial permeability transition with cyclosporin A or metformin. Cell Death Dis 2:e134

Lamarche F, Carcenac C, Gonthier B, Cottet-Rousselle C, Chauvin C et al (2013) Mitochondrial permeability transition pore inhibitors prevent ethanol-induced neuronal death in mice. Chem Res Toxicol 26(1):78–88

Li B, Chauvin C, De Paulis D, De Oliveira F, Gharib A et al (2012) Inhibition of complex I regulates the mitochondrial permeability transition through a phosphate-sensitive inhibitory site masked by cyclophilin D. Biochim Biophys Acta 1817:1628–1634

Massari S, Azzone GF (1972) The equivalent pore radius of intact and damaged mitochondria and the mechanism of active shrinkage. Biochim Biophys Acta 283:23–29

Merlini L, Angelin A, Tiepolo T, Braghetta P, Sabatelli P et al (2008) Cyclosporin A corrects mitochondrial dysfunction and muscle apoptosis in patients with collagen VI myopathies. Proc Natl Acad Sci U S A 105:5225–5229

Mitchell P (1961) Coupling of phosphorylation to electron and hydrogen transfer by a chemiosmotic type of mechanism. Nature 191:144–148

Nakagawa T, Shimizu S, Watanabe T, Yamaguchi O, Otsu K et al (2005) Cyclophilin D-dependent mitochondrial permeability transition regulates some necrotic but not apoptotic cell death. Nature 434:652–658

Palma E, Tiepolo T, Angelin A, Sabatelli P, Maraldi NM et al (2009) Genetic ablation of cyclophilin D rescues mitochondrial defects and prevents muscle apoptosis in collagen VI myopathic mice. Hum Mol Genet 18:2024–2031

Palty R, Silverman WF, Hershfinkel M, Caporale T, Sensi SL et al (2010) NCLX is an essential component of mitochondrial Na+/Ca2+ exchange. Proc Natl Acad Sci U S A 107: 436–441

Petronilli V, Szabo I, Zoratti M (1989) The inner mitochondrial membrane contains ion-conducting channels similar to those found in bacteria. FEBS Lett 259:137–143

Petronilli V, Nicolli A, Costantini P, Colonna R, Bernardi P (1994) Regulation of the permeability transition pore, a voltage-dependent mitochondrial channel inhibited by cyclosporin A. Biochim Biophys Acta 1187:255–259

Petronilli V, Miotto G, Canton M, Brini M, Colonna R et al (1999) Transient and long-lasting openings of the mitochondrial permeability transition pore can be monitored directly in intact cells by changes in mitochondrial calcein fluorescence. Biophys J 76:725–734

Petronilli V, Penzo D, Scorrano L, Bernardi P, Di Lisa F (2001) The mitochondrial permeability transition, release of cytochrome c and cell death. Correlation with the duration of pore openings in situ. J Biol Chem 276:12030–12034

Pfeiffer DR, Kuo TH, Tchen TT (1976) Some effects of Ca2+, Mg2+, and Mn2+ on the ultrastructure, light-scattering properties, and malic enzyme activity of adrenal cortex mitochondria. Arch Biochem Biophys 176:556–563

Piot C, Croisille P, Staat P, Thibault H, Rioufol G et al (2008) Effect of cyclosporine on reperfusion injury in acute myocardial infarction. N Engl J Med 359:473–481

Raaflaub J (1953a) Mechanism of adenosinetriphosphate as cofactor of isolated mitochondria. Helv Physiol Pharmacol Acta 11:157–165

Raaflaub J (1953b) Swelling of isolated mitochondria of the liver and their susceptibility to physicochemical influences. Helv Physiol Pharmacol Acta 11:142–156

Scarpa A, Azzone GF (1970) The mechanism of ion translocation in mitochondria. 4. Coupling of K+ efflux with Ca2+ uptake. Eur J Biochem 12:328–335

Schinzel AC, Takeuchi O, Huang Z, Fisher JK, Zhou Z et al (2005) Cyclophilin D is a component of mitochondrial permeability transition and mediates neuronal cell death after focal cerebral ischemia. Proc Natl Acad Sci U S A 102:12005–12010

Scorrano L, Ashiya M, Buttle K, Weiler S, Oakes SA et al (2002) A distinct pathway remodels mitochondrial cristae and mobilizes cytochrome c during apoptosis. Dev Cell 2:55–67

Szabo I, Zoratti M (1991) The giant channel of the inner mitochondrial membrane is inhibited by cyclosporin A. J Biol Chem 266:3376–3379

Tiepolo T, Angelin A, Palma E, Sabatelli P, Merlini L et al (2009) The cyclophilin inhibitor Debio 025 normalizes mitochondrial function, muscle apoptosis and ultrastructural defects in Col6a1-/- myopathic mice. Br J Pharmacol 157:1045–1052

von Stockum S, Giorgio V, Trevisan E, Lippe G, Glick GD et al (2015) F-ATPase of Drosophila melanogaster forms 53-picosiemen (53-pS) channels responsible for mitochondrial Ca2+-induced Ca2+ release. J Biol Chem 290:4537–4544

Votyakova TV, Reynolds IJ (2001) DeltaPsi(m)-Dependent and -independent production of reactive oxygen species by rat brain mitochondria. J Neurochem 79:266–277

Wang X, Carlsson Y, Basso E, Zhu C, Rousset CI et al (2009) Developmental shift of cyclophilin D contribution to hypoxic-ischemic brain injury. J Neurosci 29:2588–2596

Zoratti M, Szabo I (1995) The mitochondrial permeability transition. Biochim Biophys Acta 1241:139–176

Zorov DB, Filburn CR, Klotz LO, Zweier JL, Sollott SJ (2000) Reactive oxygen species (ROS)-induced ROS release: a new phenomenon accompanying induction of the mitochondrial permeability transition in cardiac myocytes. J Exp Med 192:1001–1014

Zorov DB, Juhaszova M, Sollott SJ (2014) Mitochondrial reactive oxygen species (ROS) and ROS-induced ROS release. Physiol Rev 94:909–950

Zulian A, Rizzo E, Schiavone M, Palma E, Tagliavini F et al (2014) NIM811, a cyclophilin inhibitor without immunosuppressive activity, is beneficial in collagen VI congenital muscular dystrophy models. Hum Mol Genet 23:5353–5363

Chapter 2
Mitochondrial Calcium and Ischemia: Reperfusion Injury in Heart

Liron Boyman, George S.B. Williams, and W. Jonathan Lederer

2.1 Introduction

Ischemia occurs when blood supply to a region of tissue is inappropriately reduced or occluded. Such ischemic conditions may lead to diminished levels of oxygen (O_2), glucose, lipids, and other energy sources needed by the cells for normal function and also the accumulation of metabolic products that need to be cleared. While the period of ischemia itself is often well-tolerated by cells, once blood flood is restored and reperfusion occurs, damage to the cells may develop. This damage is known as "Ischemia-Reperfusion" (IR) injury. One particularly dangerous example of IR injury occurs when arterial blockage takes place in the heart itself (i.e., myocardial infarction) and the resulting IR injury damages the tissue and may produce contractile dysfunction, arrhythmia, or both. Beginning in the late 1970s, the experimental model of IR injury became readily available and pioneer investigations were carried out to study IR injury and the pathophysiological underpinnings of the process.

Through the use of the experimental IR model, a critical discovery was reported by Murry and co-workers in 1986 that was called "preconditioning". At that time, it was already known that 40 min of ischemia followed by reperfusion leads to the subsequent death of cardiomyocytes and that this cell death (i.e., necrosis) begins early after the onset of the reperfusion. Murry and co-workers observed that if IR is preceded by four brief episodes of ischemia (each 5 min in duration), a procedure they termed ischemic preconditioning (IP), the IR injury was profoundly reduced.

L. Boyman, Ph.D. • G.S.B. Williams, Ph.D. • W.J. Lederer, M.D., Ph.D. (✉)
Center for Biomedical Engineering and Technology, University of Maryland School of Medicine, Baltimore, MD 21201, USA

Department of Physiology, University of Maryland School of Medicine, Baltimore, MD 21201, USA
e-mail: jlederer@umaryland.edu

D.M. Hockenbery (ed.), *Mitochondria and Cell Death*, Cell Death in Biology and Diseases, DOI 10.1007/978-1-4939-3612-0_2

X.-M. Yin and Z. Dong (Series eds.), *Cell Death in Biology and Diseases*

In fact, the IP appeared to reduce the infarct size by 75 %. The beneficial outcome of IP was further reproduced by many other investigators and in various other mammalian models (Yellon and Downey 2003; Skyschally et al. 2009). Similarly, in humans there is evidence that episodes of angina (i.e., pain associated with reduced blood flow to heart tissue) can provide a protective effect from a subsequent acute myocardial infarction (Edwards et al. 2000; Kloner and Rezkalla 2006; Rezkalla and Kloner 2004; Heusch 2013). The beneficial effects of preconditioning begin immediately, the protective window lasts a few hours, and then is lost (Yellon and Downey 2003). Surprisingly, a second protective window emerges after approximately 24 hr and persists for a few days. This protective window involves the activation of a large range of translational pathways (Baxter and Ferdinandy 2001; Das and Maulik 2006).

The beneficial effects of IP revealed that cardiomyocytes can become profoundly more resilient to IR and provoked us (and many others) to investigate three fundamental questions which will be the focus of this chapter: (1) What are the changes that occur within the cardiomyocyte during the ischemia that predispose it to cell death during reperfusion?; (2) What is the sequence of cellular events during reperfusion that ultimately lead to cell death?; and (3) What is the mechanism that underlies cellular protection via IP?

2.2 Cellular Ca^{2+} and Ischemia-Reperfusion (IR) Injury in Heart

2.2.1 Cytosolic and Mitochondrial Ca^{2+} Signals

To date, several critical changes that occur in cardiomyocytes during the ischemia have been identified:

1. Cellular content of high energy phosphates such as cytosolic ATP ($[ATP]_i$) declines.
2. Cytosolic pH (pH_i) declines (i.e., the cell becomes more acidified).
3. Cytosolic sodium ($[Na^+]_i$) rises.

These energetic and ionic imbalances develop over the time course of the ischemic phase and their extent determine the severity of the IR injury. The mechanisms that underlie these changes have been extensively studied and are discussed in Sect. 2.3. In addition to these changes, high levels of cellular reactive oxygen species ($[ROS]_i$) also play a major role and are generally thought to exacerbate the damage caused by the ionic imbalances. These conclusions are supported by a large number of investigations using different genetic murine models (see Table 2.1). The observed protective effects, as revealed by these models, are attributed either to ameliorating the ionic imbalance or to enhancing the reducing capacity of the cardiomyocytes. Note, increasing the cellular reducing capacity makes the cells less sensitive to ROS. Ischemic preconditioning (see Sect. 2.5) may also partially derive its benefit from preventing ionic imbalances. In this chapter, we will focus on the ionic and energetic imbalances during IR; for more details regarding ROS during

Table 2.1 Mediators of cell death or survival following ischemia

Protein target	Full name	Cellular localization	Role in IR	Murine model	Post IR hemodynamic recovery	Infract zone size
NCX1	Na^+/Ca^{2+} exchanger isoform 1	sarcolemmal	$[Na^+]_i$ and $[Ca^{2+}]_i$ dysregulation	NCX1 null	Improved	Reduced
NHE1	Na^+/H^+ exchanger isoform 1	sarcolemmal	$[Na^+]_i$ accumulation and rapid pH_i changes during reperfusion	NHE1 null	Improved	Reduced
CypD	Cyclophilin D	mitochondrial matrix	Regulator of mPTP formation	CypD null	ND	Reduced
CaMKII	Ca^{2+}/calmodulin-dependent protein kinase II	mitochondrial matrix ?	Disputed	Overexpressor of mitochondrial-targeted CaMKII inhibitory protein	Improved	Reduced
CuZnSOD	Copper/zinc superoxide dismutase	cytosolic	ROS scavenger	CuZnSOD overexpressor	Improved	ND
MnSOD	Manganese superoxide dismutase	mitochondrial	ROS scavenger	MnSOD overexpressor	Improved	Reduced
EC-SOD	Extracellular Superoxide Dismutase	extracellular	ROS scavenger	EC-SOD overexpressor	Improved	ND
GSHPx-1	Glutathione peroxidase	cytosolic	ROS scavenger	GSHPx-1 overexpressor	Improved	Reduced

The information included in the table about the role of different proteins in IR injury is from studies with the following genetic models: NCX1 null mouse (Imahashi et al. 2005), NHE1 null mouse (Wang et al. 2003), CypD null mouse (Baines et al. 2005; Nakagawa et al. 2005), Mitochondrial-targeted CaMKII inhibitory protein overexpressing mouse (Joiner et al. 2012), CuZnSOD overexpressing mouse (Chen et al. 2000), MnSOD overexpressing mouse (Chen et al. 1998), EC-SOD overexpressing mouse (Chen et al. 1996), GSHPx-1 overexpressing mouse (Yoshida et al. 1996). Note that the findings related to the role of CaMKII in IR injury remain in dispute, for additional details see Fieni et al. (2014) and Correll and Molkentin 2013

IR, we refer the reader to recent reviews (Chen and Zweier 2014; Baines 2011; Zweier and Talukder 2006).

A large body of evidence indicates that the ionic and energetic imbalances are adverse to cardiomyocytes because of their effects on the free Ca^{2+} in the cytosol ($[Ca^{2+}]_i$) and Ca^{2+} in the mitochondrial matrix ($[Ca^{2+}]_m$). Under normal conditions, with each heartbeat 70–80 μmol of Ca^{2+} enter a liter of cytosol during a $[Ca^{2+}]_i$ transient in ventricular myocytes and activate the contraction. Subsequently, this Ca^{2+} is cleared from the cytosol, and muscle relaxation occurs (Bassani et al. 1994; Negretti et al. 1995). A small fraction of this Ca^{2+} (less than 1 %) enters the mitochondrial matrix during the $[Ca^{2+}]_i$ transient (Bassani et al. 1994; Williams et al. 2013; Boyman et al. 2014) and becomes a critical regulator of ATP production (Glancy and Balaban 2012; Williams et al. 2015). Note, however, that some investigations have suggested that this uptake is much larger (see Maack et al. 2006; Drago et al. 2012; Isenberg et al. 1993; Cortassa et al. 2003; Gauthier et al. 2012; Pacher et al. 2000). Regardless, $[Ca^{2+}]_m$ dynamics are directly coupled to the $[Ca^{2+}]_i$ dynamics and the normal role(s) of $[Ca^{2+}]_m$ is further discussed in Sect. 2.2.2. During ischemia, however, the extrusion of Ca^{2+} out of the cell is hampered due to the ionic and energetic imbalances and net cellular accumulation of Ca^{2+} occurs. High $[Ca^{2+}]_i$ is itself detrimental to cardiomyocytes due to its effect on excitability, contractility, and $[Ca^{2+}]_m$ dynamics.

A new view has emerged which suggests that altered $[Ca^{2+}]_i$ dynamics are detrimental to the cardiomyocytes primarily because of its effects on the mitochondria. Cardiomyocytes critically depend on their mitochondria to fuel their massive ATP demand, which also make this cell type extremely vulnerable to mitochondrial dysfunction. It is suggested that mitochondrial metabolism collapses during reperfusion (possibly due to pathologically high $[Ca^{2+}]_m$) leading to acute irreversible cell injury and cell death. This collapse is attributed to the formation of a putative mega channel termed the mitochondrial permeability transition pore (mPTP). The properties of mPTP have been extensively studied in-vitro, using either isolated mitochondria in suspension or vesicles of the inner mitochondrial membrane (i.e., mitoplast). These studies reported that when the mPTP is open, it enables the passage of molecules in and out of the mitochondrial matrix that are as large as 1500 Da (Halestrap and Richardson 2014; Bernardi and von Stockum 2012; Elrod and Molkentin 2013). This results in the dissipation of virtually all ionic gradients across the IMM including loss of the proton gradient and a large positive shift of the electric potential across the IMM ($\Delta\Psi_m$). By this means, the proton motive force required for ATP generation is lost when mPTP opens. Without a significant negative $\Delta\Psi_m$, the F_1F_0-ATP synthase cannot synthesize ATP and may even operate "backwards" to hydrolyze the available ATP into ADP (Halestrap 2009; Alavian et al. 2014a, b).

In heart, deleterious openings of mPTP are thought to occur due to elevated $[Ca^{2+}]_m$. While $[Ca^{2+}]_m$ may be elevated enough during IR to open mPTP, other signals also appear to be required to trigger the opening of mPTP (e.g., pH_i, pH_m, [ADP], [ATP], and ROS). Here we discuss and critically evaluate the role of mPTP and other possible mechanisms in IR injury, and specifically focus on the conditions that may lead to altered $[Ca^{2+}]_m$ dynamics and pathological opening of mPTP (Fig. 2.2).

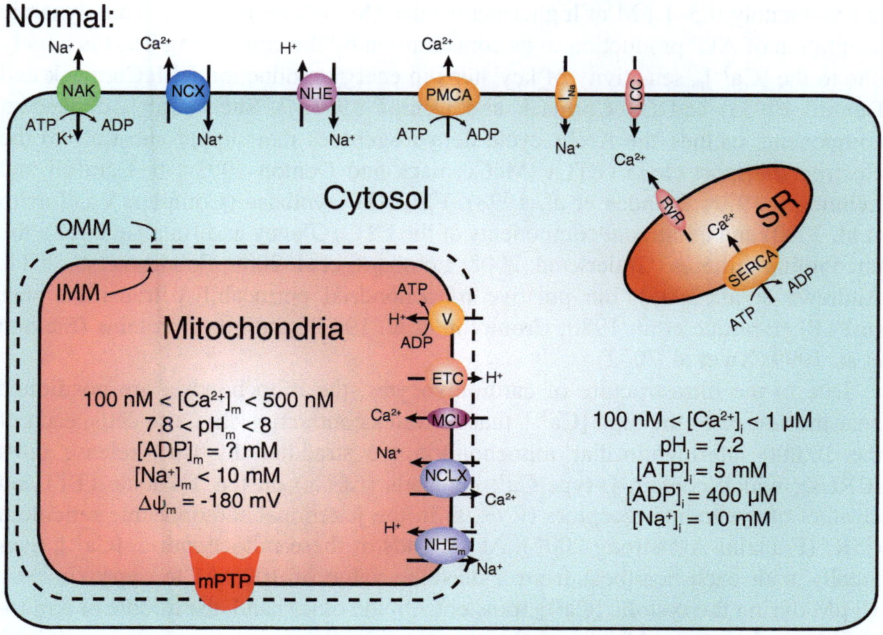

Fig. 2.1 Diagram of mitochondrial signaling under normal conditions. On the sarcolemma, the following channels and transporters are shown (*from left to right*): Na^+/K^+ ATPase (NAK), Na^+/Ca^{2+} exchanger (NCX), Na^+/H^+ exchanger (NHE), Ca^{+2} pump or the plasmalemmal Ca^{+2} ATPase (PMCA), voltage-sensitive Na^+ channels (I_{Na+}), L-type Ca^{2+} channels (LCC). On the sarcoplasmic reticulum (SR), the following channels and transporters are shown: ryanodine receptor (RyR), sarco-/endoplasmic reticulum Ca^{2+} ATPase (SERCA). The outer mitochondrial membrane (OMM), largely permeable to small ions including calcium, is shown as a dashed black line. On the inner mitochondrial membrane (IMM), the following channels and transporters are shown (*from top to bottom*): ATP synthase (complex V), electron transport chain (ETC), the mitochondrial Ca^{2+} uniporter (MCU), the mitochondrial Na^+/Ca^{2+} exchanger (NCLX), the mitochondrial Na^+/H^+ exchanger (NHE), the mitochondrial permeability transition pore (mPTP)

2.2.2 *Ca^{2+} Movement in Cardiac Mitochondria*

$[Ca^{2+}]_i$ signals regulate mitochondrial metabolism in heart. The heart relies almost entirely on mitochondrial oxidative phosphorylation as its source of ATP and to a minor extent on cytosolic glycolysis. Measurements of ATP production in the rat heart estimate that under normal conditions only 4.8 % of the total ATP produced is from glycolysis (Neely et al. 1975). The heart muscle cell has the highest volume-fraction of mitochondria of any cell in the body, about 33 % of the cell's total volume is composed of mitochondria. As the heart rate increases, the Ca^{2+} influx into the mitochondrial matrix rises; this is because the $[Ca^{2+}]_i$ transients that activate the contractions become more frequent. ATP production by the mitochondria is stimulated when $[Ca^{2+}]_m$ rises, and under normal conditions, $[Ca^{2+}]_m$ can increase from 90 to 100 nM during a prolonged diastole (Boyman et al. 2014; Lu et al. 2013) to

approximately 0.5–1 μM at higher heart rates (Miyata et al. 1992). This dynamic adaptation of ATP production to its consumption by the contracting heart is largely due to the $[Ca^{2+}]_m$ sensitivity of key steps in energy production ((McCormack and Denton 1993a) and (McCormack and Denton 1993b)). Such $[Ca^{2+}]_m$-dependent components include the Krebs cycle dehydrogenases that supply substrate to the electron transport chain (ETC) (McCormack and Denton 1993a, b; Carafoli and Lehninger 1971; Brandes et al. 1998), F_1F_0-ATP synthase (Complex V) (Territo et al. 2000) and additional components of the ETC (Glancy and Balaban 2012), the uncoupling proteins (Graier et al. 2008; Lombardi et al. 2008; Poburko et al. 2011; Andrews et al. 2005), the putative mitochondrial permeability transition pore (mPTP) (Beatrice et al. 1980; Crompton et al. 1987), and other proteins (Siemen et al. 1999; Xu et al. 2002).

Due to the ultrastructure of cardiomyocytes, the mitochondria are positioned near the source of the high $[Ca^{2+}]_i$ that activates contraction. In these cells, each of the 10,000 intermyofibrillar mitochondria are straddled by Ca^{2+} release units (CRUs), which contain L-type Ca^{2+} channels (LCCs) on the T-Tubule (TT) and clusters of ryanodine receptors (RyR2s) in the junctional sarcoplasmic reticulum (jSR) (Franzini-Armstrong 2007). At the ends of these mitochondria, $[Ca^{2+}]_i$ rises locally with each heartbeat from a diastolic value of 100 nM to approximately 10 μM during the systolic $[Ca^{2+}]_i$ transient. On the other hand, the middle of a mitochondrion (near the M-line of the sarcomere) is exposed to much less $[Ca^{2+}]_i$ (<500 nM) (Williams et al. 2013). Normally, there is a large driving force for Ca^{2+} entry from the cytosol into the mitochondrial matrix because of the electrical gradient across the IMM ($\Delta\Psi_m = -180$ mV), and yet the fluxes of Ca^{2+} in and out of the mitochondria are modest.

IMM Ca^{2+} fluxes. The main pathway for Ca^{2+} entry into the mitochondrial matrix, the mitochondrial Ca^{2+} uniporter (MCU), is a highly selective Ca^{2+} channel (Kirichok et al. 2004; Fieni et al. 2012). Each cardiac mitochondrion contains approximately 200 MCUs within its IMM; however, under normal physiological conditions, they are predominantly closed (Williams et al. 2013). $[Ca^{2+}]_m$ is set by the balance between the total Ca^{2+} influx (via MCU) and the total Ca^{2+} efflux via Ca^{2+}-extrusion/pump mechanisms. The Ca^{2+} efflux in heart muscle cells is primarily mediated by the mitochondrial Na^+/Ca^{2+} exchanger (NCLX), whose identity has only recently been identified (Palty et al. 2010). Various features of the ion exchange cycle mediated by NCLX are currently a topic of active investigations; this includes the stoichiometry (i.e., number of Na^+ ions that are exchanged per 1 Ca^{2+} ion), the electrogenicity (i.e., net transfer of electrical charge), the dependency of NCLX on cytosolic Na^+ ($[Na^+]_i$), and the magnitude of NCLX Ca^{2+} efflux [for further information on these topics see Boyman et al. (2013)].

Ischemia-induced changes in $[Ca^{2+}]_m$. It is a prevailing view that $[Ca^{2+}]_m$ rises substantially during IR in heart; however, the mechanisms that may underlie such a rise remain unclear. A large influx of Ca^{2+} via the MCUs has been suggested as a possible cause of such a rise (Joiner et al. 2012). However, the evidence implicating these high MCU fluxes in IR injury has been challenged by others (Fieni et al. 2014;

Correll and Molkentin 2013). Furthermore, the recent characterization of the MCU null mice ($MCU^{-/-}$) revealed that $MCU^{-/-}$ mice are as vulnerable to IR as their wild-type (WT) littermates (i.e., $MCU^{+/+}$) (Pan et al. 2013). It should also be noted that a large inward current of a positively charged ion such as Ca^{2+} into the matrix degrades $\Delta\Psi_m$, which not only reduces the driving force for further Ca^{2+} influx, but also diminishes the open probability (P_o) of the MCUs due to their voltage sensitivity (Kirichok et al. 2004; Michels et al. 2009). Furthermore, during ischemia $\Delta\Psi_m$ should be depolarized (i.e., less polarized) due to the lack of O_2, which limits the function of the ETC that normally generates $\Delta\Psi_m$ (Zhang et al. 2013). It is therefore unclear if MCU Ca^{2+} fluxes play any role in IR injury.

An additional mechanism that may offset normal $[Ca^{2+}]_m$ dynamics is changes in Ca^{2+} efflux from the mitochondria mediated by NCLX. In the steady-state, under normal conditions any net Ca^{2+} entry via MCU must be balanced by extrusion via NCLX before the end of diastole. During ischemia, Ca^{2+} extrusion via NCLX is most likely affected; however, conflicting factors make this difficult to predict. Such factors include the ischemic elevation of $[Na^+]_i$ and $[Ca^{2+}]_i$ and changes in $\Delta\Psi_m$. During ischemia, $[Na^+]_i$ gradually increases and can become as high as 40–45 mM (see Sect. 2.3.3), which would have promoted Ca^{2+} efflux via NCLX under normal conditions (Maack et al. 2006). However, in addition to $[Na^+]_i$, the time-average $[Ca^{2+}]_i$ also increases, potentially exerting the opposite effect on the net Ca^{2+} transport by NCLX. Reduced $\Delta\Psi_m$ is also a critical factor and the uncertainty regarding the voltage sensitivity of NCLX makes it difficult to speculate how $\Delta\Psi_m$ affects NCLX function during ischemia.

Taken together, clarifying the uncertainties regarding Ca^{2+} movement via MCU and NCLX during IR is imperative, especially given the explicit evidence that high $[Ca^{2+}]_m$ is a primary regulator of mPTP (for more details on $[Ca^{2+}]_m$ and mPTP see Sect. 2.4.2). What is more clear is that profound alterations in the $[Ca^{2+}]_i$ signals occur during IR, which changes the amount of $[Ca^{2+}]_i$ that the mitochondria are exposed to. The next sections discuss these changes and how they affect $[Ca^{2+}]_m$.

2.3 Energetic and Ionic Imbalances in Cardiomyocytes During Ischemia (Fig. 2.2)

2.3.1 ATP Depletion

Early studies using various mammalian models have shown that occlusion of coronary arteries and the subsequent lack of O_2 supply to a myocardial region causes an abrupt change in metabolism within that region. This switch from mitochondrial oxidative phosphorylation to anaerobic metabolism leads to depletion of critical metabolites and to adverse buildup of catabolites inside and outside the heart muscle cells (Braasch et al. 1968; Williamson 1966; Kübler and Spieckermann 1970; Wollenberger and Krause 1968; Neely et al. 1973; Dunn and Griggs 1975; Gudbjarnason et al. 1970).

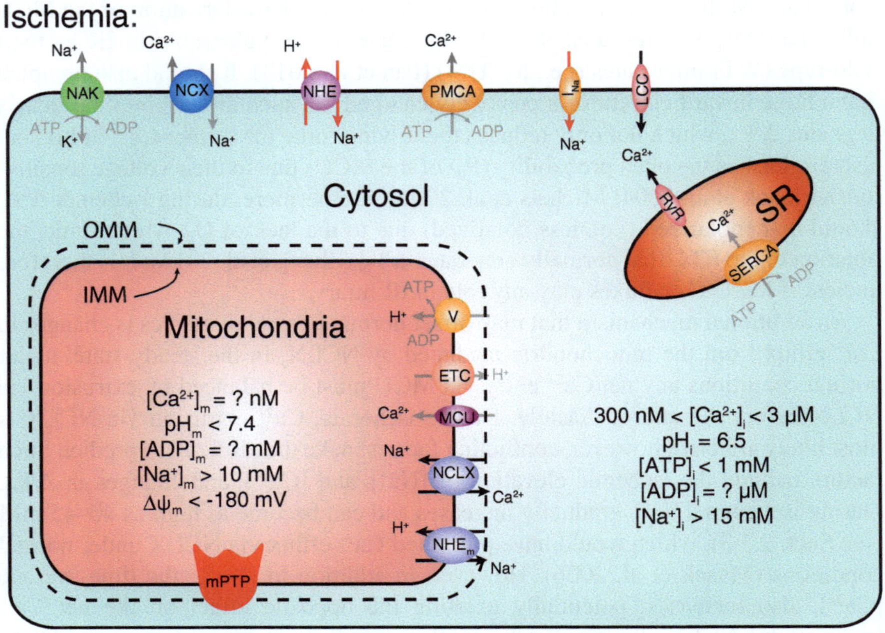

Fig. 2.2 Diagram of mitochondrial signaling during ischemia. On the sarcolemma, the following channels and transporters are shown (*from left to right*): Na^+/K^+ ATPase (NAK), Na^+/Ca^{2+} exchanger (NCX), Na^+/H^+ exchanger (NHE), Ca^{2+} pump or the plasmalemmal Ca^{2+} ATPase (PMCA), voltage-sensitive Na^+ channels (I_{Na}), L-type Ca^{2+} channels (LCC). On the sarcoplasmic reticulum (SR), the following channels and transporters are shown: ryanodine receptor (RyR), sarco-/endoplasmic reticulum Ca^{2+} ATPase (SERCA). The outer mitochondrial membrane (OMM), largely permeable to small ions including calcium, is shown as a dashed black line. On the inner mitochondrial membrane (IMM), the following channels and transporters are shown (*from top to bottom*): ATP synthase (complex V), electron transport chain (ETC), the mitochondrial Ca^{2+} uniporter (MCU), the mitochondrial Na^+/Ca^{2+} exchanger (NCLX), the mitochondrial Na^+/H^+ exchanger (NHE), the mitochondrial permeability transition pore (mPTP). *Faded gray arrows* indicate transport pathways that are either diminished or inhibited during ischemia. *Red arrows* indicate ischemically upregulated pathways

While ATP can be transiently replenished from creatine phosphate (CP), the content of CP in the myocardium declines by 70 % within 1 min of ischemia (Braasch et al. 1968; Sayen et al. 1958). Within approximately that same timeframe, the severely ischemic myocardial region ceases to contract (Sayen et al. 1958; Jennings et al. 1969); however, the rate of ATP consumption still remains higher than its production (Kübler and Spieckermann 1970; Wollenberger and Krause 1968; Scheuer 1972; Scheuer and Stezoski 1970). This is largely because anaerobic glycolysis is the only process by which high energy phosphate is now synthesized, yielding 3 mol of ATP per mol of glucose derived from breakdown of glycogen to lactate (compared to 30–32 mol when oxidative phosphorylation is fully functional) (Scheuer 1972).

Unfortunately, the rate of ATP production via glycolysis also gradually slows down. This occurs because the plasma levels of glucose in the occluded region are limited, the cellular milieu of glycogen declines (Scheuer and Stezoski 1970), and glycolysis becomes inhibited as the cytosol of the heart muscle cells acidifies (due to buildup of catabolites such as lactic acid) (Kübler and Spieckermann 1970; Neely et al. 1973). As a result, the ATP content in the ischemic myocardium is not sustained and declines by 65 % after 15 min of ischemia, by 91 % after 30 min, and by 95 % in 60 min (Jennings et al. 1978a).

2.3.2 Acidosis

One of the adverse effects early after the onset of ischemia is acidification of the myocardium, in which both the extracellular pH (pH_o) and the intracellular pH (pH_i) decline substantially and remain low throughout the entire time course of the ischemia. In humans, it has been reported that acidification of the myocardial blood plasma (i.e., acidosis) starts developing within 12 seconds following occlusion of the descending coronary artery during angioplasty (Crake et al. 1987). During ischemia, the primary source of these protons is the hydrolysis of the lactic acid that accumulates as the end product of glycolysis (Williamson 1966; Allen et al. 1985; Gevers 1977; Seelye 1980). Note that in an experimental model, where the clearance of catabolites from the myocardium is kept normal, hypoxia only leads to mild decrease of pH_o (0.02–0.07 pH units) (Neely et al. 1975; Weiss et al. 1984) and no significant change of pH_i (Neely et al. 1975). Reducing the flow rate of the perfusate through the coronary arteries leads to further pH decrease; the pH of the coronary effluent blood (venous return) declines by 0.43–0.46 pH units, and pH_i declines by 0.22–0.33 pH units (Neely et al. 1975; Steenbergen et al. 1977). During global ischemia, when the flow through the coronary arteries is completely halted, phosphorus nuclear magnetic resonance (^{31}P NMR) measurements from whole rat hearts indicated that the pH_i drops by 0.85 pH units within 13 min (Cobbe and Poole-Wilson 1980; Hirche et al. 1980; Jacobus et al. 1982; Garlick et al. 1979) and 1.3 pH units within 20 min coinciding with a pH_o drops of 1.89 pH units (Gabel et al. 1997).

In these experiments, the extracellular conditions experienced by different cells of the myocardium could vary greatly, O_2 gradients across the ventricular muscle wall should exist, and because the heart is exteriorized, cardiomyocytes at the outer surface of the heart muscle have more access to atmospheric air. To minimize these heterogeneous effects, several other studies were carried out with thinner layers of heart tissue. In addition, pH microelectrodes with faster response-time and improved accuracy are used. It has been reported that pH_o declines by 0.61–0.75 pH units after 10 min of ischemia in muscle-strips isolated from rabbit ventricular tissue (Weiss et al. 1984; Couper et al. 1984). When experimental conditions are matched to those expected to be experienced by cardiomyocytes in the severely ischemic myocardial core (i.e., midmural ventricular layers), pH_o declines by 1.1 pH units (pH_i by 0.38

pH units) within 14 min and by 1.27 pH units (pH_i by 0.47 pH units) after 18 min (Yan and Kléber 1992).

While the degree of acidosis differs depending on the location within the myocardium, all the cardiomyocytes within (and adjacent to) the ischemic region will experience some degree of acidosis within minutes following the occlusion of blood flow. The change in pH will have profound implications for numerous processes within the cells of the affected regions. This includes inhibition of ATP production (see above) and dysregulation of $[Na]_i$ and $[Ca^{2+}]_i$, which will be discussed in detail below.

2.3.3 Dysregulation of $[Na^+]_i$

Under normal physiological conditions, $[Na^+]_i$ ranges from 5 to 15 mM, depending on species (Murphy and Eisner 2009). During ischemia, $[Na]_i$ rises three to fourfold and can reach 40–45 mM (Murphy and Eisner 2009; Pike et al. 1990; Murphy et al. 1991; Anderson et al. 1996; Williams et al. 2007). Degradation of the Na^+ gradient across the sarcolemma (and possibly also the IMM) can have profound effects on the coupled transport of other ions, e.g., Ca^{2+} and K^+. In this regard, the extracellular K^+ ($[K^+]_o$) rises dramatically during ischemia reaching 11 mM (Weiss et al. 1989) and this alkalosis appears to depend on a rise in $[Na^+]_i$. It was found that while the heart has an abundance of high conductance ATP-sensitive potassium channels ($I_{K(ATP)}$), which open as the ATP levels decline (Nichols and Lederer 1990a; Ripoll et al. 1990), the loss of K^+ from the hypoxic cell is only moderately increased by $I_{K(ATP)}$ agonist cromakalim (Shivkumar et al. 1997). This suggests that increased sarcolemmal conductance to K^+ alone does not explain the increased K^+ efflux out of the cardiomyocyte. The increased K^+ efflux depends on increased Na^+ influx, and when the net $[Na^+]_i$ gain caused by hypoxia is prevented, the net K^+ loss from the hypoxic cell was almost completely eliminated (Shivkumar et al. 1997). Due to these serious implications, the mechanisms related to Na^+ accumulation by the ischemic cell have been extensively studied. To date, the evidence suggests that elevated $[Na^+]_i$ is due to the combined action of multiple Na^+ transporters as discussed below. The general consensus suggests that during ischemia the Na^+ efflux pathways (Na^+/K^+ ATPase, NAK) are reduced and influx pathways (Na^+/H^+ exchanger, NHE1 and voltage-sensitive Na^+ channels, I_{Na}) are increased, combining to significantly elevate $[Na^+]_i$.

Under normal physiological conditions, the sarcolemmal NAK sets the Na^+ gradient across the sarcolemmal membrane. It does so by using the free energy derived from hydrolysis of ATP to pump Na^+ out of the cell against its electrochemical gradient . In the steady state, during each contraction–relaxation cycle, Na^+/K^+ ATPase extrudes the same amount of sodium that enters the cell via the voltage-sensitive Na^+ channels (I_{Na}) and the sarcolemmal Na^+/Ca^{2+} exchanger (NCX) (Murphy and Eisner 2009). The decline in cellular [ATP] during ischemia reduces the overall rate of Na^+ extrusion by NAK. However, the changes in the activity of NAK cannot only be explained by the cellular [ATP] decline. The NAK remains active during the first few minutes of ischemia (Anderson et al. 1996; Cross et al. 1995) and becomes inactive before the [ATP] drops to levels that would explain this abrupt inhibition of the NAK (Cross et al. 1995). To date, the mechanism of NAK regulation during ischemia remains unclear; however, some reports suggest that NAK may be inhibited by an unknown, labile, and ROS-sensitive cytosolic compound (Fuller et al. 2003, 2004).

Two main pathways have been suggested to underlie the increased influx of Na^+ into the cardiomyocytes of the ischemic myocardium, the persistent (non-inactivating) Na^+ channels ($I_{Na,P}$) [also called the late Na^+ current, $I_{Na,L}$] and the sarcolemmal NHE1 (Williams et al. 2007; Murphy et al. 1999; Eigel and Hadley 1999). To investigate the relative contribution of each pathway, Eigel and Hadley (1999) exposed isolated guinea pig cardiomyocytes to different ischemic-related conditions (e.g., anoxia, acidosis, etc.). They reported that the rise of $[Na^+]_i$ during prolonged anoxia was almost completely blocked by HOE-642 and TTX, selective inhibitors of NHE1 and I_{Na}, respectively. However, their findings suggested that the relative contribution of each pathway is exquisitely sensitive to the conditions that prevail during ischemia. For example, I_{Na} was the predominant source of $[Na^+]_i$ gain when the prolonged anoxia was combined with acidosis (i.e., pH_o of 6.85) and NHE1 was the dominant source during simulated ischemia (e.g., hyperkalemia, acidosis, and anoxia). These findings further highlight the pathophysiological role of NHE1 during ischemia and IR injury (see below). In fact, NHE1 inhibitors have been shown to profoundly ameliorate the ischemic rise of $[Na^+]_i$ and reduce the injury to the myocardium (Murphy et al. 1991; Pike et al. 1993; Choy et al. 1997; Dizon et al. 1998; Hartmann and Decking 1999). Furthermore, it has been reported that, when compared to the WT mice, mice with a null mutation in NHE1 ($NHE1^{-/-}$) are less sensitive to IR injury (Wang et al. 2003). The post-ischemic hemodynamic performances of the $NHE1^{-/-}$ mice hearts are better preserved and [ATP] declines slower during the ischemia. It should also be noted that less-specific inhibitors of NHE1 (i.e., those suspected to also inhibit $I_{Na,P}$) have the most robust effect on $[Na^+]_i$ and are more potent in ablating the ischemic rise of $[Na^+]_i$ (Murphy and Allen 2009; Allen and Xiao 2003) than NHE1-specific inhibitors alone. Consistent with other studies (Williams et al. 2007; Butwell et al. 1993; Haigney et al. 1994), these findings suggest that Na^+ influx via $I_{Na,P}$ may increase during ischemia, further contributing to the ischemic rise in $[Na]_i$. It seems likely that $I_{Na,P}$ would contribute primarily to $[Na^+]_i$ overload early in the ischemic phase, before increased $[K^+]_o$ depolarizes the sarcolemmal membrane potential thereby reducing the electrochemical gradient for $I_{Na,P}$. Regardless of timing, it appears clear that NHE1 is the dominant $[Na^+]_i$ overload mechanism but $I_{Na,P}$ also contributes.

2.3.4 *Dysregulation of $[Ca^{2+}]_i$*

Within just a few minutes after halting the flow through the coronary arteries, the severely ischemic myocardium ceases to contract. This arrest is thought to be caused by complete metabolic failure. It has been reported that under similar extreme conditions acute suppression of the $[Ca^{2+}]_i$ transient magnitude can develop within a few minutes. This is seen following severe hypoxia (15–20 mmHg) when the heart is deprived of O_2 (Zhang et al. 2013; Kihara et al. 1989) or when isolated heart muscle cells are made anoxic (Stern et al. 1988), and also by complete metabolic inhibition (e.g., 2 mM cyanide and 10 mM 2-deoxyglucose) (Allen and Orchard 1983). Measurements of $[Ca^{2+}]_i$ from single cardiomyocytes under anoxic conditions have shown that during this early contractile failure the $[Ca^{2+}]_i$ transients that activate the contractions under normal conditions are nearly abolished. In addition, during this prolonged diastole, basal $[Ca^{2+}]_i$ does not rise (within 10 min) and remains the same as normal diastole. Since the SR Ca^{2+} remains unchanged during anoxia, the disappearance of the $[Ca^{2+}]_i$ transients has been attributed to a loss of electrical excitability (Stern et al. 1988). Later studies investigated the mechanism of this early contractile failure and found that the dissipated cellular ATP levels lead to a dramatic increase in the membrane conductance to K^+. This large outward K^+ current appeared to be caused by the opening of ATP-sensitive K^+ channels ($I_{K(ATP)}$). The massive $I_{K(ATP)}$ essentially voltage-clamps the sarcolemmal potential and keeps it close to the reversal potential of K^+ ($-70 < E_k < -80$ mV) (Lederer et al. 1989). Under these conditions, the cardiac action potential (AP) cannot be stimulated and voltage-sensitive LCC are not triggered resulting in insufficient Ca^{2+} influx to evoke a synchronized SR Ca^{2+} release (i.e., a $[Ca^{2+}]_i$ transient).

Not all cells of the myocardium experience such extreme and complete metabolic failure early at the onset of ischemia. It is likely that, as the cellular levels of ATP decline during the progression of ischemia, a more moderate increase in the P_O of the IK(ATP) occurs. A broad agreement exists, across numerous investigations and differing techniques, that a progressive rise of $[Ca^{2+}]_i$ occurs in most ischemic cells. However, it should be noted that most of these studies have carried out measurements in which the Ca^{2+} signal is averaged over large time scales (e.g., via NMR, minutes); this signal is also a global average reading from a whole heart in which heterogeneous ischemic conditions and effects may commingle and better oxygenated epicardial cells may still experience $[Ca^{2+}]_i$ transients. Yet, despite the limited spatiotemporal resolution, and some disparity in the reported results, a profound rise in $[Ca^{2+}]_i$ is clear (Ladilov et al. 1999; Allen et al. 1989; Allshire et al. 1987; Lee et al. 1987, 1988; Snowdowne et al. 1985; Murphy et al. 1985; Barry et al. 1987).

Additional studies have quantitatively estimated the extent of such an increase in the time average $[Ca^{2+}]_i$. Quantitative NMR studies have estimated that the $[Ca^{2+}]_i$ signal (averaged over 5 min) rises three to sixfold from a pre-ischemic (control) values of around 250–600 nM to as high as 3.0 μM after 15–20 min of global ischemia (Marbán et al. 1987; Steenbergen et al. 1987, 1990). Other quantitative studies used the Ca^{2+}-sensitive photoprotein Aequorin and provided some temporal improvement. When the Aequorin luminescence is measured from the surface of the left ventricle, it was reported that a brief ischemia of 5 min leads to an increase of the diastolic $[Ca^{2+}]_i$ from 330 nM to nearly 1 μM (Kihara et al. 1989). When Aequorin measurements are performed on single isolated ventricular cardiomyocytes, it was reported that $[Ca^{2+}]_i$ remains stable for the first 14 min of metabolic inhibition (i.e., 250 nM). Over the next 20 min, however, it gradually rises to approximately 3 μM (Cobbold and Bourne 1984).

2.4 Mitochondria and IR Injury

2.4.1 The Role of the Mitochondria in Irreversible IR Injury

A momentary lapse in coronary blood flow is tolerated if the myocardium is reperfused within about 10 min. Studies that investigate the mechanisms of IR injury in various mammalian models generally report that the effects of such short-lived ischemia are reversible. While some transient effects do occur (i.e., the heart experiences a "stunning" effect, see Bolli and Marbán (1999)), eventually the hemodynamic performances of the heart return to normal and necrotic cell death is averted (Bolli and Marbán 1999; Appleyard and Cohn 1993; Kloner and Jennings 2001a, b). However, if the ischemic period is extended (to over 20 min), reperfusion results in irreversible myocardial injury due to vast cellular death. Electron microscopy images from the necrotic region reveal tears in the contact sites between adjacent cardiomyocytes, hyper-contracted sarcomeres, sarcolemmal ruptures, and swollen mitochondria (Jennings et al. 1978a, b; Jennings and Reimer 1991). Because the ultrastructural deformities and infarct zone volume correlated with the severity of ATP depletion, ATP deficiency was suggested as the primarily cause of this irreversible injury (Jennings et al. 1978a; Jennings and Reimer 1981). Insufficient $[ATP]_i$ is detrimental not only because it is required for maintaining electrochemical gradients, but also because when severe ($[ATP]_i$ below 100 μM) hypercontracture develops (Nichols and Lederer 1990b). If the myocardium is not reperfused within 1 hr, rigor envelopes the entire heart muscle causing the cardiomyocytes to undergo detrimental sarcolemmal rupture (Jennings et al. 1990) (Fig. 2.3).

Reperfusion is critical to salvage the ischemic myocardium; however, the process triggers necrosis of cardiomyocytes (i.e., IR injury). The cellular events that underlie this necrosis are thought to be triggered by the abrupt changes in the myocardial pH that immediately follow reperfusion. This occurs because as soon as capillary flow resumes, the catabolites that have accumulated are cleared and pH_o rises abruptly (i.e., becomes less acidified). The sarcolemmal proton gradient reverses, favoring the efflux of protons from the cell. This H^+ efflux out of the cytosol is primarily mediated by NHE1, which extrudes H^+ in exchange for Na^+ entry, therefore exacerbating the existing $[Na^+]_i$ overload. The sarcolemmal NAK is incapable of completely extruding this accumulated Na^+ and the high $[Na^+]_i$ increases the thermodynamic force driving Ca^{2+} influx via NCX causing further accumulation of cellular Ca^{2+}. This cellular Ca^{2+} overload, while well-tolerated during ischemia, is lethal during reperfusion (Miyata et al. 1992). In fact, when Ca^{2+} overload is prevented, the IR injury is profoundly reduced. This is achieved by pharmacological blockade of NHE1 (Murphy et al. 1991; Hartmann and Decking 1999; ten Hove et al. 2007; Park et al. 1999) and is also seen in a genetic murine

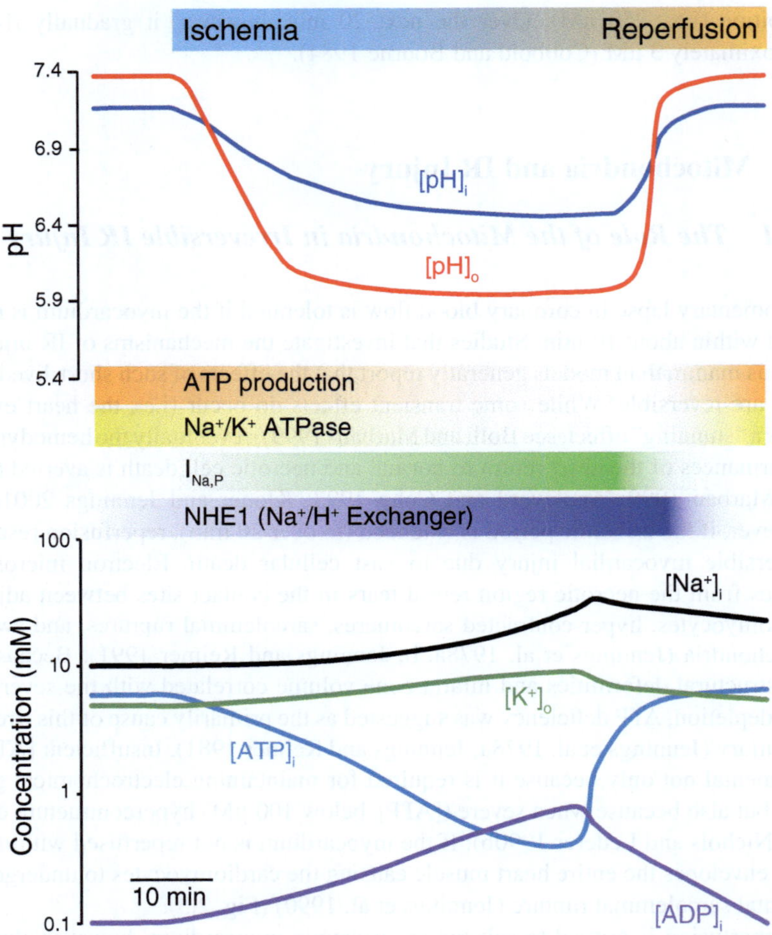

Fig. 2.3 Temporal plot of changes during IR. Plots are generated by interpolating experimental data derived from multiple sources and intended for illustrative purposes only. Additional details about the experimental measurements can be found in the Sects. 2.1–2.3

model in which NHE1 is knocked out (Wang et al. 2003). Additionally, the mouse model in which the cardiac NCX is knocked-out is also less vulnerable to IR injury and displays better hemodynamic recovery (Imahashi et al. 2005).

The sequence of intracellular events that are initiated by the high $[Ca^{2+}]_i$ and ultimately lead to necrosis of cardiomyocytes during the reperfusion phase has been at the center of many investigations. A growing body of evidence suggests, along with other deleterious effects, the high $[Ca^{2+}]_i$ leads to high $[Ca^{2+}]_m$ and mitochondrial collapse. This collapse is primarily attributed to $[Ca^{2+}]_m$-dependent formation of mPTP, which, once open, causes irreversible cell damage and cellular necrosis (Halestrap and Richardson 2014). Treatment of the heart with cyclosporin A (CsA), an immunosuppressive drug that desensitizes mPTP to Ca^{2+}, reduces the IR injury (Pan et al. 2013; Nazareth et al. 1991; Griffiths and Halestrap 1993; Hausenloy et al. 2003; Argaud et al. 2005; Gomez et al. 2007). Similarly, the murine model that lacks cyclophilin D, the mitochondrial protein target of CsA, also displays reduced IR injury (Baines et al. 2005; Nakagawa et al. 2005). The molecular aspects of mPTP (e.g., its identity, structure, protein subunits, regulation, kinetics, etc.) remain topics of great contention and are discussed more comprehensively in Sect. 2.5.1. Nevertheless, the evidence does suggest that ischemia predisposes the mitochondria to mPTP formation during reperfusion (Di Lisa et al. 2001; Javadov et al. 2003; Griffiths and Halestrap 1995). Once open, mPTP completely degrades $\Delta\Psi_m$, causing ATP production to cease. Why mPTP remains closed during ischemia and opens during reperfusion appears paradoxical. This is because the low ATP, high $[P_i]$, and high $[Ca^{2+}]_i$ conditions present during ischemia have been shown to induce mPTP opening. However, during ischemia, pH_i remains low and low pH desensitizes mPTP to Ca^{2+} (Hunter and Haworth 1979; Halestrap 1991; Bernardi et al. 1992). Following reperfusion, however, pH_i rapidly rebounds thereby relieving the "proton block" from mPTP. Additionally, reactive oxygen species (ROS) are also known to increase the sensitivity of mPTP to Ca^{2+} and the lack of O_2 during the ischemia prevents ROS production by the ETC. During reperfusion, O_2 returns causing ROS production by the ETC to increase. Note that genetically increasing the ability of the heart to scavenge ROS (see Table 2.1) has been shown to substantially reduce IR injury.

It should, however, be mentioned that other cellular mechanisms can also underlie the necrosis triggered by the reperfusion (see Fig. 2.4). As discussed above, NCX is a key player in the IR injury and is also blocked by cytosolic acidification (Boyman et al. 2011; Blaustein and Lederer 1999), but its turnover rate will rapidly rise back as the pH_i rebounds (becoming less acidified) early

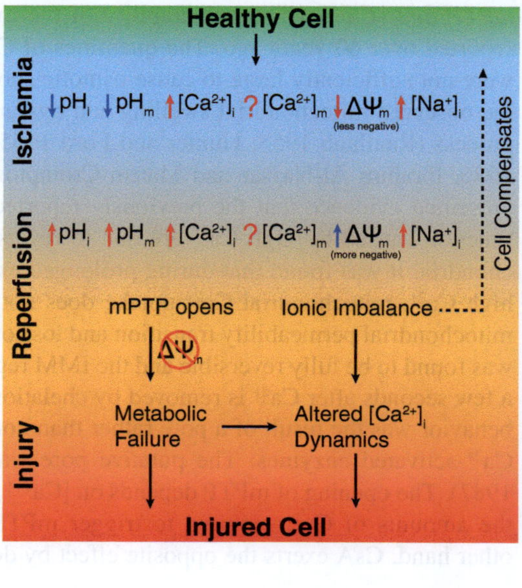

Fig. 2.4 Changes leading from a healthy cell to an IR-injured cell

after the onset of the myocardial reperfusion. Since NCX is an electrogenic exchanger, by extruding the accumulated high Ca^{2+} from the cytosol it mediates a depolarizing inward current (Blaustein and Lederer 1999). This current destabilizes the resting membrane potential and can lead to altered $[Ca^{2+}]_i$ dynamics. During ischemia, cytosolic acidification suppresses contractility because protons compete with $[Ca^{2+}]_i$ for binding to troponin c. Therefore, once pH_i rises during reperfusion, elevated $[Ca^{2+}]_i$ is more likely to cause hypercontracture and sarcolemmal rupture.

2.4.2 The Mitochondrial Permeability Transition Pore (mPTP) and Its Sensitivity to $[Ca^{2+}]_m$

Molecular Identity of mPTP. Recent publications (Alavian et al. 2014b; Giorgio et al. 2013; Bonora et al. 2013) propose a detailed molecular description to a mitochondrial process that was originally described over 60 years ago. Early on, mPTP was proposed to consist of VDAC on the OMM and ANT on the IMM (Baines 2009). In 2004, Kokoszka et al. undermined this theory by showing that mitochondria from mice lacking ANT still displayed mPTP openings (Kokoszka et al. 2004). Others have suggested that the F_1F_0 ATP synthase, or more specifically the c-subunit of F_1F_0 ATP synthase, is the pore forming unit of mPTP(Alavian et al. 2014b; Giorgio et al. 2013; Bonora et al. 2013; Carraro et al. 2014; Chinopoulos and Adam-Vizi 2012). However, many questions remain (see Karch and Molkentin (2014), Halestrap (2014)). The most compelling evidence would come from genetic testing of F_1F_0 ATP synthase as mPTP, but this would be challenging as cardiomyocytes rely heavily on F_1F_0 ATP synthase to create ATP for normal heart function.

mPTP and $[Ca^{2+}]_m$. The vulnerability of the mitochondria to high Ca^{2+} was first reported over 60 years ago. The quantities of Ca^{2+} that were used in these studies were not sufficiently large to cause osmotic shock. It was therefore suggested that the observed mitochondrial swelling and dysfunction were due to a Ca^{2+}-sensitive process (Raaflaub 1953; Hunter and Ford 1955; Nicholls and Akerman 1982). In 1986, Ibrahim Al-Nassar and Martin Crompton (Al-Nasser and Crompton 1986) presented evidence that the previously reported mitochondrial swelling and dysfunction caused by high Ca^{2+} are due to reversible permeabilization of the mitochondria. It was found that during prolonged exposure of isolated mitochondria to high Ca^{2+}, mitochondrial Ca^{2+} uptake does not cease, until the $[Ca^{2+}]_m$ triggers a mitochondrial permeability transition and loss of $\Delta\Psi_m$. This permeability transition was found to be fully reversible and the IMM regains its normal conductance within a few seconds after Ca^{2+} is removed by chelation. This suggested that the observed behavior was the result of a pore rather than non-specific digestion of the IMM by Ca^{2+}-activated enzymes. The putative pore was termed mPTP (Crompton et al. 1987). The opening of mPTP depends on $[Ca^{2+}]_m$ with high $[P_i]$ and [ROS] reducing the amounts of Ca^{2+} required to trigger mPTP (Crompton et al. 1987). On the other hand, CsA exerts the opposite effect by desensitizing mPTP opening to Ca^{2+}

(Crompton et al. 1988). Further existence of a large conductance pore(s) in the IMM comes from electrophysiological investigations. These investigations patch-clamped mitoplasts, which are vesicles whose membranes originate from the IMM (Kinnally et al. 1989; Petronilli et al. 1989). Additional investigations using this approach revealed that exposure of the matrix facing side of the IMM to 0.3 mM [Ca^{2+}] leads to opening of a large conductance pore (1–3 nS). Note this conductance exceeds the theoretical upper boundary of selective ion channels, suggesting that the IMM contains a non-specific pore, likely mPTP (Bernardi et al. 1992; Szabo and Zoratti 1991, 1992).

2.5 Ischemic Preconditioning and the Mitochondria

2.5.1 Ischemic Preconditioning and mPTP

The effects of ischemic preconditioning (IP) are pronounced; preconditioned hearts are less susceptible to the reperfusion injury, the extent of necrosis is lessened by approximately 75 %, and the hemodynamic performances recover more easily. Since the discovery of IP protection, the beneficial effects of IP were attributed to changes in energy metabolism. This was based on investigations that showed the size of the myocardial injury correlates with the ATP content of the heart and the preconditioned hearts display slower ATP decline during ischemia. The cellular changes that are triggered by IP and ultimately make the cardiomyocyte more resilient to IR injury remain controversial. We discuss some of these unexplored aspects of IP in Sect. 2.5.2. In short, there is a general agreement that [Ca^{2+}]$_i$-induced mPTP transitions are less likely in hearts subjected to IP than in control hearts (Halestrap 2009; Murphy and Steenbergen 2011). Preventing or reducing the extent of mitochondrial collapse, via mPTP, should increase the likelihood of a cell-surviving IR. For example, agents that desensitize mPTP to [Ca^{2+}]$_i$, such as CsA or sanglifehrin A, significantly reduce IR injury (Pan et al. 2013; Nazareth et al. 1991; Griffiths and Halestrap 1993; Hausenloy et al. 2003; Argaud et al. 2005; Javadov et al. 2003).

Early evidence that IP alters mPTP in-situ arises from the study by Halestrap and colleges in 2003 (Javadov et al. 2003) where isolated mitochondria were taken from hearts at different time points during and after IR. By measuring the uptake of radiolabeled 2-deoxy-D-glucose ([^{3}H]DOG), which is unable to cross the IMM, Halestrap and co-workers were able to indirectly measure mPTP activity. These results showed that IP reduces the opening of mPTP early after the onset of reperfusion and also at later times (Javadov et al. 2003). Other investigations suggest that these effects of IP on mPTP are by altering the inherent properties of mPTP, creating a form of sustained inhibition of mPTP. Supporting this are experiments conducted on mitochondria isolated from the hearts subjected to an IR model. These mitochondria require higher [Ca^{2+}]$_i$ to trigger mPTP than control hearts (Argaud et al. 2004; Khaliulin et al. 2004; Clarke et al. 2008). Furthermore, single cell studies have suggested that simulated hypoxic-preconditioning reduces the sensitivity of mPTP to ROS. In fact, these effects were just as robust as those observed after treating the cardiomyocytes with CsA (Hausenloy et al. 2004). However, other studies have undermined these conclusions, reporting that IP does not change the $[Ca^{2+}]_i$ sensitivity of mPTP (Javadov et al. 2003). These later studies suggest that it is more likely that IP acts to attenuate the factors that promote mPTP opening (e.g., ROS and high $[Ca^{2+}]_m$).

2.5.2 Possible Mechanism(s) Behind IP Protection

To date, it is still not clear how IP affects $[Ca^{2+}]_i$ dynamics during ischemia or during the reperfusion that follows. These critical questions remain open primarily due to the lack of quantitative reports where the measurements of Ca^{2+} are of high enough spatiotemporal resolution to permit measurement of $[Ca^{2+}]_i$ signaling events such as Ca^{2+} sparks and $[Ca^{2+}]_i$ transients. The available information about $[Ca^{2+}]_i$ and IP is derived from NMR measurements or fluorescence measurements of $[Ca^{2+}]_i$ from whole hearts. These studies report that IP alters the time-averaged Ca^{2+} signal during IR by attenuating $[Ca^{2+}]_i$ rise during the ischemia and reperfusion (Steenbergen et al. 1993; Ylitalo et al. 2000). It has been suggested that $[Ca^{2+}]_i$ does not rise as much in preconditioned hearts because the ischemic pH_i that follows IP is less acidified (Steenbergen et al. 1993). As discussed in Sect. 2.5.1, this acidification and the extent of reperfusion-induced pH_i rebound are tightly linked to $[Ca^{2+}]_i$ dysregulation (Murphy et al. 1991; Steenbergen et al. 1993; Kitakaze et al. 1997). Additionally, there is some evidence that pharmacologically attenuating pH_i swings reduce the opening of mPTP (Toda et al. 2007; Javadov et al. 2008).

Lingering in the background are more straightforward questions regarding $[Ca^{2+}]_m$ dynamics during IR and IP. Measurements of $[Ca^{2+}]_m$ during ischemia have yet to be reported, likely due to technical difficulties in carrying out such measurements. Even so, it is widely asserted that $[Ca^{2+}]_i$ accumulation during ischemia leads to elevated $[Ca^{2+}]_m$ (Joiner et al. 2012; Halestrap 2006; García-Rivas Gde et al. 2006). However, steady-state $[Ca^{2+}]_m$ depends not on $[Ca^{2+}]_i$ alone, but on the balance of Ca^{2+} fluxes across the IMM. Many of the conditions associated with ischemia (e.g., degraded $\Delta\Psi_m$, elevated $[Na^+]_i$, etc.) might in fact shift the balance of IMM fluxes toward Ca^{2+} extrusion via NCLX.

Taken together, this evidence led to a current, provocative proposal which suggests that the benefits of preconditioning depend directly on reductions of $[Ca^{2+}]_m$ and $[ROS]_m$ or on the sensitivity of the mPTP complex to $[Ca^{2+}]_m$, and $[ROS]_m$, Nevertheless, quantitative investigations of $[Ca^{2+}]_m$, $[ROS]_m$, and the molecular character of mPTP are needed for a compelling mechanistic investigation of preconditioning, its benefits, and therapeutic investigations of IP-styled small molecule treatments. With the new tools that we (Boyman et al. 2014) and others (Lu et al. 2013) have developed, such quantitative investigations are now possible.

2.6 Summary

Ischemia-reperfusion injury in heart and the benefits of ischemic preconditioning have raised important questions that appear to center on the role of $[Ca^{2+}]_i$ and $[Ca^{2+}]_m$ in heart cell function. However, diverse critical features of mitochondrial function still need to be studied quantitatively and mechanistically. How $[Ca^{2+}]_m$ is related to mitochondrial energy (e.g., ATP) production, ROS balance, cellular reducing capacity, and cellular contractile behavior are exciting areas of active research that will inform such investigations. These areas of active research should lay the foundation for a broad new understating of heart cell and mitochondrial behavior at the molecular level and provide diverse therapeutic opportunities.

References

Alavian KN, Dworetzky SI, Bonanni L, Zhang P, Sacchetti S, Li H, Signore AP, Smith PJS, Gribkoff VK, Jonas EA (2014a) The mitochondrial complex V-associated large-conductance inner membrane current is regulated by cyclosporine and dexpramipexole. Mol Pharmacol. doi:10.1124/mol.114.095661

Alavian KN, Beutner G, Lazrove E, Sacchetti S, Park H-A, Licznerski P, Li H, Nabili P, Hockensmith K, Graham M, Porter GA, Jonas EA (2014b) An uncoupling channel within the c-subunit ring of the F1FO ATP synthase is the mitochondrial permeability transition pore. Proc Natl Acad Sci U S A 111:10580–10585. doi:10.1073/pnas.1401591111

Allen DG, Orchard CH (1983) Intracellular calcium concentration during hypoxia and metabolic inhibition in mammalian ventricular muscle. J Physiol Lond 339:107–122

Allen DG, Xiao X-H (2003) Role of the cardiac Na+/H+ exchanger during ischemia and reperfusion. Cardiovasc Res 57:934–941

Allen DG, Morris PG, Orchard CH, Pirolo JS (1985) A nuclear magnetic resonance study of metabolism in the ferret heart during hypoxia and inhibition of glycolysis. J Physiol Lond 361:185–204

Allen DG, Lee JA, Smith GL (1989) The consequences of simulated ischaemia on intracellular Ca2+ and tension in isolated ferret ventricular muscle. J Physiol Lond 410:297–323

Allshire A, Piper HM, Cuthbertson KS, Cobbold PH (1987) Cytosolic free Ca2+ in single rat heart cells during anoxia and reoxygenation. Biochem J 244:381–385

Al-Nasser I, Crompton M (1986) The reversible Ca2+-induced permeabilization of rat liver mitochondria. Biochem J 239:19–29

Anderson SE, Dickinson CZ, Liu H, Cala PM (1996) Effects of Na-K-2Cl cotransport inhibition on myocardial Na and Ca during ischemia and reperfusion. Am J Physiol 270:C608–C618

Andrews ZB, Diano S, Horvath TL (2005) Mitochondrial uncoupling proteins in the CNS: in support of function and survival. Nat Rev Neurosci 6:829–840. doi:10.1038/nrn1767

Appleyard RF, Cohn LH (1993) Myocardial stunning and reperfusion injury in cardiac surgery. J Card Surg 8:316–324

Argaud L, Gateau-Roesch O, Chalabreysse L, Gomez L, Loufouat J, Thivolet-Béjui F, Robert D, Ovize M (2004) Preconditioning delays Ca2+-induced mitochondrial permeability transition. Cardiovasc Res 61:115–122

Argaud L, Gateau-Roesch O, Muntean D, Chalabreysse L, Loufouat J, Robert D, Ovize M (2005) Specific inhibition of the mitochondrial permeability transition prevents lethal reperfusion injury. J Mol Cell Cardiol 38:367–374. doi:10.1016/j.yjmcc.2004.12.001

Baines CP (2009) The molecular composition of the mitochondrial permeability transition pore. J Mol Cell Cardiol 46:850–857. doi:10.1016/j.yjmcc.2009.02.007

Baines CP (2011) How and when do myocytes die during ischemia and reperfusion: the late phase. J Cardiovasc Pharmacol Ther 16:239–243. doi:10.1177/1074248411407769

Baines CP, Kaiser RA, Purcell NH, Blair NS, Osinska H, Hambleton MA, Brunskill EW, Sayen MR, Gottlieb RA, Dorn GW, Robbins J, Molkentin JD (2005) Loss of cyclophilin D reveals a critical role for mitochondrial permeability transition in cell death. Nature 434:658–662. doi:10.1038/nature03434

Barry WH, Peeters GA, Rasmussen CA, Cunningham MJ (1987) Role of changes in [Ca2+]i in energy deprivation contracture. Circ Res 61:726–734

Bassani JW, Bassani RA, Bers DM (1994) Relaxation in rabbit and rat cardiac cells: species-dependent differences in cellular mechanisms. J Physiol Lond 476:279–293

Baxter GF, Ferdinandy P (2001) Delayed preconditioning of myocardium: current perspectives. Basic Res Cardiol 96:329–344

Beatrice MC, Palmer JW, Pfeiffer DR (1980) The relationship between mitochondrial membrane permeability, membrane potential, and the retention of Ca2+ by mitochondria. J Biol Chem 255:8663–8671

Bernardi P, von Stockum S (2012) The permeability transition pore as a Ca(2+) release channel: new answers to an old question. Cell Calcium 52:22–27. doi:10.1016/j.ceca.2012.03.004

Bernardi P, Vassanelli S, Veronese P, Colonna R, Szabo I, Zoratti M (1992) Modulation of the mitochondrial permeability transition pore. Effect of protons and divalent cations. J Biol Chem 267:2934–2939

Blaustein MP, Lederer WJ (1999) Sodium/calcium exchange: its physiological implications. Physiol Rev 79:763–854

Bolli R, Marbán E (1999) Molecular and cellular mechanisms of myocardial stunning. Physiol Rev 79:609–634

Bonora M, Bononi A, De Marchi E, Giorgi C, Lebiedzinska M, Marchi S, Patergnani S, Rimessi A, Suski JM, Wojtala A, Wieckowski MR, Kroemer G, Galluzzi L, Pinton P (2013) Role of the c subunit of the FO ATP synthase in mitochondrial permeability transition. Cell Cycle 12:674–683. doi:10.4161/cc.23599

Boyman L, Hagen BM, Giladi M, Hiller R, Lederer WJ, Khananshvili D (2011) Proton-sensing Ca2+ binding domains regulate the cardiac Na+/Ca2+ exchanger. J Biol Chem 286:28811–28820. doi:10.1074/jbc.M110.214106

Boyman L, Williams GSB, Khananshvili D, Sekler I, Lederer WJ (2013) NCLX: the mitochondrial sodium calcium exchanger. J Mol Cell Cardiol 59:205–213. doi:10.1016/j.yjmcc.2013.03.012

Boyman L, Chikando AC, Williams GSB, Khairallah RJ, Kettlewell S, Ward CW, Smith GL, Kao JPY, Lederer WJ (2014) Calcium movement in cardiac mitochondria. Biophys J 107:1289–1301. doi:10.1016/j.bpj.2014.07.045

Braasch W, Gudbjarnason S, Puri PS, Ravens KG, Bing RJ (1968) Early changes in energy metabolism in the myocardium following acute coronary artery occlusion in anesthetized dogs. Circ Res 23:429–438

Brandes R, Maier LS, Bers DM (1998) Regulation of mitochondrial [NADH] by cytosolic [Ca2+] and work in trabeculae from hypertrophic and normal rat hearts. Circ Res 82:1189–1198

Butwell NB, Ramasamy R, Lazar I, Sherry AD, Malloy CR (1993) Effect of lidocaine on contracture, intracellular sodium, and pH in ischemic rat hearts. Am J Physiol 264:H1884–H1889

Carafoli E, Lehninger AL (1971) A survey of the interaction of calcium ions with mitochondria from different tissues and species. Biochem J 122:681–690

Carraro M, Giorgio V, Sileikyte J, Sartori G, Forte M, Lippe G, Zoratti M, Szabò I, Bernardi P (2014) Channel formation by yeast F-ATP synthase and the role of dimerization in the mitochondrial permeability transition. J Biol Chem. doi:10.1074/jbc.C114.559633

Chen Y-R, Zweier JL (2014) Cardiac mitochondria and reactive oxygen species generation. Circ Res 114:524–537. doi:10.1161/CIRCRESAHA.114.300559

Chen EP, Bittner HB, Davis RD, Folz RJ, Van Trigt P (1996) Extracellular superoxide dismutase transgene overexpression preserves postischemic myocardial function in isolated murine hearts. Circulation 94:II412–II417

Chen Z, Siu B, Ho YS, Vincent R, Chua CC, Hamdy RC, Chua BH (1998) Overexpression of MnSOD protects against myocardial ischemia/reperfusion injury in transgenic mice. J Mol Cell Cardiol 30:2281–2289. doi:10.1006/jmcc.1998.0789

Chen Z, Oberley TD, Ho Y, Chua CC, Siu B, Hamdy RC, Epstein CJ, Chua BH (2000) Overexpression of CuZnSOD in coronary vascular cells attenuates myocardial ischemia/reperfusion injury. Free Radic Biol Med 29:589–596

Chinopoulos C, Adam-Vizi V (2012) Modulation of the mitochondrial permeability transition by cyclophilin D: moving closer to F(0)-F(1) ATP synthase? Mitochondrion 12:41–45. doi:10.1016/j.mito.2011.04.007

Choy IO, Schepkin VD, Budinger TF, Obayashi DY, Young JN, DeCampli WM (1997) Effects of specific sodium/hydrogen exchange inhibitor during cardioplegic arrest. Ann Thorac Surg 64:94–99

Clarke SJ, Khaliulin I, Das M, Parker JE, Heesom KJ, Halestrap AP (2008) Inhibition of mitochondrial permeability transition pore opening by ischemic preconditioning is probably mediated by reduction of oxidative stress rather than mitochondrial protein phosphorylation. Circ Res 102:1082–1090. doi:10.1161/CIRCRESAHA.107.167072

Cobbe SM, Poole-Wilson PA (1980) The time of onset and severity of acidosis in myocardial ischaemia. J Mol Cell Cardiol 12:745–760

Cobbold PH, Bourne PK (1984) Aequorin measurements of free calcium in single heart cells. Nature 312:444–446

Correll RN, Molkentin JD (2013) CaMKII does it again: even the mitochondria cannot escape its influence. Circ Res 112:1208–1211. doi:10.1161/CIRCRESAHA.113.301263

Cortassa S, Aon MA, Marbán E, Winslow RL, O'Rourke B (2003) An integrated model of cardiac mitochondrial energy metabolism and calcium dynamics. Biophys J 84:2734–2755. doi:10.1016/S0006-3495(03)75079-6

Couper GS, Weiss J, Hiltbrand B, Shine KI (1984) Extracellular pH and tension during ischemia in the isolated rabbit ventricle. Am J Physiol 247:H916–H927

Crake T, Crean PA, Shapiro LM, Rickards AF, Poole-Wilson PA (1987) Coronary sinus pH during percutaneous transluminal coronary angioplasty: early development of acidosis during myocardial ischaemia in man. Br Heart J 58:110–115

Crompton M, Costi A, Hayat L (1987) Evidence for the presence of a reversible Ca2+-dependent pore activated by oxidative stress in heart mitochondria. Biochem J 245:915–918

Crompton M, Ellinger H, Costi A (1988) Inhibition by cyclosporin A of a Ca2+-dependent pore in heart mitochondria activated by inorganic phosphate and oxidative stress. Biochem J 255:357–360

Cross HR, RADDA GK, Clarke K (1995) The role of Na+/K+ ATPase activity during low flow ischemia in preventing myocardial injury: a 31P, 23Na and 87Rb NMR spectroscopic study. Magn Reson Med 34:673–685

Das DK, Maulik N (2006) Cardiac genomic response following preconditioning stimulus. Cardiovasc Res 70:254–263. doi:10.1016/j.cardiores.2006.02.023

Di Lisa F, Menabò R, Canton M, Barile M, Bernardi P (2001) Opening of the mitochondrial permeability transition pore causes depletion of mitochondrial and cytosolic NAD+ and is a causative event in the death of myocytes in postischemic reperfusion of the heart. J Biol Chem 276:2571–2575. doi:10.1074/jbc.M006825200

Dizon J, Burkhoff D, Tauskela J, Whang J, Cannon P, Katz J (1998) Metabolic inhibition in the perfused rat heart: evidence for glycolytic requirement for normal sodium homeostasis. Am J Physiol 274:H1082–H1089

Drago I, De Stefani D, Rizzuto R, Pozzan T (2012) Mitochondrial Ca2+ uptake contributes to buffering cytoplasmic Ca2+ peaks in cardiomyocytes. Proc Natl Acad Sci U S A 109:12986–12991. doi:10.1073/pnas.1210718109

Dunn RB, Griggs DM (1975) Transmural gradients in ventricular tissue metabolites produced by stopping coronary blood flow in the dog. Circ Res 37:438–445

Edwards RJ, Saurin AT, Rakhit RD, Marber MS (2000) Therapeutic potential of ischaemic preconditioning. Br J Clin Pharmacol 50:87–97

Eigel BN, Hadley RW (1999) Contribution of the Na(+) channel and Na(+)/H(+) exchanger to the anoxic rise of [Na(+)] in ventricular myocytes. Am J Physiol 277:H1817–H1822

Elrod JW, Molkentin JD (2013) Physiologic functions of cyclophilin D and the mitochondrial permeability transition pore. Circ J 77:1111–1122

Fieni F, Lee SB, Jan YN, Kirichok Y (2012) Activity of the mitochondrial calcium uniporter varies greatly between tissues. Nat Commun 3:1317. doi:10.1038/ncomms2325

Fieni F, Johnson DE, Hudmon A, Kirichok Y (2014) Mitochondrial Ca2+ uniporter and CaMKII in heart. Nature 513:E1–E2. doi:10.1038/nature13626

Franzini-Armstrong C (2007) ER-mitochondria communication. How privileged? Physiology (Bethesda) 22:261–268. doi:10.1152/physiol.00017.2007

Fuller W, Parmar V, Eaton P, Bell JR, Shattock MJ (2003) Cardiac ischemia causes inhibition of the Na/K ATPase by a labile cytosolic compound whose production is linked to oxidant stress. Cardiovasc Res 57:1044–1051

Fuller W, Eaton P, Bell JR, Shattock MJ (2004) Ischemia-induced phosphorylation of phospholemman directly activates rat cardiac Na/K-ATPase. FASEB J 18:197–199. doi:10.1096/fj.03-0213fje

Gabel SA, Cross HR, London RE, Steenbergen C, Murphy E (1997) Decreased intracellular pH is not due to increased H+ extrusion in preconditioned rat hearts. Am J Physiol 273:H2257–H2262

García-Rivas Gde J, Carvajal K, Correa F, Zazueta C (2006) Ru360, a specific mitochondrial calcium uptake inhibitor, improves cardiac post-ischaemic functional recovery in rats in vivo. Br J Pharmacol 149:829–837. doi:10.1038/sj.bjp.0706932

Garlick PB, RADDA GK, Seeley PJ (1979) Studies of acidosis in the ischaemic heart by phosphorus nuclear magnetic resonance. Biochem J 184:547–554

Gauthier LD, Greenstein JL, Winslow RL (2012) Toward an integrative computational model of the Guinea pig cardiac myocyte. Front Physiol 3:244. doi:10.3389/fphys.2012.00244

Gevers W (1977) Generation of protons by metabolic processes in heart cells. J Mol Cell Cardiol 9:867–874

Giorgio V, von Stockum S, Antoniel M, Fabbro A, Fogolari F, Forte M, Glick GD, Petronilli V, Zoratti M, Szabò I, Lippe G, Bernardi P (2013) Dimers of mitochondrial ATP synthase form the permeability transition pore. Proc Natl Acad Sci U S A 110:5887–5892. doi:10.1073/pnas.1217823110

Glancy B, Balaban RS (2012) Role of mitochondrial Ca2+ in the regulation of cellular energetics. Biochemistry 51:2959–2973. doi:10.1021/bi2018909

Gomez L, Thibault H, Gharib A, Dumont J-M, Vuagniaux G, Scalfaro P, Derumeaux G, Ovize M (2007) Inhibition of mitochondrial permeability transition improves functional recovery and reduces mortality following acute myocardial infarction in mice. Am J Physiol Heart Circ Physiol 293:H1654–H1661. doi:10.1152/ajpheart.01378.2006

Graier WF, Trenker M, Malli R (2008) Mitochondrial Ca2+, the secret behind the function of uncoupling proteins 2 and 3? Cell Calcium 44:36–50. doi:10.1016/j.ceca.2008.01.001

Griffiths EJ, Halestrap AP (1993) Protection by Cyclosporin A of ischemia/reperfusion-induced damage in isolated rat hearts. J Mol Cell Cardiol 25:1461–1469. doi:10.1006/jmcc.1993.1162

Griffiths EJ, Halestrap AP (1995) Mitochondrial non-specific pores remain closed during cardiac ischaemia, but open upon reperfusion. Biochem J 307(Pt 1):93–98

Gudbjarnason S, Mathes P, Ravens KG (1970) Functional compartmentation of ATP and creatine phosphate in heart muscle. J Mol Cell Cardiol 1:325–339

Haigney MC, Lakatta EG, Stern MD, Silverman HS (1994) Sodium channel blockade reduces hypoxic sodium loading and sodium-dependent calcium loading. Circulation 90:391–399

Halestrap AP (1991) Calcium-dependent opening of a non-specific pore in the mitochondrial inner membrane is inhibited at pH values below 7. Implications for the protective effect of low pH against chemical and hypoxic cell damage. Biochem J 278(Pt 3):715–719

Halestrap AP (2006) Calcium, mitochondria and reperfusion injury: a pore way to die. Biochem Soc Trans 34:232–237. doi:10.1042/BST20060232
Halestrap AP (2009) Mitochondria and reperfusion injury of the heart--a holey death but not beyond salvation. J Bioenerg Biomembr 41:113–121. doi:10.1007/s10863-009-9206-x
Halestrap AP (2014) The C ring of the F1Fo ATP synthase forms the mitochondrial permeability transition pore: a critical appraisal. Mol Cell Oncol. doi:10.3389/fonc.2014.00234
Halestrap AP, Richardson AP (2014) The mitochondrial permeability transition: a current perspective on its identity and role in ischaemia/reperfusion injury. J Mol Cell Cardiol. doi:10.1016/j.yjmcc.2014.08.018
Hartmann M, Decking UK (1999) Blocking Na(+)-H+ exchange by cariporide reduces Na(+)-overload in ischemia and is cardioprotective. J Mol Cell Cardiol 31:1985–1995. doi:10.1006/jmcc.1999.1029
Hausenloy DJ, Duchen MR, Yellon DM (2003) Inhibiting mitochondrial permeability transition pore opening at reperfusion protects against ischaemia-reperfusion injury. Cardiovasc Res 60:617–625
Hausenloy DJ, Yellon DM, Mani-Babu S, Duchen MR (2004) Preconditioning protects by inhibiting the mitochondrial permeability transition. Am J Physiol Heart Circ Physiol 287:H841–H849. doi:10.1152/ajpheart.00678.2003
Heusch G (2013) Cardioprotection: chances and challenges of its translation to the clinic. Lancet 381:166–175. doi:10.1016/S0140-6736(12)60916-7
Hirche H, Franz C, Bös L, Bissig R, Lang R, Schramm M (1980) Myocardial extracellular K+ and H+ increase and noradrenaline release as possible cause of early arrhythmias following acute coronary artery occlusion in pigs. J Mol Cell Cardiol 12:579–593
Hunter FE, Ford L (1955) Inactivation of oxidative and phosphorylative systems in mitochondria by preincubation with phosphate and other ions. J Biol Chem 216:357–369
Hunter DR, Haworth RA (1979) The Ca2+-induced membrane transition in mitochondria. I. The protective mechanisms. Arch Biochem Biophys 195:453–459
Imahashi K, Pott C, Goldhaber JI, Steenbergen C, Philipson KD, Murphy E (2005) Cardiac-specific ablation of the Na+-Ca2+ exchanger confers protection against ischemia/reperfusion injury. Circ Res 97:916–921. doi:10.1161/01.RES.0000187456.06162.cb
Isenberg G, Han S, Schiefer A, Wendt-Gallitelli MF (1993) Changes in mitochondrial calcium concentration during the cardiac contraction cycle. Cardiovasc Res 27:1800–1809
Jacobus WE, Pores IH, Lucas SK, Weisfeldt ML, Flaherty JT (1982) Intracellular acidosis and contractility in the normal and ischemic heart as examined by 31P NMR. J Mol Cell Cardiol 14(Suppl 3):13–20
Javadov SA, Clarke S, Das M, Griffiths EJ, Lim KHH, Halestrap AP (2003) Ischaemic preconditioning inhibits opening of mitochondrial permeability transition pores in the reperfused rat heart. J Physiol Lond 549:513–524. doi:10.1113/jphysiol.2003.034231
Javadov S, Choi A, Rajapurohitam V, Zeidan A, Basnakian AG, Karmazyn M (2008) NHE-1 inhibition-induced cardioprotection against ischaemia/reperfusion is associated with attenuation of the mitochondrial permeability transition. Cardiovasc Res 77:416–424. doi:10.1093/cvr/cvm039
Jennings RB, Reimer KA (1981) Lethal myocardial ischemic injury. Am J Pathol 102:241–255
Jennings RB, Reimer KA (1991) The cell biology of acute myocardial ischemia. Annu Rev Med 42:225–246. doi:10.1146/annurev.me.42.020191.001301
Jennings RB, Sommers HM, Herdson PB, Kaltenbach JP (1969) Ischemic injury of myocardium. Ann N Y Acad Sci 156:61–78
Jennings RB, Hawkins HK, Lowe JE, Hill ML, Klotman S, Reimer KA (1978a) Relation between high energy phosphate and lethal injury in myocardial ischemia in the dog. Am J Pathol 92:187–214
Jennings RB, Shen AC, Hill ML, Ganote CE, Herdson PB (1978b) Mitochondrial matrix densities in myocardial ischemia and autolysis. Exp Mol Pathol 29:55–65
Jennings RB, Murry CE, Steenbergen C, Reimer KA (1990) Development of cell injury in sustained acute ischemia. Circulation 82:II2–II12

Joiner M-LA, Koval OM, Li J, He BJ, Allamargot C, Gao Z, Luczak ED, Hall DD, Fink BD, Chen B, Yang J, Moore SA, Scholz TD, Strack S, Mohler PJ, Sivitz WI, Song L-S, Anderson ME (2012) CaMKII determines mitochondrial stress responses in heart. Nature 491:269–273. doi:10.1038/nature11444

Karch J, Molkentin JD (2014) Identifying the components of the elusive mitochondrial permeability transition pore. Proc Natl Acad Sci U S A 111:10396–10397. doi:10.1073/pnas.1410104111

Khaliulin I, Schwalb H, Wang P, Houminer E, Grinberg L, Katzeff H, Borman JB, Powell SR (2004) Preconditioning improves postischemic mitochondrial function and diminishes oxidation of mitochondrial proteins. Free Radic Biol Med 37:1–9. doi:10.1016/j.freeradbiomed.2004.04.017

Kihara Y, Grossman W, Morgan JP (1989) Direct measurement of changes in intracellular calcium transients during hypoxia, ischemia, and reperfusion of the intact mammalian heart. Circ Res 65:1029–1044

Kinnally KW, Campo ML, Tedeschi H (1989) Mitochondrial channel activity studied by patch-clamping mitoplasts. J Bioenerg Biomembr 21:497–506

Kirichok Y, Krapivinsky G, Clapham DE (2004) The mitochondrial calcium uniporter is a highly selective ion channel. Nature 427:360–364. doi:10.1038/nature02246

Kitakaze M, Takashima S, Funaya H, Minamino T, Node K, Shinozaki Y, Mori H, Hori M (1997) Temporary acidosis during reperfusion limits myocardial infarct size in dogs. Am J Physiol 272:H2071–H2078

Kloner RA, Jennings RB (2001a) Consequences of brief ischemia: stunning, preconditioning, and their clinical implications: part 1. Circulation 104:2981–2989

Kloner RA, Jennings RB (2001b) Consequences of brief ischemia: stunning, preconditioning, and their clinical implications: part 2. Circulation 104:3158–3167

Kloner RA, Rezkalla SH (2006) Preconditioning, postconditioning and their application to clinical cardiology. Cardiovasc Res 70:297–307. doi:10.1016/j.cardiores.2006.01.012

Kokoszka JE, Waymire KG, Levy SE, Sligh JE, Cai J, Jones DP, MacGregor GR, Wallace DC (2004) The ADP/ATP translocator is not essential for the mitochondrial permeability transition pore. Nature 427:461–465. doi:10.1038/nature02229

Kübler W, Spieckermann PG (1970) Regulation of glycolysis in the ischemic and the anoxic myocardium. J Mol Cell Cardiol 1:351–377

Ladilov YV, Balser-Schäfer C, Haffner S, Maxeiner H, Piper HM (1999) Pretreatment with PKC activator protects cardiomyocytes against reoxygenation-induced hypercontracture independently of Ca2+ overload. Cardiovasc Res 43:408–416

Lederer WJ, Nichols CG, Smith GL (1989) The mechanism of early contractile failure of isolated rat ventricular myocytes subjected to complete metabolic inhibition. J Physiol Lond 413:329–349

Lee HC, Smith N, Mohabir R, Clusin WT (1987) Cytosolic calcium transients from the beating mammalian heart. Proc Natl Acad Sci U S A 84:7793–7797

Lee HC, Mohabir R, Smith N, Franz MR, Clusin WT (1988) Effect of ischemia on calcium-dependent fluorescence transients in rabbit hearts containing indo 1. Correlation with monophasic action potentials and contraction. Circulation 78:1047–1059

Lombardi A, Grasso P, Moreno M, de Lange P, Silvestri E, Lanni A, Goglia F (2008) Interrelated influence of superoxides and free fatty acids over mitochondrial uncoupling in skeletal muscle. Biochim Biophys Acta 1777:826–833. doi:10.1016/j.bbabio.2008.04.019

Lu X, Ginsburg KS, Kettlewell S, Bossuyt J, Smith GL, Bers DM (2013) Measuring local gradients of intramitochondrial [Ca(2+)] in cardiac myocytes during sarcoplasmic reticulum Ca(2+) release. Circ Res 112:424–431. doi:10.1161/CIRCRESAHA.111.300501

Maack C, Cortassa S, Aon MA, Ganesan AN, Liu T, O'Rourke B (2006) Elevated cytosolic Na+ decreases mitochondrial Ca2+ uptake during excitation-contraction coupling and impairs energetic adaptation in cardiac myocytes. Circ Res 99:172–182. doi:10.1161/01.RES.0000232546.92777.05

Marbán E, Kitakaze M, Kusuoka H, Porterfield JK, Yue DT, Chacko VP (1987) Intracellular free calcium concentration measured with 19F NMR spectroscopy in intact ferret hearts. Proc Natl Acad Sci U S A 84:6005–6009

McCormack JG, Denton RM (1993a) The role of intramitochondrial Ca2+ in the regulation of oxidative phosphorylation in mammalian tissues. Biochem Soc Trans 21(Pt 3):793–799

McCormack JG, Denton RM (1993b) Mitochondrial Ca2+ transport and the role of intramitochondrial Ca2+ in the regulation of energy metabolism. Dev Neurosci 15:165–173

Michels G, Khan IF, Endres-Becker J, Rottlaender D, Herzig S, Ruhparwar A, Wahlers T, Hoppe UC (2009) Regulation of the human cardiac mitochondrial Ca2+ uptake by 2 different voltage-gated Ca2+ channels. Circulation 119:2435–2443. doi:10.1161/CIRCULATIONAHA.108.835389

Miyata H, Lakatta EG, Stern MD, Silverman HS (1992) Relation of mitochondrial and cytosolic free calcium to cardiac myocyte recovery after exposure to anoxia. Circ Res 71:605–613

Murphy E, Allen DG (2009) Why did the NHE inhibitor clinical trials fail? J Mol Cell Cardiol 46:137–141. doi:10.1016/j.yjmcc.2008.09.715

Murphy E, Eisner DA (2009) Regulation of intracellular and mitochondrial sodium in health and disease. Circ Res 104:292–303. doi:10.1161/CIRCRESAHA.108.189050

Murphy E, Steenbergen C (2011) What makes the mitochondria a killer? Can we condition them to be less destructive? Biochim Biophys Acta 1813:1302–1308. doi:10.1016/j.bbamcr.2010.09.003

Murphy E, Jacob R, Lieberman M (1985) Cytosolic free calcium in chick heart cells. Its role in cell injury. J Mol Cell Cardiol 17:221–231

Murphy E, Perlman M, London RE, Steenbergen C (1991) Amiloride delays the ischemia-induced rise in cytosolic free calcium. Circ Res 68:1250–1258

Murphy E, Cross H, Steenbergen C (1999) Sodium regulation during ischemia versus reperfusion and its role in injury. Circ Res 84:1469–1470

Murry CE, Jennings RB, Reimer KA (1986) Preconditioning with ischemia: a delay of lethal cell injury in ischemic myocardium. Circulation 74:1124–1136

Nakagawa T, Shimizu S, Watanabe T, Yamaguchi O, Otsu K, Yamagata H, Inohara H, Kubo T, Tsujimoto Y (2005) Cyclophilin D-dependent mitochondrial permeability transition regulates some necrotic but not apoptotic cell death. Nature 434:652–658. doi:10.1038/nature03317

Nazareth W, Yafei N, Crompton M (1991) Inhibition of anoxia-induced injury in heart myocytes by cyclosporin A. J Mol Cell Cardiol 23:1351–1354

Neely JR, Rovetto MJ, Whitmer JT, Morgan HE (1973) Effects of ischemia on function and metabolism of the isolated working rat heart. Am J Physiol 225:651–658

Neely JR, Whitmer JT, Rovetto MJ (1975) Effect of coronary blood flow on glycolytic flux and intracellular pH in isolated rat hearts. Circ Res 37:733–741

Negretti N, Varro A, Eisner DA (1995) Estimate of net calcium fluxes and sarcoplasmic reticulum calcium content during systole in rat ventricular myocytes. J Physiol Lond 486(Pt 3):581–591

Nicholls D, Akerman K (1982) Mitochondrial calcium transport. Biochim Biophys Acta 683:57–88

Nichols CG, Lederer WJ (1990a) The regulation of ATP-sensitive K+ channel activity in intact and permeabilized rat ventricular myocytes. J Physiol Lond 423:91–110

Nichols CG, Lederer WJ (1990b) The role of ATP in energy-deprivation contractures in unloaded rat ventricular myocytes. Can J Physiol Pharmacol 68:183–194

Pacher P, Csordás P, Schneider T, Hajnóczky G (2000) Quantification of calcium signal transmission from sarco-endoplasmic reticulum to the mitochondria. J Physiol Lond 529(Pt 3):553–564

Palty R, Silverman WF, Hershfinkel M, Caporale T, Sensi SL, Parnis J, Nolte C, Fishman D, Shoshan-Barmatz V, Herrmann S, Khananshvili D, Sekler I (2010) NCLX is an essential component of mitochondrial Na+/Ca2+ exchange. Proc Natl Acad Sci U S A 107:436–441. doi:10.1073/pnas.0908099107

Pan X, Liu J, Nguyen T, Liu C, Sun J, Teng Y, Fergusson MM, Rovira II, Allen M, Springer DA, Aponte AM, Gucek M, Balaban RS, Murphy E, Finkel T (2013) The physiological role of mitochondrial calcium revealed by mice lacking the mitochondrial calcium uniporter. Nat Cell Biol 15:1464–1472. doi:10.1038/ncb2868

Park CO, Xiao XH, Allen DG (1999) Changes in intracellular Na+ and pH in rat heart during ischemia: role of Na+/H+ exchanger. Am J Physiol 276:H1581–H1590

Petronilli V, Szabo I, Zoratti M (1989) The inner mitochondrial membrane contains ion-conducting channels similar to those found in bacteria. FEBS Lett 259:137–143

Pike MM, Kitakaze M, Marbán E (1990) 23Na-NMR measurements of intracellular sodium in intact perfused ferret hearts during ischemia and reperfusion. Am J Physiol 259:H1767–H1773

Pike MM, Luo CS, Clark MD, Kirk KA, Kitakaze M, Madden MC, Cragoe EJ, Pohost GM (1993) NMR measurements of Na+ and cellular energy in ischemic rat heart: role of Na(+)-H+ exchange. Am J Physiol 265:H2017–H2026

Poburko D, Santo-Domingo J, Demaurex N (2011) Dynamic regulation of the mitochondrial proton gradient during cytosolic calcium elevations. J Biol Chem 286:11672–11684. doi:10.1074/jbc.M110.159962

Raaflaub J (1953) Swelling of isolated mitochondria of the liver and their susceptibility to physicochemical influences. Helv Physiol Pharmacol Acta 11:142–156

Rezkalla SH, Kloner RA (2004) Ischemic preconditioning and preinfarction angina in the clinical arena. Nat Clin Pract Cardiovasc Med 1:96–102. doi:10.1038/ncpcardio0047

Ripoll C, Lederer WJ, Nichols CG (1990) Modulation of ATP-sensitive K+ channel activity and contractile behavior in mammalian ventricle by the potassium channel openers cromakalim and RP49356. J Pharmacol Exp Ther 255:429–435

Sayen JJ, Sheldon WF, Peirce G, Kuo PT (1958) Polarographic oxygen, the epicardial electrocardiogram and muscle contraction in experimental acute regional ischemia of the left ventricle. Circ Res 6:779–798

Scheuer J (1972) The effect of hypoxia on glycolytic ATP production. J Mol Cell Cardiol 4:689–692

Scheuer J, Stezoski SW (1970) Protective role of increased myocardial glycogen stores in cardiac anoxia in the rat. Circ Res 27:835–849

Seelye RN (1980) Proton generation and control during anaerobic glycolysis in heart cells. J Mol Cell Cardiol 12:1483–1486

Shivkumar K, Deutsch NA, Lamp ST, Khuu K, Goldhaber JI, Weiss JN (1997) Mechanism of hypoxic K loss in rabbit ventricle. J Clin Invest 100:1782–1788. doi:10.1172/JCI119705

Siemen D, Loupatatzis C, Borecky J, Gulbins E, Lang F (1999) Ca2+-activated K channel of the BK-type in the inner mitochondrial membrane of a human glioma cell line. Biochem Biophys Res Commun 257:549–554. doi:10.1006/bbrc.1999.0496

Skyschally A, van Caster P, Iliodromitis EK, Schulz R, Kremastinos DT, Heusch G (2009) Ischemic postconditioning: experimental models and protocol algorithms. Basic Res Cardiol 104:469–483. doi:10.1007/s00395-009-0040-4

Snowdowne KW, Ertel RJ, Borle AB (1985) Measurement of cytosolic calcium with aequorin in dispersed rat ventricular cells. J Mol Cell Cardiol 17:233–241

Steenbergen C, Deleeuw G, Rich T, Williamson JR (1977) Effects of acidosis and ischemia on contractility and intracellular pH of rat heart. Circ Res 41:849–858

Steenbergen C, Murphy E, Levy L, London RE (1987) Elevation in cytosolic free calcium concentration early in myocardial ischemia in perfused rat heart. Circ Res 60:700–707

Steenbergen C, Murphy E, Watts JA, London RE (1990) Correlation between cytosolic free calcium, contracture, ATP, and irreversible ischemic injury in perfused rat heart. Circ Res 66:135–146

Steenbergen C, Perlman ME, London RE, Murphy E (1993) Mechanism of preconditioning. Ionic alterations. Circ Res 72:112–125

Stern MD, Silverman HS, Houser SR, Josephson RA, Capogrossi MC, Nichols CG, Lederer WJ, Lakatta EG (1988) Anoxic contractile failure in rat heart myocytes is caused by failure of intracellular calcium release due to alteration of the action potential. Proc Natl Acad Sci U S A 85:6954–6958

Szabo I, Zoratti M (1991) The giant channel of the inner mitochondrial membrane is inhibited by cyclosporin A. J Biol Chem 266:3376–3379

Szabo I, Zoratti M (1992) The mitochondrial megachannel is the permeability transition pore. J Bioenerg Biomembr 24:111–117

ten Hove M, Jansen MA, Nederhoff MGJ, Van Echteld CJA (2007) Combined blockade of the Na+ channel and the Na+/H+ exchanger virtually prevents ischemic Na+ overload in rat hearts. Mol Cell Biochem 297:101–110. doi:10.1007/s11010-006-9334-0

Territo PR, Mootha VK, French SA, Balaban RS (2000) Ca(2+) activation of heart mitochondrial oxidative phosphorylation: role of the F(0)/F(1)-ATPase. Am J Physiol Cell Physiol 278:C423–C435

Toda T, Kadono T, Hoshiai M, Eguchi Y, Nakazawa S, Nakazawa H, Higashijima N, Ishida H (2007) Na+/H+ exchanger inhibitor cariporide attenuates the mitochondrial Ca2+ overload and PTP opening. Am J Physiol Heart Circ Physiol 293:H3517–H3523. doi:10.1152/ajpheart.00483.2006

Wang Y, Meyer JW, Ashraf M, Shull GE (2003) Mice with a null mutation in the NHE1 Na+-H+ exchanger are resistant to cardiac ischemia-reperfusion injury. Circ Res 93:776–782. doi:10.1161/01.RES.0000094746.24774.DC

Weiss J, Couper GS, Hiltbrand B, Shine KI (1984) Role of acidosis in early contractile dysfunction during ischemia: evidence from pHo measurements. Am J Physiol 247:H760–H767

Weiss JN, Lamp ST, Shine KI (1989) Cellular K+ loss and anion efflux during myocardial ischemia and metabolic inhibition. Am J Physiol 256:H1165–H1175

Williams IA, Xiao X-H, Ju Y-K, Allen DG (2007) The rise of [Na(+)] (i) during ischemia and reperfusion in the rat heart-underlying mechanisms. Pflugers Arch 454:903–912. doi:10.1007/s00424-007-0241-3

Williams GSB, Boyman L, Chikando AC, Khairallah RJ, Lederer WJ (2013) Mitochondrial calcium uptake. Proc Natl Acad Sci U S A 110:10479–10486. doi:10.1073/pnas.1300410110

Williams GSB, Boyman L, Lederer WJ (2015) Mitochondrial calcium and the regulation of metabolism in the heart. J Mol Cell Cardiol 78:35–45. doi:10.1016/j.yjmcc.2014.10.019

Williamson JR (1966) Glycolytic control mechanisms II. Kinetics of intermediate changes during the aerobic-anoxic transition in perfused rat heart. J Biol Chem 241:5026–5036

Wollenberger A, Krause EG (1968) Metabolic control characteristics of the acutely ischemic myocardium. Am J Cardiol 22:349–359

Xu W, Liu Y, Wang S, McDonald T, Van Eyk JE, Sidor A, O'Rourke B (2002) Cytoprotective role of Ca2+- activated K+ channels in the cardiac inner mitochondrial membrane. Science 298:1029–1033. doi:10.1126/science.1074360

Yan GX, Kléber AG (1992) Changes in extracellular and intracellular pH in ischemic rabbit papillary muscle. Circ Res 71:460–470

Yellon DM, Downey JM (2003) Preconditioning the myocardium: from cellular physiology to clinical cardiology. Physiol Rev 83:1113–1151. doi:10.1152/physrev.00009.2003

Ylitalo KV, Ala-Rämi A, Liimatta EV, Peuhkurinen KJ, Hassinen IE (2000) Intracellular free calcium and mitochondrial membrane potential in ischemia/reperfusion and preconditioning. J Mol Cell Cardiol 32:1223–1238. doi:10.1006/jmcc.2000.1157

Yoshida T, Watanabe M, Engelman DT, Engelman RM, Schley JA, Maulik N, Ho YS, Oberley TD, Das DK (1996) Transgenic mice overexpressing glutathione peroxidase are resistant to myocardial ischemia reperfusion injury. J Mol Cell Cardiol 28:1759–1767. doi:10.1006/jmcc.1996.0165

Zhang H, Shang W, Zhang X, Gu J, Wang X, Zheng M, Wang Y, Zhou Z, Cao J-M, Ji G, Zhang R, Cheng H (2013) B-adrenergic-stimulated L-type channel Ca^2+ entry mediates hypoxic Ca^2+ overload in intact heart. J Mol Cell Cardiol 65:51–58. doi:10.1016/j.yjmcc.2013.09.002

Zweier JL, Talukder MAH (2006) The role of oxidants and free radicals in reperfusion injury. Cardiovasc Res 70:181–190. doi:10.1016/j.cardiores.2006.02.025

Chapter 3
Mislocalization of Mitochondrial Intermembrane Space Proteins

Lorna Flanagan*, Federico Lucantoni*, and J.H.M. Prehn

Abbreviations

Apaf-1	Apoptotic protease-activating factor 1
ATP	Adenosine 5′-(tetrahydrogen triphosphate)
Bcl-2	B-cell lymphoma-2
CARD	Caspase recruitment domain
CASPASE	Cysteinyl aspartate-specific proteinase
cFLIP	Cellular FLICE-like inhibitory protein
CHX	Cycloheximide
c-IAP1	Cellular-inhibitor of apoptosis 1
c-IAP2	Cellular-inhibitor of apoptosis 2
Cyt-c	Cytochrome-c
dATP	Deoxyadenosine 5′-(tetrahydrogen triphosphate)
DD	Death domain
DED	Death effector domin
DISC	Death-inducing signaling complex
FADD	Fas-associated protein with death domain
FasL	Fas ligand
FITC	Fluorescein isothiocyanate
FLIP	FLICE-like inhibitory protein
GFP	Green fluorescent protein
IAP	Inhibitor of apoptosis protein
ICAD	Inhibitor of caspase-activated DNase
IMS	Intermembrane space
kDa	Kilodaltons
MOMP	Mitochondrial outer membrane permeabilization

Author contributed equally with all other contributors.

L. Flanagan • F. Lucantoni • J.H.M. Prehn (✉)
Centre for Systems Medicine, Department of Physiology and Medical Physics, Royal College of Surgeons in Ireland, 123 St. Stephen's Green, Dublin 2, Ireland
e-mail: prehn@rcsi.ie

D.M. Hockenbery (ed.), *Mitochondria and Cell Death*, Cell Death in Biology and Diseases, DOI 10.1007/978-1-4939-3612-0_3

X.-M. Yin and Z. Dong (Series eds.), *Cell Death in Biology and Diseases*

PARP	Poly(ADP-ribose) polymerase
siRNA	Small interfering RNA
Smac	Second mitochondria-derived activator of caspases
TNF-α	Tumor necrosis factor-alpha
TNFR	Tumor necrosis factor receptor
TRADD	TNF receptor-associated death domain protein
TRAIL	Tumor necrosis factor-related apoptosis-inducing ligand
XIAP	x-Linked inhibitor of apoptosis protein
YFP	Yellow fluorescent protein
Z-VAD	Benzyloxycarbonyl-Val-Ala-Asp(*O*-methyl)-fmk

3.1 The Intermembrane Space of Mitochondria

During apoptosis, proteins are released from the mitochondrial intermembrane space (IMS). Mitochondria have a complex structure which allows this organelle to perform a wide range of physiological functions important for cellular homeostasis. The IMS is a sub-compartment of this organelle with a wide range of physiological properties. The IMS is delimited by the outer mitochondrial membrane (OMM) and the internal mitochondrial membrane (IMM). These two membranes differ in their lipid composition: the outer mitochondrial membrane is enriched in phosphatidylcholine (PC), phosphatidylethanolamine (PE), and phosphatidylinositol (PI) and, in smaller amounts, phosphatidylserine (PS) and cardiolipin (CL), while the internal mitochondrial membrane is constituted mainly of CL and PE (Colbeau et al. 1971; de Kroon et al. 1997; Hovius et al. 1993). The IMM presents with invaginations called cristae that show a pleiomorphic structure linked through narrow tubular segments (cristae junctions) of varying length to the peripheral surface of the inner membrane, but also to one another, creating the 'intracristae' space (Mannella et al. 2001). Cristae morphology is regulated by several proteins including prohibitins (Tatsuta et al. 2005), mitofilin (John et al. 2005), and OPA1 (Frezza et al. 2006); however, the IMM topology is considered to be highly dynamic and results from ongoing membrane fusion and fission processes (Mannella 2006). It was proposed that cristae have different physiological properties compared to the peripheral IMM and are enriched in complexes of oxidative phosphorylation (Vogel et al. 2006).

3.2 The IMS as a Controlled Environment

The IMS is the interface between the OMM and IMM, and therefore the molecular composition of this environment is highly regulated and complex. It has been shown that the pH of the IMS is more acidic than the cytosol. By using targeted pH-sensitive GFP probes, studies in isolated mitochondria and intact cultured mammalian cells have measured pH values for the IMS of 6.88, lower than the cytosol by a margin of 0.7–0.9 pH units (Porcelli et al. 2005). On the other hand, the ionic strength of the

IMS is similar to the cytosol with a value of 100–150 mM (Cortese et al. 1991). This is physiologically relevant, especially for proteins such as cytochrome c; at this ionic strength, the diffusion rate of cytochrome c is highest, while its affinity for the inner membrane is lowest and its electron transport rate highest (Cortese et al. 1991). The IMS redox potential has been examined as well and was shown to be considerably more oxidizing than the cytosol or the matrix suggesting a redox control that is independently regulated within this cellular compartment (Hu et al. 2008).

The exchange of ions, metabolites, and small molecules between the cytosol and the matrix is regulated by different channel proteins. The main transporter that influences metabolite flux through a gating system is the voltage-dependent anion channel (VDAC), which may also interact with other outer membrane proteins to further alter the permeability of the OMM (Colombini 1979). This 30 kDa protein forms a β-barrel structure that possesses different transport properties based on the open/closed conformation. The open state favors the transport of anions over cations, while the closed state permits the passage of small cations (such as Ca^{2+} and K^{+}) and blocks anionic metabolites such as ATP or succinate (Colombini 2012). On the other hand, the transport across the IMM is regulated by the mitochondrial carrier family (MCF) whose major function is to transport anions (ADP/ATP, P_i, oxoglutarate, aspartate/glutamate, pyruvate, citrate, dicarboxylates), and in lesser amount cations and zwitterions. These transporters have a tripartite structure of three tandem repeated domains of 100 amino acids (Ferramosca and Zara 2013; Palmieri et al. 2011). Another class of transporter is constituted by a ubiquitous class of protein termed ATP-binding cassette (ABC) transporters. These transporters are present in the OMM and IMM where they are involved in the transport of ions, metabolites, cofactors, and biosynthetic precursors (Zutz et al. 2009).

3.3 Functions of the IMS

The IMS can act as a sorting hub for lipid, proteins, metal ions, and metabolites. The lipid composition in the mitochondrial membranes is highly regulated. Three proteins from the PRELI family, Ups1, Ups2, and Ups3, have been identified in the regulation and control of CL levels in the IMM (Osman et al. 2009; Tamura et al. 2009). Metal ion transport is also finely regulated since several enzymes residing in the matrix or at the interface of the IMM use metal ions as cofactors to carry out enzymatic reactions. Iron transport is regulated by mitoferrin 1 at the IMM, while copper and zinc import can be regulated by different proteins such as MIA40/TIMM40, metallothioneins, and COX17 (Palumaa et al. 2004; Terziyska et al. 2005; Ye et al. 2001).

The IMS has also a sorting function for nuclear-encoded proteins that function in the matrix or IMM. These proteins use different protein translocases localized in the OMM and IMM, which facilitate the targeting of these proteins to different sub-mitochondrial compartments. The translocase of the outer membrane (TOM) complex is a prominent complex which is used as an 'entry' gate, while in the IMM two translocases of the inner membrane (TIM complexes) have been identified (Rehling et al. 2004).

The IMS is also important for the detoxification of reactive oxygen species (ROS), mainly produced by complexes I and III of the respiratory chain. In the IMS, the superoxide radical is transformed to hydrogen peroxide by enzyme reactions. This disproportionation reaction is mainly catalyzed by superoxide dismutase, a cytosolic enzyme which is also IMS-localized (Kawamata and Manfredi 2010). However, ROS species are also used as signalling molecules which can activate certain cellular responses for oxidative stress control (Finkel 2011). This regulation can occur in the IMS where cysteines in key proteins, involved in processes like respiration and protein import, may undergo oxidation (Fischer and Riemer 2013; Riemer et al. 2009).

Due to the high diversity of cellular processes occurring in the IMS, its proteome was analyzed. Using mass spectrometry, 127 proteins were identified in human mitochondria IMS (Hung et al. 2014), while 51 proteins were found in yeast mitochondria (Vogtle et al. 2012). Both proteomes were highly specific (>90 %) and revealed new mitochondrial proteins. It was also found that the yeast IMS had a total of 35 human orthologs.

3.4 Mechanisms of IMS Proteins Release

IMS proteins also function in the control of apoptosis. Apoptosis has an essential role in regulating development and tissue homeostasis in multicellular organisms. Dysfunction or deregulation in apoptotic pathways is implicated in a variety of pathological conditions including neurodegenerative diseases, ischemic damage, autoimmune disorders, and many types of cancer (Portt et al. 2011). Several apoptotic signalling pathways lead to the release of proteins from the intermembrane space to the cytosol: the intrinsic pathway and the extrinsic pathway which are activated by internal or external stressors and the granzyme B (GrB)-induced pathway which is triggered to kill infected cells.

3.5 The Intrinsic Pathway

The intrinsic pathway, also named the mitochondria-mediated pathway, leads to the process of mitochondrial outer membrane permeabilization (MOMP) as a result of uncompensated cell stress signalling. The permeabilization process is highly regulated by proteins in the B-cell CLL/lymphoma 2 (BCL2) family of proteins, but also involves mitochondrial lipids (Martinou and Youle Richard 2011). The BCL2 family of proteins is constituted by a large number of components with either pro- or anti-apoptotic function and is divided into three different subclasses: The anti-apoptotic BCL2 subfamily including BCL2, BCLXL, BCLW, MCL1, and A1; the pro-apoptotic

BAX family including BAX and BAK (and potentially BOK); and the pro-apoptotic BH3 subfamily that includes BIM, PUMA, BAD, tBID, BIK, HRK, and BNIP3.

The balance between the pro- and anti-apoptotic proteins and the regulation of their interactions determine if the cell commits to apoptosis or aborts the process (Youle and Strasser 2008). BCL2 family proteins share up to four conserved BCL2 homology (BH) domains known as BH1, BH2, BH3, and BH4, which correspond to α-helical-structured segments involved in the formation of a complex interaction network (Chao and Korsmeyer 1998). The C-terminal amino acids encode for a stretch of hydrophobic amino acids, followed by two charged residues, which participate in membrane docking, targeting, and lipid interaction. Several of these proteins can be found in different subcellular localization such as the cytoplasmic face of the mitochondrial outer membrane, the nuclear envelope, and the endoplasmic reticulum (Gross et al. 1999). Pro- and anti-apoptotic family members can hetero-dimerize: the BH1, BH2, and BH3 domains of an anti-apoptotic member form a hydrophobic cleft to which a BH3 amphipathic α-helix can bind. This BH3 cleft has a key role for all dimerization within the family and the integrity of this hydrophobic groove is required both for anti-apoptotic activity and for the binding of anti-apoptotic proteins to their pro-apoptotic partners (Sattler et al. 1997; Yin et al. 1994).

The original notion dictates that, once activated, all BH3-only proteins trigger apoptosis through a common mode of action. However, several recent reports have demonstrated that different classes of BH3-only proteins can be distinguished (Chen et al. 2005); to date three different models were proposed (Czabotar et al. 2014). In the direct activation model, only a few proteins (BIM, tBID, and possibly PUMA) have the capacity to directly bind and activate BAX and BAK. In this model, BAX and BAK cannot be activated in the absence of a "direct activator" of the BIM/tBID/PUMA class. The BH3-only proteins that lack the capacity to directly activate BAX and BAK (such as BAD, BIK, HRK, NOXA), termed "sensitizers", are able to induce apoptosis through the binding to anti-apoptotic proteins and displacing direct activators, in order to activate BAX and BAK indirectly (Certo et al. 2006; Kuwana et al. 2005; Letai et al. 2002). On the other hand, the indirect activation model proposes that activated BAX and BAK proteins are sequestered by anti-apoptotic proteins, and in order to permeabilize the OMM, the anti-apoptotic proteins need to be neutralized by BH3-only proteins (Fletcher et al. 2008; Willis et al. 2005, 2007). The third model, known as the unified model, takes into account both modes of interaction: in this scenario anti-apoptotic proteins sequester BAX, BAK, and BH3-only protein activators and sensitizers (Llambi et al. 2011). Impairments and altered expression in pro- and anti-apoptotic BCL2 proteins can lead to cancer; more recently, a systematic analysis of BCL2 proteins by using the direct activation model was able to predict the response of cell lines and patient tumors to genotoxic drug treatments (Lindner et al. 2013). Once activated, BAX and BAK are able to oligomerize (as a result of an apoptotic stimuli), leading to the formation of a pore structure on the OMM with the release of cytochrome c and other apoptotic proteins from the IMS to the cytosol (Green and Kroemer 2004).

3.6 The Extrinsic Pathway

The extrinsic pathway is regulated by a complex series of transmembrane receptors located in the plasma membrane of the cell. These membrane proteins are members of the tumor necrosis factor (TNF) receptor gene superfamily and are called death receptors. The most characterized receptors are CD95 (APO-1/Fas), TNF receptor 1, TNF-related apoptosis-inducing ligand-receptor 1 (TRAILR1), and TRAILR2 (Fulda and Debatin 2006). CD95 is a type I transmembrane receptor present at the cell surface. This 48 kDa protein is expressed on activated lymphocytes, on several lymphoid and non-lymphoid tissues and tumor cells (Krammer 2000). The binding of the ligand to the receptor (CD95 or TRAILR1 and R2) causes the trimerization of the receptor and the clustering of the receptors' death domains. Through this domain, there is the recruitment of the adaptor molecule Fas-associated DD-containing protein (FADD) (Walczak and Krammer 2000). This adaptor protein contains another protein–protein interaction domain called the death effector domain (DED) that is able to recruit caspase 8 to the activated receptor forming the death-inducing signalling complex (DISC) (Kischkel et al. 1995). This finally leads to caspase 8 activation. Once activated, these cysteine proteases activate downstream effector caspases such as caspase 3. For death receptor-induced apoptosis, two principal types of cells have been identified: in the first one (type I cells) DISC formation is able to activate sufficient quantities of caspase 8, while in type II cells an amplification loop is necessary for complete activation of caspases (Scaffidi et al. 1998). It is at this step that crosstalk occurs between the Type II extrinsic pathway and the intrinsic pathway, as t-BID activates BAK and BAX proteins and then induces pore formation in the mitochondria (Wei et al. 2001). MOMP subsequently occurs, leading to activation of the caspase cascade and cell death.

3.7 Granzyme B-Induced Apoptosis

Granzyme B (GRB) is a serine protease which is involved in cytotoxic granule-mediated apoptosis, an important pathway used by cytotoxic T lymphocytes (CTLs) and natural killer (NK) cells to remove and clear the tissue environment of allogeneic, virally infected and tumor cells (Rousalova and Krepela 2010). After the recognition of the harmful cell by CTLs and NK cells, the secretory granules are transported and released towards an intercellular cleft termed the immunological synapse (Jenkins and Griffiths 2010). These granules are rich in GRB and the pore-forming protein perforin, but also contain smaller amounts of different types of granzymes such as A, H, K, and M, which trigger different death signals (Cullen and Martin 2008; Mahrus and Craik 2005). GRB structure is characterized by two six-stranded ß-barrel folds, linked by three trans-domain segments which allow a peptide bond cleavage immediately adjacent to Asp residues (Rotonda et al. 2001); among the substrates, there are at least three hundred potential targets identified (Rousalova and

Krepela 2010). GRB can proteolytically cleave and activate the executioner procaspases 3 and 7 (Adrain et al. 2005; Martin et al. 1996; Metkar et al. 2003; Quan et al. 1996) and can cleave, but not activate, procaspases 8, 10, 9, and 2 (Adrain et al. 2005; Medema et al. 1997; Srinivasula et al. 1996; Talanian et al. 1997). This protease can also have a role in MOMP by cleaving the BH3-only protein BId, forming truncated p15 BID (tBID) which translocates to the mitochondria and subsequently recruits cytosolic BAX to the mitochondria, resulting in MOMP and release of cytochrome c and other apoptotic proteins from the IMS (Barry et al. 2000; Heibein et al. 2000; Sutton et al. 2000). GRB can also disrupt the complex formed by MCL1 and BID as a second mechanism to induce MOMP (Han et al. 2004).

3.8 Mitochondrial Structure Changes During MOMP

During apoptosis, various changes in the mitochondrial structure occur to allow the release of apoptotic factors from the IMS. These alterations are triggered by proteins in the IMS and by changes in lipid compositions between mitochondrial membranes. One of the major events is the translocation of CL from the IMM to the OMM (Garcia Fernandez et al. 2002; Kagan et al. 2005). At the OMM, CL can participate in the recruitment of BCL2 proteins through a lipid–protein interaction mediated by the transmembrane domain of these proteins (Andreu-Fernandez et al. 2014; Gonzalvez et al. 2005; Lutter et al. 2000). However, CL is not the only lipid that changes with the induction of MOMP; several lipid species such as lysophosphatidyl-choline and phosphatidylinositol can increase or decrease depending on the apoptotic stimulus (Crimi and Esposti 2011). Another important change is cristae remodeling triggered by OPA1. This protein resides in the inner membrane as a hetero-oligomeric complex where it regulates IMM fusion and maintains cristae junctions (Delettre et al. 2000; Olichon et al. 2006). During MOMP, OPA1 promotes cristae remodeling and cristae junction opening and allows the release of cytochrome *c* which is bound to the cristae (Ow et al. 2008; Scorrano et al. 2002), due to different electrostatic interactions between lysine residues and CL (Kostrzewa et al. 2000; Rytomaa and Kinnunen 1994; Sinibaldi et al. 2013). Finally, a change in the permeability of the OMM is necessary for the release of apoptotic proteins from the IMS. In normal conditions, the OMM allows the passage of small molecules (less than 5 kDa) using VDAC channels; BCL2 proteins are able to control and alter the mitochondrial permeability by forming a complex interaction network (Czabotar et al. 2014). The two pro-apoptotic members BAX and BAK mediate the mitochondrial permeabilization by oligomerization (Dewson and Kluck 2009; Mikhailov et al. 2003). In healthy conditions, both proteins are globular monomers; however, BAX is primarily cytosolic while BAK is anchored at the OMM through its c-terminal transmembrane domain (Suzuki et al. 2000). Furthermore, inactive BAX is retrotranslocated between the cytosol and the OMM by interacting with BCLXL (Edlich et al. 2011; Schellenberg et al. 2013). However, the exact mechanism used by these two proteins to disrupt the membrane integrity is not well-known. One mechanism proposes that α-5 and -6 helices of BAX and BAK insert and span the OMM (Fuertes et al. 2010; Garcia-Saez et al. 2006). Another model shows how these two helices lie on the bilayer membrane and alter its integrity by applying a surface tension and a curvature (Westphal et al. 2014).

Once MOMP is activated, the release of apoptotic proteins is triggered. Cytochrome *c* release occurs in one step and can be initiated at different points in a cell and propagated across the cell; this process is rapid and completed within minutes (Bhola et al. 2009; Goldstein et al. 2000; Rehm et al. 2009). During apoptosis, mitochondria undergo fission. This event might occur simultaneously, or prior to the release of cytochrome *c* (Youle and Karbowski 2005). Drp1 is one of the components of the fission machinery, and during apoptosis, is translocated to the OMM where it colocalizes with BAX foci (Karbowski et al. 2002) . DRP1 is also able to promote membrane remodeling and stimulate BAX oligomerization and has been shown to trigger a hemifusion intermediary between the IMM and the OMM, which again depends on the presence of CL (Montessuit et al. 2010).

3.9 Cytochrome *c*

Cytochrome *c* is a water-soluble 13 kDa heme-containing protein that is encoded by a nuclear gene. This protein plays an important role in mitochondrial bioenergetics, but is also paramount to the intrinsic apoptosis pathway (Liu et al. 1996; Ow et al. 2008). It is synthesized in the cytosol as an apoprotein and then transported to the mitochondria where it normally resides within the intracristae and mitochondrial intermembrane spaces (IMS). Upon its localization to the IMS, a heme group is added to form holocytochrome *c*, a mature functional protein (Mavridou et al. 2013). This heme addition is vital for cytochrome *c* function in the apoptotic pathway as the precursor protein, apocytochrome *c*, cannot induce apoptosis (Martin and Fearnhead 2002). The mature protein is also required for its bioenergetics functions, as the heme group is responsible for shuttling electrons from complex III to complex IV in the electron transport chain, an essential step in ATP production. The fact that cytochrome *c* knockout mice die in utero at midgestation highlights the importance of cytochrome *c* in the ETC (Li et al. 2000).

A significant proportion of cytochrome *c* in the mitochondria seems to be associated with the membrane phospholipid, cardiolipin (Hanske et al. 2012). During MOMP, cytochrome *c* detaches from the inner mitochondrial membrane, dissociates from cardiolipin, and is released into the cytosol. MOMP is a rapid process that allows for cytochrome *c* release within 5 min (Goldstein et al. 2000). Single cell imaging studies suggest that MOMP results in the simultaneous release of IMS proteins into the cytosol (Rehm et al. 2003). Fluorescent fusion IMS proteins, shown to reliably mimic the behavior of the native proteins, were used for these studies (Munoz-Pinedo et al. 2006; Rehm et al. 2003). Another technique involves fusing a small tag (10–15 aa residues) containing a tetracysteine motif to the protein of interest. These small tags are specifically recognized by a cell-permeant fluorescein- or resorufin-based fluorescent dye (Munoz-Pinedo et al. 2006). Time-lapse confocal microscopy showed that release of cytochrome *c*-EGFP from mitochondria could occur as a rapid, one- or multiple-step process, or slowly over several hours, depending on cell death stimulus (Luetjens et al. 2001). More recent studies, using high-speed cellular imaging in intact living cells, observed that MOMP-induced cytochrome *c*-GFP release started in subcellular mitochondrial regions and proceeded in a wave-like manner throughout the cell body (Rehm et al. 2009). Some mitochondria in the cell may not undergo MOMP during apoptosis. It was found that incomplete MOMP occurred in a subset of cells, and in these cells, release of cytochrome *c*-GFP can also be observed to be incomplete (Tait et al. 2010).

While the initiation of release of IMS proteins is synchronous, the duration of release can vary for individual IMS proteins. Activation of caspases was shown to occur ~10 min after cytochrome *c* –GFP release (Flanagan et al. 2010; Munoz-Pinedo et al. 2006; Rehm et al. 2003). Following release, cytochrome *c* binds to apoptotic protease-activating factor 1 (APAF-1), inducing its oligomerization. Seven APAF-1 molecules form a heptameric wheel structure known as the apoptosome (Chandra et al. 2006; Kim et al. 2005). The apoptosome recruits procaspase 9, leading to dimerization and activation of procaspase 9 to active caspase 9 (Würstle and Rehm 2014). Active caspase 9 then subsequently activates downstream executioner caspases, which are responsible for the hallmarks of apoptosis. MOMP and subsequent release of IMS proteins occurs independently of caspase activity (Düßmann et al. 2003; Rehm et al. 2003).

Cytochrome *c* plays a key role in the electron transport chain, where it is responsible for transporting electrons from respiratory complex III to complex IV. Cytochrome *c* release following MOMP leads to a bioenergetic crisis for the cell due to breakdown of the electron transport chain, characterized by ATP depletion, loss of ionic homeostasis, increased osmotic pressure, and necrotic cell death. Cancer cells evade these processes by altering their metabolism, via enhanced glucose utilization and caspase activation pathways. Studies show that if caspase activation is blocked or defective, some cells can survive apoptotic cytochrome *c* release. For example, autophagy and glyceraldehyde-3-phosphate dehydrogenase preserve survival after cytochrome *c* release in the absence of caspase activation (Colell et al. 2007). It has been reported that following MOMP, mitochondria can use cytoplasmic cytochrome *c* to maintain mitochondrial transmembrane potential and ATP synthesis, provided caspases are inhibited (Waterhouse et al. 2001). The level of accessible cytochrome *c* in the cytosol after release and the cell's glycolytic capacity are associated with levels of ATP synthase activity following MOMP (Huber et al. 2011).

3.10 Second Mitochondria-Derived Activator of Caspases (SMAC)

SMAC is encoded by nuclear genes, but is shuttled to the mitochondria after cytoplasmic translation via classical mitochondrial targeting sequences in the amino terminus (Du et al. 2000; Verhagen et al. 2000). The nuclear-encoded SMAC protein is 239 amino acids in length. Upon its import into the mitochondria, the 55 amino acid mitochondrial targeting sequence is proteolytically removed, leaving the mature SMAC protein (Du et al. 2000). Mature SMAC exists as a homodimer of two 21 kDa triple α helical bundle monomers (Chai et al. 2000). The SMAC dimeric structure is very stable and has the highest dissociation constant reported for any dimer (Gonçalves et al. 2008). SMAC is expressed in most tissues, but the liver, kidney, lung, pancreas, and testes exhibit the highest levels of the protein (Tikoo et al. 2002).

Import of SMAC into the mitochondria and subsequent cleavage of the mitochondrial targeting sequence leaves the mature protein. As a consequence of this cleavage at the N-terminus, four hydrophobic amino acids (AVPI) are exposed. This tetrapeptide represents the IAP-binding motif (IBM) and its integrity is essential for SMAC antagonism of XIAP (Shi 2002). Once MOMP is initiated, SMAC is released into the cytosol. Real-time single cell analysis of a SMAC fusion protein demonstrated that the protein is released in one step, and release is complete within 10 min (Rehm et al. 2003). Although SMAC and cytochrome *c* release is initiated at the same time, the duration of SMAC release can be prolonged when compared to cytochrome *c* release (Flanagan et al. 2010; Munoz-Pinedo et al. 2006; Rehm et al. 2003). Early studies suggested SMAC release was dependent on caspase activation (Adrain et al. 2001). However, real-time single cell analysis demonstrated SMAC release occurred independently of caspase activation and also independently of BAX (Rehm et al. 2003).

Cytochrome *c* release activates initiator caspase 9 and executioner caspases 3 and 7. Executioner caspase activation proceeds in a rapid all-or-none fashion (Rehm et al. 2002), but X-linked inhibitor of apoptosis protein (XIAP) can bind to and inhibit these proteins, provided it is present in concentrations above a critical threshold (Deveraux et al. 1998; Rehm et al. 2006). Dimeric SMAC sterically and/or competitively antagonizes XIAP, thereby relieving XIAP-mediated capase inhibition and allowing caspases to continue in the apoptotic pathway (Liu et al. 2000; Wu et al. 2000). SMAC can also bind and inhibit other IAPs, such as livin/ML-IAP, Bruce/Apollon, CIAP1, and 2 (Bartke et al. 2004; Du et al. 2000; Hao et al. 2004; Verhagen et al. 2000; Vucic et al. 2002), which antagonize caspases in cell death pathways and/or play additional roles in proliferation and pro-survival signaling. In cells overexpressing XIAP, dimeric SMAC is retained within the mitochondria, despite MOMP occurring. This suggests that XIAP not only protects the cell by inhibiting caspases, but may also preserve mitochondrial function (Flanagan et al. 2011). Recently, an inherited mutation was reported in *SMAC* that can lead to progressive hearing loss. Overexpression of this mutant SMAC has been shown to increase mitochondrial sensitivity to calcium, leading to mitochondrial dysfunction (Cheng et al. 2011). Mice deficient in *Smac* are viable, show no obvious histological abnormalities, and respond normally to apoptotic stimuli (Okada et al. 2002). However, low SMAC levels may be associated with resistance to chemotherapy, possibly due to disturbed SMAC/XIAP balance in these cells (Mizutani et al. 2010).

3.11 OMI/HTRA2

The serine protease OMI/HTRA2 is encoded as a 49 kDa proenzyme in the nucleus and then targeted to the mitochondria. In the intermembrane space, the 133 amino acid targeting peptide is removed, generating the mature 36 kDa protein (Martins et al. 2002; Suzuki et al. 2001). Once MOMP occurs, mature OMI/HTRA2 is released and interacts with XIAP in a manner similar to SMAC via its first four N-terminal amino acids; Ala-Val-Pro-Ser. This interaction results in catalytic cleavage of XIAP, which relieves caspase inhibition and restores apoptosis (Martins et al. 2002; Suzuki et al. 2001). OMI/HTRA2 also cleaves structural proteins upon its release (Vande Walle et al. 2007), thereby contributing to both caspase-dependent and -independent apoptosis. tBID was found to be essential for the release of OMI from mitochondria, while BCL2 inhibits this release (van Loo et al. 2002). Time-lapse imaging studies utilizing a small tetracysteine tag fused to OMI demonstrated that release of this protein from the mitochondria was caspase-independent. The onset of OMI release was found to correspond with cytochrome *c* and SMAC release. Although its released kinetics were slightly slower, OMI was still completely released in ~ 5 min (Munoz-Pinedo et al. 2006).

Additionally, OMI/HTRA2 has been shown to have a link with mitochondrial integrity. Mnd2 (motor neuron degeneration 2) mice, which exhibit early-onset neurodegeneration associated with Parkinsonism and juvenile lethality, were found to contain a missense mutation, Ser276Cys, in the protease domain of OMI/HTRA2 resulting in loss of protease activity (Jones et al. 2003). This loss of activity leads to mitochondrial dysfunction with changes in mitochondrial morphology and biogenesis, suggesting OMI/HTRA2 is required in striatal neurons to protect against stress-related damage (Abou-Sleiman et al. 2006; Jones et al. 2003; Strauss et al. 2005). A recent study examining the role of OMI/HTRA2 as a regulator of mitochondrial biogenesis indicates that this may be through the GSK3β/PGC1α pathway (Xu et al. 2014). PGC1α is an important factor for mitochondrial biogenesis (Austin and St-Pierre 2012) and is negatively regulated by glycogen synthase kinase 3β (GSK3β). OMI/HTRA2 has recently been shown to cleave GSK3β (Xu et al. 2014). GSK3β was found to be increased in mnd2 mice, while PGC1α is decreased significantly, suggesting loss of the protease activity of OMI/HTRA2 is associated with the reduction of PGC1α. Inhibition of GSK3β or overexpression of PGC1α can restore mitochondrial biogenesis in these mice. Inhibition of GSK3β also results in a significant improvement of the movement ability of mnd2 mice. Thus, OMI/HTRA2 seems to have a vital role as a novel regulator of mitochondrial biogenesis, and dysfunction of biogenesis may be associated with OMI/HTRA2 protease deficiency-induced neurodegeneration (Xu et al. 2014).

3.12 Apoptosis-Inducing Factor (AIF)

AIF is nuclearly encoded as a 67 kDa protein and shuttled to the mitochondria via its N-terminal mitochondrial localization signal. Once imported into the mitochondria, this localization signal is removed to leave a 62 kDa mature protein (Otera et al. 2005). This protein is anchored to the inner mitochondrial membrane where it displays NADH oxidase activity. When the cell receives an apoptotic stimulus, proteases, such as calpains and/ or cathepsins (Polster et al. 2005; Yuste et al. 2005), cleave AIF from its membrane anchor to generate truncated AIF (tAIF), a 57 kD protein (Otera et al. 2005; Susin et al. 1999). tAIF is shuttled to the nucleus upon its release from the mitochondria via two nuclear localization sequences. Studies using both a GFP-tagged AIF and AIF fused to a small tetracysteine tag indicated that the release of AIF from the mitochondria is slow, incomplete, and caspase-independent, occurring up to 5 h after cytochrome *c* release. In the presence of a caspase inhibitor, AIF release was slow and not complete until as long as 10 h after cytochrome *c* release (Munoz-Pinedo et al. 2006). AIF release seems to occur in two steps. Initially upon MOMP, a small fraction of AIF preexisting as a soluble form in the intermembrane space is released. The second release event occurs between 2 and 5 h after cytochrome *c* release and corresponds with plasma membrane permeabilization. As AIF is a transmembrane protein, it is feasible that a secondary event is required in order for complete release to occur during apoptosis (Munoz-Pinedo et al. 2006). Overexpression of BCL2 can prevent AIF release (Otera et al. 2005; Susin et al. 1999; Susin et al. 1996), while BAX was found to be insufficient for AIF release, likely due to the association of AIF with the mitochondrial inner membrane (Arnoult et al. 2002).

Once in the nucleus, AIF causes chromatin condensation and DNA fragmentation, leading to caspase-independent cell death and necrosis (Artus et al. 2010). Studies elucidating the exact role of AIF in the nucleus reveal that AIF forms a complex with histone H2AX and cyclophilin A, generating an active DNA-degrading complex that regulates chromatinolysis and programmed necrosis (Artus et al. 2010; Baritaud et al. 2010). In 2008, Wilkinson and colleagues identified AIF as a XIAP-binding protein. Both the mature 62 kDa AIF protein and the 57 kDa tAIF are capable of binding to XIAP, indicating that AIF may also have a role in caspase-dependent cell death (Wilkinson et al. 2008).

To examine the role of AIF in the mitochondria, knockout mice were generated. However, these mice do not survive past embryonic day 12 (Brown et al. 2006). Consequently, tissue-specific knockout mice have been developed. In cardiac and skeletal muscle, AIF knock out resulted in decreased ETC complex I activity, leading to severe dilated cardiomyopathy and skeletal muscle atrophy (Joza et al. 2005). In Harlequin mutant mice, where AIF expression is reduced to 10–20 % of the normal level, animals survive into adulthood but suffer progressive degeneration of terminally differentiated cerebellar and retinal neurons (Klein et al. 2002). Expression of mitochondrially anchored AIF was sufficient to rescue these neurons and enhance survival (Cheung et al. 2006; Klein et al. 2002). Knockdown of AIF in

HeLa cells was associated with a 40–50 % reduction in complex I activity, which led to reduced oxidative phosphorylation and cells becoming more reliant on glycolysis (Vahsen et al. 2004). These studies indicate that AIF plays an important role in mitochondrial bioenergetics alongside its role in cell death.

3.13 Endonuclease G

Endonuclease G is encoded in the nucleus as a 32 kDa precursor protein and shuttled to the mitochondria via a 48 amino acid N-terminal mitochondrial targeting sequence (Cote and Ruiz-Carrillo 1993). Once in the intermembrane space of the mitochondria, the mitochondrial targeting sequence is removed leaving the mature 27 kDa protein. Reports suggest that this protein is not soluble in the intermembrane space and may be tightly bound to the IMM (Arnoult et al. 2003; Uren et al. 2005). In the intermembrane space, endonuclease G has been shown to participate in the replication of mitochondrial DNA (mtDNA) via generation of primers required by DNA polymerase γ in the initiation of mtDNA replication. Overexpression of endonuclease G in cardiac tissue causes an increase in mitochondrial mass and respiration rates. Additionally, endonuclease G was shown to regulate genes controlling the ETC, including complex I, IV, and V subunits (Cote and Ruiz-Carrillo 1993). These data indicate that endonuclease G plays an important role in mitochondrial function and bioenergetics.

Upon MOMP, endonuclease G is released from the mitochondria. The BCL2 family of proteins are pivotal to this release. Absence of tBID and also overexpression of BCL2 completely block endonuclease G release (Li et al. 2001; van Loo et al. 2001). BAX is not sufficient to induce any detectable endonuclease G release (Arnoult et al. 2003). Endonuclease G release was observed in tBID-treated mitochondria after 15 min, and release was complete by 90 min (Li et al. 2001).

Recent studies examining the role of endonuclease G in Parkinson's disease also found that BNA3 and ECM33 may be involved in Endonuclease G release upon alpha-synuclein expression (Buttner et al. 2013). Once it is released from the mitochondria, endonuclease G localizes to the nucleus. Here, it first induces large scale DNA fragmentation, followed by oligonucleosomal DNA fragmentation in a caspase-independent manner (Lee et al. 2005; Li et al. 2001; Zanna et al. 2005). A link also exists between endonuclease G nuclear localization and delayed neuronal death after ischemic stroke (Lee et al. 2005). Endonuclease G was detected in the nucleus 4–24 h after transient focal cerebral ischemia. As the protein builds up in the nucleus, its levels in the mitochondria correspondingly drop. This group also noted that AIF and endonuclease G migrate to the nucleus in a simultaneous manner, suggesting AIF and endonuclease G work together in a caspase-independent cell death pathway (Cho and Toledo-Pereyra 2008; Lee et al. 2005), but their exact relationship has not been elucidated.

3.14 Cell Fate After MOMP and Release of IMS Proteins

Once MOMP occurs, it is often viewed as a point of no return for cells. However, characteristics of cancer cells promote survival following MOMP in certain situations. As caspases are the main executioners in the cell, it may be inferred that caspase inhibition may be the most efficient way to promote cell survival. However, even in the absence of caspase activation, cell death can still occur, albeit at a slower rate, in a process termed Caspase-Independent Cell Death (CICD) (Colell et al. 2007; Green and Kroemer 2004; Haraguchi et al. 2000; Tait and Green 2008). Mitochondrial IMS proteins such as AIF, endonuclease G, and OMI/HTRA2 contribute to CICD (Green and Kroemer 2004; Tait and Green 2008). While CICD occurs upon overexpression of these proteins, the process will still occur in their absence, indicating other mechanisms are involved in CICD. CICD was shown to correlate with a decline in ATP generation and mitochondrial bioenergetics, suggesting that loss of mitochondrial function is an important factor in CICD (Colell et al. 2007). In situations where caspases are inhibited, GAPDH promotes cell survival. GAPDH not only saved cells from CICD, it also promoted clonogenic outgrowths (Colell et al. 2007). This effect is mediated by GAPDH on two levels. Firstly, GAPDH increases the rate of glycolysis, thus providing ATP for mitochondrial regeneration. Secondly, it enhances autophagy, which protects the cell from CICD (Colell et al. 2007). Interestingly, many tumors have defects in APAF1, leading to impaired caspase activation, in combination with elevated GAPDH levels, which may contribute to survival in these cells (Liu et al. 2002; Revillion et al. 2000). Another mechanism by which cells can evade death is via incomplete MOMP of a subset of mitochondria. Live cell imaging revealed that not all mitochondria undergo MOMP during apoptosis, independent of caspase activity. Incomplete MOMP is associated with cellular recovery, but only in the absence of caspase activity (Tait et al. 2010). The development of selective BCL2 subfamily antagonists, which can be titrated to induce partial MOMP, will in the future allow a more detailed analysis of such cell survival and cell death decisions.

Acknowledgements Research in the Prehn laboratory is supported by Science Foundation Ireland (13/IA/1881).

References

Abou-Sleiman PM, Muqit MM, Wood NW (2006) Expanding insights of mitochondrial dysfunction in Parkinson's disease. Nat Rev Neurosci 7:207–219

Adrain C, Creagh EM, Martin SJ (2001) Apoptosis-associated release of Smac/DIABLO from mitochondria requires active caspases and is blocked by Bcl-2. EMBO J 20:6627–6636

Adrain C, Murphy BM, Martin SJ (2005) Molecular ordering of the caspase activation cascade initiated by the cytotoxic T lymphocyte/natural killer (CTL/NK) protease granzyme B. J Biol Chem 280:4663–4673

Andreu-Fernandez V, Genoves A, Lee TH, Stellato M, Lucantoni F, Orzaez M, Mingarro I, Aguilar MI, Perez-Paya E (2014) Peptides derived from the transmembrane domain of Bcl-2 proteins as potential mitochondrial priming tools. ACS Chem Biol 9(8):1799–1811

Arnoult D, Gaume B, Karbowski M, Sharpe JC, Cecconi F, Youle RJ (2003) Mitochondrial release of AIF and EndoG requires caspase activation downstream of Bax/Bak-mediated permeabilization. EMBO J 22:4385–4399

Arnoult D, Parone P, Martinou JC, Antonsson B, Estaquier J, Ameisen JC (2002) Mitochondrial release of apoptosis-inducing factor occurs downstream of cytochrome c release in response to several proapoptotic stimuli. J Cell Biol 159:923–929

Artus C, Boujrad H, Bouharrour A, Brunelle MN, Hoos S, Yuste VJ, Lenormand P, Rousselle JC, Namane A, England P, Lorenzo HK, Susin SA (2010) AIF promotes chromatinolysis and caspase-independent programmed necrosis by interacting with histone H2AX. EMBO J 29:1585–1599

Austin S, St-Pierre J (2012) PGC1alpha and mitochondrial metabolism--emerging concepts and relevance in ageing and neurodegenerative disorders. J Cell Sci 125:4963–4971

Baritaud M, Boujrad H, Lorenzo HK, Krantic S, Susin SA (2010) Histone H2AX: The missing link in AIF-mediated caspase-independent programmed necrosis. Cell Cycle 9:3166–3173

Barry M, Heibein JA, Pinkoski MJ, Lee SF, Moyer RW, Green DR, Bleackley RC (2000) Granzyme B short-circuits the need for caspase 8 activity during granule-mediated cytotoxic T-lymphocyte killing by directly cleaving Bid. Mol Cell Biol 20:3781–3794

Bartke T, Pohl C, Pyrowolakis G, Jentsch S (2004) Dual role of BRUCE as an antiapoptotic IAP and a chimeric E2/E3 ubiquitin ligase. Mol Cell 14:801–811

Bhola PD, Mattheyses AL, Simon SM (2009) Spatial and temporal dynamics of mitochondrial membrane permeability waves during apoptosis. Biophys J 97:2222–2231

Brown D, Yu BD, Joza N, Benit P, Meneses J, Firpo M, Rustin P, Penninger JM, Martin GR (2006) Loss of Aif function causes cell death in the mouse embryo, but the temporal progression of patterning is normal. Proc Natl Acad Sci U S A 103:9918–9923

Buttner S, Habernig L, Broeskamp F, Ruli D, Vogtle FN, Vlachos M, Macchi F, Kuttner V, Carmona-Gutierrez D, Eisenberg T, Ring J, Markaki M, Taskin AA, Benke S, Ruckenstuhl C, Braun R, Van den Haute C, Bammens T, van der Perren A, Frohlich KU, Winderickx J, Kroemer G, Baekelandt V, Tavernarakis N, Kovacs GG, Dengjel J, Meisinger C, Sigrist SJ, Madeo F (2013) Endonuclease G mediates alpha-synuclein cytotoxicity during Parkinson's disease. EMBO J 32:3041–3054

Certo M, Del Gaizo MV, Nishino M, Wei G, Korsmeyer S, Armstrong SA, Letai A (2006) Mitochondria primed by death signals determine cellular addiction to antiapoptotic BCL-2 family members. Cancer Cell 9:351–365

Chai J, Du C, Wu J-W, Kyin S, Wang X, Shi Y (2000) Structural and biochemical basis of apoptotic activation by Smac/DIABLO. Nature 406:855

Chandra D, Bratton SB, Person MD, Tian Y, Martin AG, Ayres M, Fearnhead HO, Gandhi V, Tang DG (2006) Intracellular nucleotides act as critical prosurvival factors by binding to cytochrome c and inhibiting apoptosome. Cell 125:1333–1346

Chao DT, Korsmeyer SJ (1998) BCL-2 family: regulators of cell death. Annu Rev Immunol 16:395–419

Chen L, Willis SN, Wei A, Smith BJ, Fletcher JI, Hinds MG, Colman PM, Day CL, Adams JM, Huang DC (2005) Differential targeting of prosurvival Bcl-2 proteins by their BH3-only ligands allows complementary apoptotic function. Mol Cell 17:393–403

Cheng J, Zhu Y, He S, Lu Y, Chen J, Han B, Petrillo M, Wrzeszczynski KO, Yang S, Dai P, Zhai S, Han D, Zhang MQ, Li W, Liu X, Li H, Chen ZY, Yuan H (2011) Functional mutation of SMAC/DIABLO, encoding a mitochondrial proapoptotic protein, causes human progressive hearing loss DFNA64. Am J Hum Genet 89:56–66

Cheung EC, Joza N, Steenaart NA, McClellan KA, Neuspiel M, McNamara S, MacLaurin JG, Rippstein P, Park DS, Shore GC, McBride HM, Penninger JM, Slack RS (2006) Dissociating the dual roles of apoptosis-inducing factor in maintaining mitochondrial structure and apoptosis. EMBO J 25:4061–4073

Cho BB, Toledo-Pereyra LH (2008) Caspase-independent programmed cell death following ischemic stroke. J Invest Surg 21:141–147

Colbeau A, Nachbaur J, Vignais PM (1971) Enzymic characterization and lipid composition of rat liver subcellular membranes. Biochim Biophys Acta 249:462–492

Colell A, Ricci J-E, Tait S, Milasta S, Maurer U, Bouchier-Hayes L, Fitzgerald P, Guio-Carrion A, Waterhouse NJ, Li CW, Mari B, Barbry P, Newmeyer DD, Beere HM, Green DR (2007) GAPDH and autophagy preserve survival after apoptotic cytochrome c release in the absence of caspase activation. Cell 129:983–997

Colombini M (1979) A candidate for the permeability pathway of the outer mitochondrial membrane. Nature 279:643–645

Colombini M (2012) VDAC structure, selectivity, and dynamics. Biochim Biophys Acta 1818:1457–1465

Cortese JD, Voglino AL, Hackenbrock CR (1991) Ionic strength of the intermembrane space of intact mitochondria as estimated with fluorescein-BSA delivered by low pH fusion. J Cell Biol 113:1331–1340

Cote J, Ruiz-Carrillo A (1993) Primers for mitochondrial DNA replication generated by endonuclease G. Science 261:765–769

Crimi M, Esposti MD (2011) Apoptosis-induced changes in mitochondrial lipids. Biochim Biophys Acta 1813:551–557

Cullen SP, Martin SJ (2008) Mechanisms of granule-dependent killing. Cell Death Differ 15:251–262

Czabotar PE, Lessene G, Strasser A, Adams JM (2014) Control of apoptosis by the BCL-2 protein family: implications for physiology and therapy. Nat Rev Mol Cell Biol 15:49–63

de Kroon AI, Dolis D, Mayer A, Lill R, de Kruijff B (1997) Phospholipid composition of highly purified mitochondrial outer membranes of rat liver and Neurospora crassa. Is cardiolipin present in the mitochondrial outer membrane? Biochim Biophys Acta 1325:108–116

Delettre C, Lenaers G, Griffoin JM, Gigarel N, Lorenzo C, Belenguer P, Pelloquin L, Grosgeorge J, Turc-Carel C, Perret E, Astarie-Dequeker C, Lasquellec L, Arnaud B, Ducommun B, Kaplan J, Hamel CP (2000) Nuclear gene OPA1, encoding a mitochondrial dynamin-related protein, is mutated in dominant optic atrophy. Nat Genet 26:207–210

Deveraux QL, Roy N, Stennicke HR, Van Arsdale T, Zhou Q, Srinivasula SM, Alnemri ES, Salvesen GS, Reed JC (1998) IAPs block apoptotic events induced by caspase-8 and cytochrome c by direct inhibition of distinct caspases. EMBO J 17:2215–2223

Dewson G, Kluck RM (2009) Mechanisms by which Bak and Bax permeabilise mitochondria during apoptosis. J Cell Sci 122:2801–2808

Du C, Fang M, Li Y, Li L, Wang X (2000) Smac, a mitochondrial protein that promotes cytochrome c-dependent caspase activation by eliminating IAP inhibition. Cell 102:33

Düßmann H, Rehm M, Kögel D, Prehn JHM (2003) Outer mitochondrial membrane permeabilization during apoptosis triggers caspase-independent mitochondrial and caspase-dependent plasma membrane potential depolarization: a single-cell analysis. J Cell Sci 116:525–536

Edlich F, Banerjee S, Suzuki M, Cleland MM, Arnoult D, Wang C, Neutzner A, Tjandra N, Youle RJ (2011) Bcl-x(L) retrotranslocates Bax from the mitochondria into the cytosol. Cell 145:104–116

Ferramosca A, Zara V (2013) Biogenesis of mitochondrial carrier proteins: molecular mechanisms of import into mitochondria. Biochim Biophys Acta 1833:494–502

Finkel T (2011) Signal transduction by reactive oxygen species. J Cell Biol 194:7–15

Fischer M, Riemer J (2013) The mitochondrial disulfide relay system: roles in oxidative protein folding and beyond. Int J Cell Biol 2013:12

Flanagan L, Sebastia J, Delgado ME, Lennon JC, Rehm M (2011) Dimerization of Smac is crucial for its mitochondrial retention by XIAP subsequent to mitochondrial outer membrane permeabilization. Biochim Biophys Acta 1813:819–826

Flanagan L, Sebastia J, Tuffy LP, Spring A, Lichawska A, Devocelle M, Prehn JHM, Rehm M (2010) XIAP impairs Smac release from the mitochondria during apoptosis. Cell Death Dis 1, e49

Fletcher JI, Meusburger S, Hawkins CJ, Riglar DT, Lee EF, Fairlie WD, Huang DC, Adams JM (2008) Apoptosis is triggered when prosurvival Bcl-2 proteins cannot restrain Bax. Proc Natl Acad Sci U S A 105:18081–18087

Frezza C, Cipolat S, Martins de Brito O, Micaroni M, Beznoussenko GV, Rudka T, Bartoli D, Polishuck RS, Danial NN, De Strooper B, Scorrano L (2006) OPA1 controls apoptotic cristae remodeling independently from mitochondrial fusion. Cell 126:177–189

Fuertes G, Garcia-Saez AJ, Esteban-Martin S, Gimenez D, Sanchez-Munoz OL, Schwille P, Salgado J (2010) Pores formed by Baxalpha5 relax to a smaller size and keep at equilibrium. Biophys J 99:2917–2925

Fulda S, Debatin KM (2006) Extrinsic versus intrinsic apoptosis pathways in anticancer chemotherapy. Oncogene 25:4798–4811

Garcia-Saez AJ, Coraiola M, Serra MD, Mingarro I, Muller P, Salgado J (2006) Peptides corresponding to helices 5 and 6 of Bax can independently form large lipid pores. FEBS J 273:971–981

Garcia Fernandez M, Troiano L, Moretti L, Nasi M, Pinti M, Salvioli S, Dobrucki J, Cossarizza A (2002) Early changes in intramitochondrial cardiolipin distribution during apoptosis. Cell Growth Differ 13:449–455

Goldstein JC, Waterhouse NJ, Juin P, Evan GI, Green DR (2000) The coordinate release of cytochrome c during apoptosis is rapid, complete and kinetically invariant. Nat Cell Biol 2:156

Gonçalves RB, Sanches D, Souza TLF, Silva JL, Oliveira AC (2008) The proapoptotic protein smac/DIABLO dimer Has the highest stability as measured by pressure and urea denaturation†. Biochemistry 47:3832–3841

Gonzalvez F, Pariselli F, Dupaigne P, Budihardjo I, Lutter M, Antonsson B, Diolez P, Manon S, Martinou JC, Goubern M, Wang X, Bernard S, Petit PX (2005) tBid interaction with cardiolipin primarily orchestrates mitochondrial dysfunctions and subsequently activates Bax and Bak. Cell Death Differ 12:614–626

Green DR, Kroemer G (2004) The pathophysiology of mitochondrial cell death. Science (New York, NY) 305:626–629

Gross A, McDonnell JM, Korsmeyer SJ (1999) BCL-2 family members and the mitochondria in apoptosis. Genes Dev 13:1899–1911

Han J, Goldstein LA, Gastman BR, Froelich CJ, Yin XM, Rabinowich H (2004) Degradation of Mcl-1 by granzyme B: implications for Bim-mediated mitochondrial apoptotic events. J Biol Chem 279:22020–22029

Hanske J, Toffey JR, Morenz AM, Bonilla AJ, Schiavoni KH, Pletneva EV (2012) Conformational properties of cardiolipin-bound cytochrome c. Proc Natl Acad Sci 109:125–130

Hao Y, Sekine K, Kawabata A, Nakamura H, Ishioka T, Ohata H, Katayama R, Hashimoto C, Zhang X, Noda T, Tsuruo T, Naito M (2004) Apollon ubiquitinates SMAC and caspase-9, and has an essential cytoprotection function. Nat Cell Biol 6:849–860

Haraguchi M, Torii S, Matsuzawa S, Xie Z, Kitada S, Krajewski S, Yoshida H, Mak TW, Reed JC (2000) Apoptotic protease activating factor 1 (Apaf-1)-independent cell death suppression by Bcl-2. J Exp Med 191:1709–1720

Heibein JA, Goping IS, Barry M, Pinkoski MJ, Shore GC, Green DR, Bleackley RC (2000) Granzyme B-mediated cytochrome c release is regulated by the Bcl-2 family members bid and Bax. J Exp Med 192:1391–1402

Hovius R, Thijssen J, van der Linden P, Nicolay K, de Kruijff B (1993) Phospholipid asymmetry of the outer membrane of rat liver mitochondria. Evidence for the presence of cardiolipin on the outside of the outer membrane. FEBS Lett 330:71–76

Hu J, Dong L, Outten CE (2008) The redox environment in the mitochondrial intermembrane space is maintained separately from the cytosol and matrix. J Biol Chem 283:29126–29134

Huber HJ, Dussmann H, Kilbride SM, Rehm M, Prehn JH (2011) Glucose metabolism determines resistance of cancer cells to bioenergetic crisis after cytochrome-c release. Mol Syst Biol 7:470

Hung V, Zou P, Rhee HW, Udeshi ND, Cracan V, Svinkina T, Carr SA, Mootha VK, Ting AY (2014) Proteomic mapping of the human mitochondrial intermembrane space in live cells via ratiometric APEX tagging. Mol Cell 55:332–341

Jenkins MR, Griffiths GM (2010) The synapse and cytolytic machinery of cytotoxic T cells. Curr Opin Immunol 22:308–313

John GB, Shang Y, Li L, Renken C, Mannella CA, Selker JM, Rangell L, Bennett MJ, Zha J (2005) The mitochondrial inner membrane protein mitofilin controls cristae morphology. Mol Biol Cell 16:1543–1554

Jones JM, Datta P, Srinivasula SM, Ji W, Gupta S, Zhang Z, Davies E, Hajnoczky G, Saunders TL, Van Keuren ML, Fernandes-Alnemri T, Meisler MH, Alnemri ES (2003) Loss of Omi

mitochondrial protease activity causes the neuromuscular disorder of mnd2 mutant mice. Nature 425:721–727
Joza N, Oudit GY, Brown D, Benit P, Kassiri Z, Vahsen N, Benoit L, Patel MM, Nowikovsky K, Vassault A, Backx PH, Wada T, Kroemer G, Rustin P, Penninger JM (2005) Muscle-specific loss of apoptosis-inducing factor leads to mitochondrial dysfunction, skeletal muscle atrophy, and dilated cardiomyopathy. Mol Cell Biol 25:10261–10272
Kagan VE, Tyurin VA, Jiang J, Tyurina YY, Ritov VB, Amoscato AA, Osipov AN, Belikova NA, Kapralov AA, Kini V, Vlasova II, Zhao Q, Zou M, Di P, Svistunenko DA, Kurnikov IV, Borisenko GG (2005) Cytochrome c acts as a cardiolipin oxygenase required for release of proapoptotic factors. Nat Chem Biol 1:223–232
Karbowski M, Lee YJ, Gaume B, Jeong SY, Frank S, Nechushtan A, Santel A, Fuller M, Smith CL, Youle RJ (2002) Spatial and temporal association of Bax with mitochondrial fission sites, Drp1, and Mfn2 during apoptosis. J Cell Biol 159:931–938
Kawamata H, Manfredi G (2010) Import, maturation, and function of SOD1 and its copper chaperone CCS in the mitochondrial intermembrane space. Antioxid Redox Signal 13:1375–1384
Kim H-E, Du F, Fang M, Wang X (2005) Formation of apoptosome is initiated by cytochrome c-induced dATP hydrolysis and subsequent nucleotide exchange on Apaf-1. Proc Natl Acad Sci U S A 102:17545–17550
Kischkel FC, Hellbardt S, Behrmann I, Germer M, Pawlita M, Krammer PH, Peter ME (1995) Cytotoxicity-dependent APO-1 (Fas/CD95)-associated proteins form a death-inducing signaling complex (DISC) with the receptor. EMBO J 14:5579–5588
Klein JA, Longo-Guess CM, Rossmann MP, Seburn KL, Hurd RE, Frankel WN, Bronson RT, Ackerman SL (2002) The harlequin mouse mutation downregulates apoptosis-inducing factor. Nature 419:367–374
Kostrzewa A, Pali T, Froncisz W, Marsh D (2000) Membrane location of spin-labeled cytochrome c determined by paramagnetic relaxation agents. Biochemistry 39:6066–6074
Krammer PH (2000) CD95's deadly mission in the immune system. Nature 407:789–795
Kuwana T, Bouchier-Hayes L, Chipuk JE, Bonzon C, Sullivan BA, Green DR, Newmeyer DD (2005) BH3 domains of BH3-only proteins differentially regulate Bax-mediated mitochondrial membrane permeabilization both directly and indirectly. Mol Cell 17:525–535
Lee BI, Lee DJ, Cho KJ, Kim GW (2005) Early nuclear translocation of endonuclease G and subsequent DNA fragmentation after transient focal cerebral ischemia in mice. Neurosci Lett 386:23–27
Letai A, Bassik MC, Walensky LD, Sorcinelli MD, Weiler S, Korsmeyer SJ (2002) Distinct BH3 domains either sensitize or activate mitochondrial apoptosis, serving as prototype cancer therapeutics. Cancer Cell 2:183–192
Li K, Li Y, Shelton JM, Richardson JA, Spencer E, Chen ZJ, Wang X, Williams RS (2000) Cytochrome c deficiency causes embryonic lethality and attenuates stress-induced apoptosis. Cell 101:389–399
Li LY, Luo X, Wang X (2001) Endonuclease G is an apoptotic DNase when released from mitochondria. Nature 412:95–99
Lindner AU, Concannon CG, Boukes GJ, Cannon MD, Llambi F, Ryan D, Boland K, Kehoe J, McNamara DA, Murray F, Kay EW, Hector S, Green DR, Huber HJ, Prehn JH (2013) Systems analysis of BCL2 protein family interactions establishes a model to predict responses to chemotherapy. Cancer Res 73:519–528
Liu JR, Opipari AW, Tan L, Jiang Y, Zhang Y, Tang H, Nunez G (2002) Dysfunctional apoptosome activation in ovarian cancer: implications for chemoresistance. Cancer Res 62:924–931
Liu X, Kim CN, Yang J, Jemmerson R, Wang X (1996) Induction of apoptotic program in cell-free extracts: requirement for dATP and cytochrome c. Cell 86:147–157
Liu Z, Sun C, Olejniczak ET, Meadows RP, Betz SF, Oost T, Herrmann J, Wu JC, Fesik SW (2000) Structural basis for binding of Smac/DIABLO to the XIAP BIR3 domain. Nature 408:1004–1008
Llambi F, Moldoveanu T, Tait SW, Bouchier-Hayes L, Temirov J, McCormick LL, Dillon CP, Green DR (2011) A unified model of mammalian BCL-2 protein family interactions at the mitochondria. Mol Cell 44:517–531

Luetjens CM, Kögel D, Reimertz C, Düßmann H, Renz A, Schulze-Osthoff K, Nieminen A-L, Poppe M, Prehn JHM (2001) Multiple kinetics of mitochondrial cytochrome cRelease in drug-induced apoptosis. Mol Pharmacol 60:1008–1019
Lutter M, Fang M, Luo X, Nishijima M, Xie X, Wang X (2000) Cardiolipin provides specificity for targeting of tBid to mitochondria. Nat Cell Biol 2:754–761
Mahrus S, Craik CS (2005) Selective chemical functional probes of granzymes A and B reveal granzyme B is a major effector of natural killer cell-mediated lysis of target cells. Chem Biol 12:567–577
Mannella CA (2006) Structure and dynamics of the mitochondrial inner membrane cristae. Biochim Biophys Acta 1763:542–548
Mannella CA, Pfeiffer DR, Bradshaw PC, Moraru II, Slepchenko B, Loew LM, Hsieh CE, Buttle K, Marko M (2001) Topology of the mitochondrial inner membrane: dynamics and bioenergetic implications. IUBMB Life 52:93–100
Martin AG, Fearnhead HO (2002) Apocytochrome c blocks caspase-9 activation and Bax-induced apoptosis. J Biol Chem 277:50834–50841
Martin SJ, Amarante-Mendes GP, Shi L, Chuang TH, Casiano CA, O'Brien GA, Fitzgerald P, Tan EM, Bokoch GM, Greenberg AH, Green DR (1996) The cytotoxic cell protease granzyme B initiates apoptosis in a cell-free system by proteolytic processing and activation of the ICE/CED-3 family protease, CPP32, via a novel two-step mechanism. EMBO J 15:2407–2416
Martinou J-C, Youle Richard J (2011) Mitochondria in apoptosis: Bcl-2 family members and mitochondrial dynamics. Dev Cell 21:92–101
Martins LM, Iaccarino I, Tenev T, Gschmeissner S, Totty NF, Lemoine NR, Savopoulos J, Gray CW, Creasy CL, Dingwall C, Downward J (2002) The serine protease Omi/HtrA2 regulates apoptosis by binding XIAP through a reaper-like motif. J Biol Chem 277:439–444
Mavridou DA, Ferguson SJ, Stevens JM (2013) Cytochrome c assembly. IUBMB Life 65:209–216
Medema JP, Toes RE, Scaffidi C, Zheng TS, Flavell RA, Melief CJ, Peter ME, Offringa R, Krammer PH (1997) Cleavage of FLICE (caspase-8) by granzyme B during cytotoxic T lymphocyte-induced apoptosis. Eur J Immunol 27:3492–3498
Metkar SS, Wang B, Ebbs ML, Kim JH, Lee YJ, Raja SM, Froelich CJ (2003) Granzyme B activates procaspase-3 which signals a mitochondrial amplification loop for maximal apoptosis. J Cell Biol 160:875–885
Mikhailov V, Mikhailova M, Degenhardt K, Venkatachalam MA, White E, Saikumar P (2003) Association of Bax and Bak homo-oligomers in mitochondria. Bax requirement for Bak reorganization and cytochrome c release. J Biol Chem 278:5367–5376
Mizutani Y, Katsuoka Y, Bonavida B (2010) Prognostic significance of second mitochondria-derived activator of caspase (Smac/DIABLO) expression in bladder cancer and target for therapy. Int J Oncol 37:503–508
Montessuit S, Somasekharan SP, Terrones O, Lucken-Ardjomande S, Herzig S, Schwarzenbacher R, Manstein DJ, Bossy-Wetzel E, Basanez G, Meda P, Martinou JC (2010) Membrane remodeling induced by the dynamin-related protein Drp1 stimulates Bax oligomerization. Cell 142:889–901
Munoz-Pinedo C, Guio-Carrion A, Goldstein JC, Fitzgerald P, Newmeyer DD, Green DR (2006) Different mitochondrial intermembrane space proteins are released during apoptosis in a manner that is coordinately initiated but can vary in duration. Proc Natl Acad Sci 103:11573–11578
Okada H, Suh WK, Jin J, Woo M, Du C, Elia A, Duncan GS, Wakeham A, Itie A, Lowe SW, Wang X, Mak TW (2002) Generation and characterization of Smac/DIABLO-deficient mice. Mol Cell Biol 22:3509–3517
Olichon A, Guillou E, Delettre C, Landes T, Arnaune-Pelloquin L, Emorine LJ, Mils V, Daloyau M, Hamel C, Amati-Bonneau P, Bonneau D, Reynier P, Lenaers G, Belenguer P (2006) Mitochondrial dynamics and disease, OPA1. Biochim Biophys Acta 1763:500–509
Osman C, Haag M, Potting C, Rodenfels J, Dip PV, Wieland FT, Brugger B, Westermann B, Langer T (2009) The genetic interactome of prohibitins: coordinated control of cardiolipin and phosphatidylethanolamine by conserved regulators in mitochondria. J Cell Biol 184:583–596

Otera H, Ohsakaya S, Nagaura Z, Ishihara N, Mihara K (2005) Export of mitochondrial AIF in response to proapoptotic stimuli depends on processing at the intermembrane space. EMBO J 24:1375–1386

Ow Y-LP, Green DR, Hao Z, Mak TW (2008) Cytochrome c: functions beyond respiration. Nat Rev Mol Cell Biol 9:532–542

Palmieri F, Pierri CL, De Grassi A, Nunes-Nesi A, Fernie AR (2011) Evolution, structure and function of mitochondrial carriers: a review with new insights. Plant J 66:161–181

Palumaa P, Kangur L, Voronova A, Sillard R (2004) Metal-binding mechanism of Cox17, a copper chaperone for cytochrome c oxidase. Biochem J 382:307–314

Polster BM, Basanez G, Etxebarria A, Hardwick JM, Nicholls DG (2005) Calpain I induces cleavage and release of apoptosis-inducing factor from isolated mitochondria. J Biol Chem 280:6447–6454

Porcelli AM, Ghelli A, Zanna C, Pinton P, Rizzuto R, Rugolo M (2005) pH difference across the outer mitochondrial membrane measured with a green fluorescent protein mutant. Biochem Biophys Res Commun 326:799–804

Portt L, Norman G, Clapp C, Greenwood M, Greenwood MT (2011) Anti-apoptosis and cell survival: a review. Biochim Biophys Acta 1813:238–259

Quan LT, Tewari M, O'Rourke K, Dixit V, Snipas SJ, Poirier GG, Ray C, Pickup DJ, Salvesen GS (1996) Proteolytic activation of the cell death protease Yama/CPP32 by granzyme B. Proc Natl Acad Sci U S A 93:1972–1976

Rehling P, Brandner K, Pfanner N (2004) Mitochondrial import and the twin-pore translocase. Nat Rev Mol Cell Biol 5:519–530

Rehm M, Dussmann H, Janicke RU, Tavare JM, Kogel D, Prehn JH (2002) Single-cell fluorescence resonance energy transfer analysis demonstrates that caspase activation during apoptosis is a rapid process. Role of caspase-3. J Biol Chem 277:24506–24514

Rehm M, Dussmann H, Prehn JHM (2003) Real-time single cell analysis of Smac/DIABLO release during apoptosis. J Cell Biol 162:1031–1043

Rehm M, Huber HJ, Dussmann H, Prehn JH (2006) Systems analysis of effector caspase activation and its control by X-linked inhibitor of apoptosis protein. EMBO J 25:4338–4349

Rehm M, Huber HJ, Hellwig CT, Anguissola S, Dussmann H, Prehn JH (2009) Dynamics of outer mitochondrial membrane permeabilization during apoptosis. Cell Death Differ 16:613–623

Revillion F, Pawlowski V, Hornez L, Peyrat JP (2000) Glyceraldehyde-3-phosphate dehydrogenase gene expression in human breast cancer. Eur J Cancer 36:1038–1042

Riemer J, Bulleid N, Herrmann JM (2009) Disulfide formation in the ER and mitochondria: Two solutions to a common process. Science (New York, NY) 324:1284–1287

Rotonda J, Garcia-Calvo M, Bull HG, Geissler WM, McKeever BM, Willoughby CA, Thornberry NA, Becker JW (2001) The three-dimensional structure of human granzyme B compared to caspase-3, key mediators of cell death with cleavage specificity for aspartic acid in P1. Chem Biol 8:357–368

Rousalova I, Krepela E (2010) Granzyme B-induced apoptosis in cancer cells and its regulation (review). Int J Oncol 37:1361–1378

Rytomaa M, Kinnunen PK (1994) Evidence for two distinct acidic phospholipid-binding sites in cytochrome c. J Biol Chem 269:1770–1774

Sattler M, Liang H, Nettesheim D, Meadows RP, Harlan JE, Eberstadt M, Yoon HS, Shuker SB, Chang BS, Minn AJ, Thompson CB, Fesik SW (1997) Structure of Bcl-xL-Bak peptide complex: recognition between regulators of apoptosis. Science (New York, NY) 275:983–986

Scaffidi C, Fulda S, Srinivasan A, Friesen C, Li F, Tomaselli KJ, Debatin KM, Krammer PH, Peter ME (1998) Two CD95 (APO-1/Fas) signaling pathways. EMBO J 17:1675–1687

Schellenberg B, Wang P, Keeble JA, Rodriguez-Enriquez R, Walker S, Owens TW, Foster F, Tanianis-Hughes J, Brennan K, Streuli CH, Gilmore AP (2013) Bax exists in a dynamic equilibrium between the cytosol and mitochondria to control apoptotic priming. Mol Cell 49:959–971

Scorrano L, Ashiya M, Buttle K, Weiler S, Oakes SA, Mannella CA, Korsmeyer SJ (2002) A distinct pathway remodels mitochondrial cristae and mobilizes cytochrome c during apoptosis. Dev Cell 2:55–67

Shi Y (2002) A conserved tetrapeptide motif: potentiating apoptosis through IAP-binding. Cell Death Differ 9:93–95

Sinibaldi F, Howes BD, Droghetti E, Polticelli F, Piro MC, Di Pierro D, Fiorucci L, Coletta M, Smulevich G, Santucci R (2013) Role of lysines in cytochrome c-cardiolipin interaction. Biochemistry 52:4578–4588

Srinivasula SM, Fernandes-Alnemri T, Zangrilli J, Robertson N, Armstrong RC, Wang L, Trapani JA, Tomaselli KJ, Litwack G, Alnemri ES (1996) The Ced-3/interleukin 1beta converting enzyme-like homolog Mch6 and the lamin-cleaving enzyme Mch2alpha are substrates for the apoptotic mediator CPP32. J Biol Chem 271:27099–27106

Strauss KM, Martins LM, Plun-Favreau H, Marx FP, Kautzmann S, Berg D, Gasser T, Wszolek Z, Muller T, Bornemann A, Wolburg H, Downward J, Riess O, Schulz JB, Kruger R (2005) Loss of function mutations in the gene encoding Omi/HtrA2 in Parkinson's disease. Hum Mol Genet 14:2099–2111

Susin SA, Lorenzo HK, Zamzami N, Marzo I, Snow BE, Brothers GM, Mangion J, Jacotot E, Costantini P, Loeffler M, Larochette N, Goodlett DR, Aebersold R, Siderovski DP, Penninger JM, Kroemer G (1999) Molecular characterization of mitochondrial apoptosis-inducing factor. Nature 397:441–446

Susin SA, Zamzami N, Castedo M, Hirsch T, Marchetti P, Macho A, Daugas E, Geuskens M, Kroemer G (1996) Bcl-2 inhibits the mitochondrial release of an apoptogenic protease. J Exp Med 184:1331–1341

Sutton VR, Davis JE, Cancilla M, Johnstone RW, Ruefli AA, Sedelies K, Browne KA, Trapani JA (2000) Initiation of apoptosis by granzyme B requires direct cleavage of bid, but not direct granzyme B-mediated caspase activation. J Exp Med 192:1403–1414

Suzuki M, Youle RJ, Tjandra N (2000) Structure of Bax: coregulation of dimer formation and intracellular localization. Cell 103:645–654

Suzuki Y, Imai Y, Nakayama H, Takahashi K, Takio K, Takahashi R (2001) A serine protease, HtrA2, is released from the mitochondria and interacts with XIAP, inducing cell death. Mol Cell 8:613–621

Tait SW, Green DR (2008) Caspase-independent cell death: leaving the set without the final cut. Oncogene 27:6452–6461

Tait SW, Parsons MJ, Llambi F, Bouchier-Hayes L, Connell S, Munoz-Pinedo C, Green DR (2010) Resistance to caspase-independent cell death requires persistence of intact mitochondria. Dev Cell 18:802–813

Talanian RV, Yang X, Turbov J, Seth P, Ghayur T, Casiano CA, Orth K, Froelich CJ (1997) Granule-mediated killing: pathways for granzyme B-initiated apoptosis. J Exp Med 186:1323–1331

Tamura Y, Endo T, Iijima M, Sesaki H (2009) Ups1p and Ups2p antagonistically regulate cardiolipin metabolism in mitochondria. J Cell Biol 185:1029–1045

Tatsuta T, Model K, Langer T (2005) Formation of membrane-bound ring complexes by prohibitins in mitochondria. Mol Biol Cell 16:248–259

Terziyska N, Lutz T, Kozany C, Mokranjac D, Mesecke N, Neupert W, Herrmann JM, Hell K (2005) Mia40, a novel factor for protein import into the intermembrane space of mitochondria is able to bind metal ions. FEBS Lett 579:179–184

Tikoo A, O'Reilly L, Day CL, Verhagen AM, Pakusch M, Vaux DL (2002) Tissue distribution of Diablo/Smac revealed by monoclonal antibodies. Cell Death Differ 9:710–716

Uren RT, Dewson G, Bonzon C, Lithgow T, Newmeyer DD, Kluck RM (2005) Mitochondrial release of pro-apoptotic proteins: ELECTROSTATIC INTERACTIONS CAN HOLD Cytochrome c BUT NOT Smac/DIABLO TO MITOCHONDRIAL MEMBRANES. J Biol Chem 280:2266–2274

Vahsen N, Cande C, Briere JJ, Benit P, Joza N, Larochette N, Mastroberardino PG, Pequignot MO, Casares N, Lazar V, Feraud O, Debili N, Wissing S, Engelhardt S, Madeo F, Piacentini M, Penninger JM, Schagger H, Rustin P, Kroemer G (2004) AIF deficiency compromises oxidative phosphorylation. EMBO J 23:4679–4689 van Loo G, Schotte P, van Gurp M, Demol H, Hoorelbeke B, Gevaert K, Rodriguez I, Ruiz-Carrillo A, Vandekerckhove J, Declercq W, Beyaert R, Vandenabeele P (2001) Endonuclease G: a mitochondrial protein released in apoptosis and involved in caspase-independent DNA degradation. Cell Death Differ 8:1136–1142

van Loo G, van Gurp M, Depuydt B, Srinivasula SM, Rodriguez I, Alnemri ES, Gevaert K, Vandekerckhove J, Declercq W, Vandenabeele P (2002) The serine protease Omi/HtrA2 is released from mitochondria during apoptosis. Omi interacts with caspase-inhibitor XIAP and induces enhanced caspase activity. Cell Death Differ 9:20–26

Vande Walle L, Van Damme P, Lamkanfi M, Saelens X, Vandekerckhove J, Gevaert K, Vandenabeele P (2007) Proteome-wide Identification of HtrA2/Omi Substrates. J Proteome Res 6:1006–1015

Verhagen AM, Ekert PG, Pakusch M, Silke J, Connolly LM, Reid GE, Moritz RL, Simpson RJ, Vaux DL (2000) Identification of DIABLO, a mammalian protein that promotes apoptosis by binding to and antagonizing IAP proteins. Cell 102:43–53

Vogel F, Bornhovd C, Neupert W, Reichert AS (2006) Dynamic subcompartmentalization of the mitochondrial inner membrane. J Cell Biol 175:237–247

Vogtle FN, Burkhart JM, Rao S, Gerbeth C, Hinrichs J, Martinou JC, Chacinska A, Sickmann A, Zahedi RP, Meisinger C (2012) Intermembrane space proteome of yeast mitochondria. Mol Cell Proteomics 11:1840–1852

Vucic D, Deshayes K, Ackerly H, Pisabarro MT, Kadkhodayan S, Fairbrother WJ, Dixit VM (2002) SMAC negatively regulates the anti-apoptotic activity of melanoma inhibitor of apoptosis (ML-IAP). J Biol Chem 277:12275–12279

Walczak H, Krammer PH (2000) The CD95 (APO-1/Fas) and the TRAIL (APO-2L) apoptosis systems. Exp Cell Res 256:58–66

Waterhouse NJ, Goldstein JC, von Ahsen O, Schuler M, Newmeyer DD, Green DR (2001) Cytochrome c maintains mitochondrial transmembrane potential and ATP generation after outer mitochondrial membrane permeabilization during the apoptotic process. J Cell Biol 153:319–328

Wei MC, Zong W-X, Cheng EHY, Lindsten T, Panoutsakopoulou V, Ross AJ, Roth KA, MacGregor GR, Thompson CB, Korsmeyer SJ (2001) Proapoptotic BAX and BAK: a requisite gateway to mitochondrial dysfunction and death. Science 292:727–730

Westphal D, Dewson G, Menard M, Frederick P, Iyer S, Bartolo R, Gibson L, Czabotar PE, Smith BJ, Adams JM, Kluck RM (2014) Apoptotic pore formation is associated with in-plane insertion of Bak or Bax central helices into the mitochondrial outer membrane. Proc Natl Acad Sci U S A 111:E4076–E4085

Wilkinson JC, Wilkinson AS, Galban S, Csomos RA, Duckett CS (2008) Apoptosis-inducing factor is a target for ubiquitination through interaction with XIAP. Mol Cell Biol 28:237–247

Willis SN, Chen L, Dewson G, Wei A, Naik E, Fletcher JI, Adams JM, Huang DC (2005) Proapoptotic Bak is sequestered by Mcl-1 and Bcl-xL, but not Bcl-2, until displaced by BH3-only proteins. Genes Dev 19:1294–1305

Willis SN, Fletcher JI, Kaufmann T, van Delft MF, Chen L, Czabotar PE, Ierino H, Lee EF, Fairlie WD, Bouillet P, Strasser A, Kluck RM, Adams JM, Huang DC (2007) Apoptosis initiated when BH3 ligands engage multiple Bcl-2 homologs, not Bax or Bak. Science (New York, NY) 315:856–859

Wu G, Chai J, Suber TL, Wu JW, Du C, Wang X, Shi Y (2000) Structural basis of IAP recognition by Smac/DIABLO. Nature 408:1008–1012

Würstle ML, Rehm M (2014) A systems biology analysis of apoptosome formation and apoptosis execution supports allosteric procaspase-9 activation. J Biol Chem 289:26277–26289

Xu R, Hu Q, Ma Q, Liu C, Wang G (2014) The protease Omi regulates mitochondrial biogenesis through the GSK3[beta]/PGC-1[alpha] pathway. Cell Death Dis 5, e1373

Ye B, Maret W, Vallee BL (2001) Zinc metallothionein imported into liver mitochondria modulates respiration. Proc Natl Acad Sci U S A 98:2317–2322

Yin XM, Oltvai ZN, Korsmeyer SJ (1994) BH1 and BH2 domains of Bcl-2 are required for inhibition of apoptosis and heterodimerization with Bax. Nature 369:321–323

Youle RJ, Karbowski M (2005) Mitochondrial fission in apoptosis. Nat Rev Mol Cell Biol 6:657–663

Youle RJ, Strasser A (2008) The BCL-2 protein family: opposing activities that mediate cell death. Nat Rev Mol Cell Biol 9:47–59

Yuste VJ, Moubarak RS, Delettre C, Bras M, Sancho P, Robert N, d'Alayer J, Susin SA (2005) Cysteine protease inhibition prevents mitochondrial apoptosis-inducing factor (AIF) release. Cell Death Differ 12:1445–1448

Zanna C, Ghelli A, Porcelli AM, Martinuzzi A, Carelli V, Rugolo M (2005) Caspase-independent death of Leber's hereditary optic neuropathy cybrids is driven by energetic failure and mediated by AIF and Endonuclease G. Apoptosis 10:997–1007

Zutz A, Gompf S, Schagger H, Tampe R (2009) Mitochondrial ABC proteins in health and disease. Biochim Biophys Acta 1787:681–690

Chapter 4
Bcl-2 Protein Interplay on the Outer Mitochondrial Membrane

Frank Edlich and Jean-Claude Martinou

Apoptosis is a fundamental process in all multicellular organisms; it removes cells that are less fit, superfluous, infected, or damaged (Clavería et al. 2013; Sulston and Horvitz 1977; Zychlinsky et al. 1992) and it provides the most prominent defense against tumor development. Cells that undergo apoptosis become fragmented into membrane-bound vesicles termed "apoptotic bodies," which are then completely eliminated by phagocytosis (Kerr et al. 1972). This form of cell death protects the neighboring cells from potentially harmful contents of damaged cells and is essential for tissue homeostasis, immunity, and development.

Mitochondrial apoptosis signaling is the most common form of programmed death (Hotchkiss et al. 2009) and involves proteins of the B-cell lymphoma-2 (Bcl-2) family. Bcl-2 has been found to be overexpressed following the common t(14; 18) chromosome translocation in B-cell follicular lymphoma cells (Tsujimoto et al. 1985) and later has been shown to protect cells from programmed cell death (Vaux et al. 1988). Bcl-2 proteins have been classified into two functional classes according to their activities in programmed cell death (Youle and Strasser 2008). The pro-apoptotic family members, Bcl-2-associated X protein (Bax) and Bcl-2 antagonist killer 1 (Bak), can commit cells to apoptosis (Oltvai et al. 1993; Shimizu et al. 1999). Their activation leads to the permeabilization of the outer mitochondrial

F. Edlich (✉)
Institute for Biochemistry and Molecular Biology, University of Freiburg, Stefan-Meier-Str.17, Freiburg 79104, Germany

BIOSS, Centre for Biological Signaling Studies, University of Freiburg, Freiburg 79104, Germany
e-mail: frank.edlich@biochemie.uni-freiburg.de

J.-C. Martinou
Department of Cell Biology, University of Geneva, 30 quai Ernest-Ansermet, 1211, Genève 4, Switzerland
e-mail: jean-claude.martinou@unige.ch

D.M. Hockenbery (ed.), *Mitochondria and Cell Death*, Cell Death in Biology and Diseases, DOI 10.1007/978-1-4939-3612-0_4

X.-M. Yin and Z. Dong (Series eds.), *Cell Death in Biology and Diseases*

membrane (OMM) and subsequent release of proteins such as cytochrome c from the intermembrane space (IMS) into the cytoplasm (Eskes et al. 1998). Release of IMS proteins by Bax and/or Bak results in mitochondrial dysfunction and initiates a cascade of cysteinyl aspartate proteases (caspases) that ultimately leads to complete dismantling of the cell (Bratton and Cohen 2001; Green and Kroemer 2004). Therefore, the activation of Bax and Bak is the first irreversible step in intrinsic apoptosis signaling, leading to full activation of caspases. However, some neuronal cells and cardiomyocytes can survive mitochondrial outer membrane permeabilization (MOMP), when caspase activation is inhibited or reduced (Martinou et al. 1999; Potts et al. 2005). Moreover, overexpression of glyceraldehyde-3-phosphate dehydrogenase has been found to protect cells downstream of MOMP, through stimulation of autophagy (Colell and Green 2009). Therefore, some cell types could have additional layers of protection from regulated suicide.

Bax and Bak share a high degree of functional redundancy (Lindsten et al. 2000). While each single gene knockout has only a mild phenotype in the mouse, the double knockout of both pro-apoptotic genes results in high embryonic lethality. Recent observations suggest that the Bcl-2-related ovarian killer, Bok, could have a similar function to Bax and Bak (D.R. Green and A. Strasser, personal communications). However, the function of this protein remains controversial, in particular because Bok-deficient mice are normal and deletion of Bok in a Bax and Bak double KO strain does not further modify the phenotype (Ke et al. 2013). In addition, Bok has been shown to localize to the ER and the GOLGI (Echeverry et al. 2013), which may argue against a role of Bok in mitochondrial apoptosis signaling.

Pro-apoptotic Bcl-2 proteins are antagonized by their pro-survival counterparts, namely Bcl-2, B-cell lymphoma-extra large (Bcl-x_L), Bcl-2-like protein 2 (Bcl-w), Bcl-2-like protein 10 (Bcl-B), myeloid cell leukemia 1 (Mcl-1), and Bcl-2-related gene A1 (A1). All Bcl-2 family proteins contain four conserved motifs known as Bcl-2 homology (BH) domains 1-4 (BH1-4). They also share a globular protein fold with bundled α helices surrounding a central hydrophobic α helix (Fig. 4.1; Muchmore et al. 1996). This fold generates a hydrophobic surface groove that is occupied in the cytosolic forms of Bax, Bcl-w, and Bcl-x_L by the hydrophobic C-terminal membrane anchor (MA) domain (Hinds et al. 2003; Suzuki et al. 2000). Of note, other Bcl-2 protein structures have been obtained using C-terminal protein truncations and thus intramolecular interactions between MA and hydrophobic groove in these proteins have not been characterized. The hydrophobic groove can also mediate intermolecular interactions with the BH3 domain of other Bcl-2 proteins, as originally shown for the binding of the Bak BH3 peptide to Bcl-x_L (Sattler et al. 1997). The central importance of this interaction in the regulation of pro-apoptotic and pro-survival Bcl-2 proteins and thus mitochondrial apoptosis signaling has been demonstrated by single amino acid substitutions in both the hydrophobic groove and the BH3 domains and by the use of low molecular weight mimetics of the BH3 domain (van Delft et al. 2006; Desagher et al. 1999; Dlugosz et al. 2006; Fletcher et al. 2008; Kvansakul et al. 2007; Sedlak et al. 1995). From these studies, the structural insights into Bcl-2 protein complexes have been deduced from interactions between short peptides and truncated Bcl-2 proteins, and it would be inter-

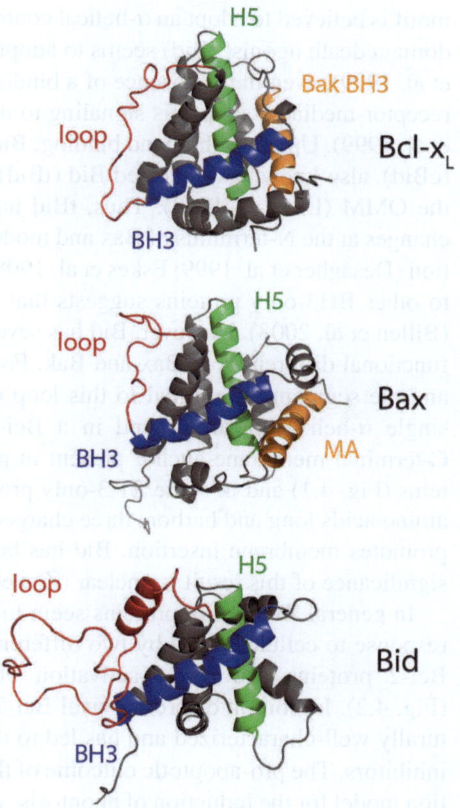

Fig. 4.1 Representation of the three-dimensional structure of Bcl-xL complexed with the Bak BH3 peptide (*top*, PDB, 1BXL), Bax (*center*, PDB, 1F16), and Bid (*bottom*, PDB, 2BID) in gray using Pymol software (DeLano Scientific LLC). The BH3 domains (*blue*), the central α helix 5 (H5, *green*), and the extended loop (*red*) between α helices 1 and 2 are displayed in all three structures. In addition, the Bax MA and the Bak BH3 peptide binding the hydrophobic groove of Bcl-xL are shown in orange

esting to compare these findings with studies of the full-length proteins. However, a prerequisite for interaction between Bcl-2 proteins is the presence of a lipid membrane (Roucou et al. 2002). This has led to the so-called "embedded together" model, suggesting that intermolecular interactions between Bcl-2 family proteins form homo- and hetero-complexes and require membrane insertion (Leber et al. 2007; Lovell et al. 2008). Within cells, Bcl-2 proteins interact predominantly in the OMM and localization to this compartment is mediated by the C-terminal membrane anchor (Jeong et al. 2004).

4.1 Bcl-2 Protein Regulation by BH3-Only Proteins

As the name suggests, the unifying feature among BH3-only proteins is the presence of a single Bcl-2 homology motif (BH3). In contrast to pro-apoptotic and pro-survival Bcl-2 proteins, many BH3-only proteins, such as the Bcl-2 interacting mediator of cell death (Bim), Bcl-2-associated death promoter (Bad), and Bcl-2-modifying factor (Bmf), are believed to be intrinsically unstructured (Hinds et al. 2007), and the BH3 motif is believed to adopt an α-helical configuration. In contrast, the BH3-interacting domain death agonist (Bid) seems to adopt a Bcl-2 protein-like fold (Fig. 4.1, Chou et al. 1999), even in the absence of a binding partner. Bid plays a key role in linking receptor-mediated apoptosis signaling to the mitochondrial apoptosis pathway (Yin et al. 1999). Upon death ligand binding, Bid is cleaved by caspase-8 and cleaved Bid (cBid), also known as truncated Bid (tBid), can efficiently activate Bax and Bak on the OMM (Li et al. 1998). Thus, tBid has been shown to induce conformational changes at the N-terminus of Bax and modulate its OMM insertion and oligomerization (Desagher et al. 1999; Eskes et al. 1998). The unusual structure of Bid compared to other BH3-only proteins suggests that it may share similar functions with Bax (Billen et al. 2008). However, Bid has several unique structural features, suggesting functional differences to Bax and Bak. For instance, Bid features a very long loop, and the segment N-terminal to this loop consists of two α-helices rather than the single α-helix typically found in a Bcl-2-like protein fold. Bid also lacks the C-terminal membrane anchor present in pro-apoptotic and pro-survival Bcl-2 proteins (Fig. 4.1) and in some BH3-only proteins. The C-terminal Bid helix is only 9 amino acids long and harbors three charged residues, which suggests that it unlikely promotes membrane insertion. Bid has been reported to be myristoylated, but the significance of this result is unclear (Zha et al. 2000).

In general, BH3-only proteins seem to carry out their pro-apoptotic function in response to cellular stress by two different mechanisms: inhibition of pro-survival Bcl-2 proteins and direct activation of pro-apoptotic Bcl-2 family members (Fig. 4.2). Inhibition of pro-survival Bcl-2 proteins by BH3-only proteins is structurally well-characterized and has led to the development of low molecular weight inhibitors. The pro-apoptotic outcome of this interaction suggests a simple competition model for the induction of apoptosis, whereby BH3-only proteins compete with Bax or Bak for pro-survival Bcl-2 protein binding Bax and Bak (Willis et al. 2005, 2007). This model suggests that Bax or Bak become active upon release from pro-survival Bcl-2 proteins and are thus free to commit the cell to apoptosis. However, stable complexes between pro-survival and pro-apoptotic Bcl-2 proteins have not been detected in cells (Hsu and Youle 1997) and an alternative direct activation model has been proposed, which suggests that Bax or Bak can be activated directly by binding to "activator" BH3-only proteins (Kim et al. 2006; Kuwana et al. 2005; Letai et al. 2002). In this model, pro-survival Bcl-2 proteins are suggested to prevent Bax/Bak activation by binding to and sequestering the putative activator BH3-only proteins. Stress-induced signaling pathways release the activating BH3-only proteins from the hydrophobic groove of pro-survival Bcl-2 proteins through competition with other BH3-only proteins. The displaced activator BH3-only proteins are then able to activate Bax or Bak by direct interaction. The detailed mechanism of this activation remains unknown, although based on the fact that BH3-only proteins do not remain associated with activated pro-apoptotic Bcl-2 proteins, a "hit-and-run" scenario has been proposed (Wei et al. 2000). Clearly, these two proposed activities of BH3-only proteins on either the pro-apoptotic or pro-survival members of the Bcl-2 family are not mutually exclusive (Llambi et al. 2011) and the predominant mode of action may vary depending on the cell type and stimulus.

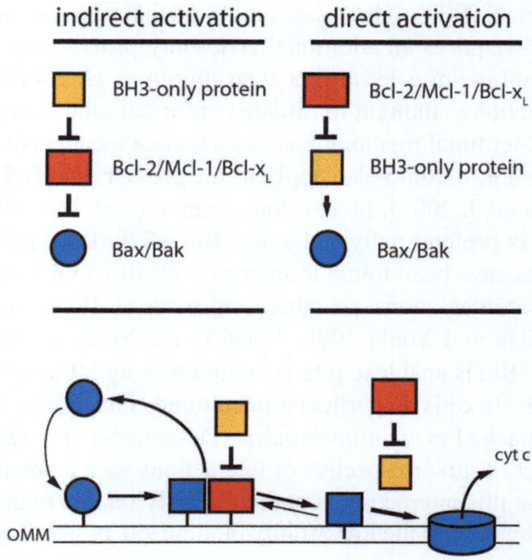

Fig. 4.2 Bax (*blue*) is regulated by the competitive interplay between pro-survival Bcl-2 proteins (*red*) and BH3-only proteins (*orange*), specifically by interactions leading to indirect (*left*) or direct (*right*) Bax activation. In cells, cytosolic Bax translocates constantly to the OMM and undergoes a conformational change (*bottom*). In the absence of apoptotic signaling, mitochondrial Bax is retrotranslocated into the cytosol by pro-survival Bcl-2 proteins and is thereby inhibited. BH3-only proteins can inhibit the retrotranslocation activity of pro-survival Bcl-2 proteins thus leading indirectly to Bax activation (*left top* and *bottom*). Bax activation, on the other hand, does not result from mitochondrial Bax accumulation per se, but at least in some situations seems to require direct interactions with BH3-only proteins (*right top* and *bottom*). Direct Bax activation can be counteracted by pro-survival Bcl-2 proteins

According to the "embedded together" model, BH3-only proteins possess a third role in apoptosis signaling. Experiments using liposomes have shown the potential of BH3-only proteins to recruit pro-survival Bcl-2 family members to the OMM (Lovell et al. 2008). Since this recruitment into the OMM is thought to activate pro-survival Bcl-2 proteins, this function of BH3-only proteins would be anti-apoptotic and thus contradict other reported BH3-only protein activities.

4.2 Retrotranslocation of Bcl-2 Proteins Controls Commitment to Apoptosis

Extensive studies have investigated the mechanism by which pro-survival Bcl-2 family members control the activity of Bax and Bak, but our understanding of this central process in cell physiology is still far from complete. One issue is an apparent difference in the subcellular localizations of the pro-apoptotic proteins: while Bak is found largely on the OMM, Bax resides primarily in the cytoplasm (Hsu and

Youle 1998; Wei et al. 2000). It has been suggested that translocation of Bax, but not Bak, to the OMM requires an additional BH3-only protein-dependent interaction with a low affinity binding site on Bax (Gavathiotis et al. 2008; Leshchiner et al. 2013). This interaction is thought to initiate conformational changes leading to the exposure of the C-terminal membrane anchor. If such mechanism is able to expose the membrane anchor, it could also apply to the pro-survival Bcl-2 proteins Bcl-w and Bcl-x_L (Hinds et al. 2003). In addition, it remains unclear whether some death stimuli activate Bax preferentially and others Bak (Geissler et al. 2013).

Several proteins have been found to interact with Bax in the cytosol, but the significance of these interactions remains controversial (Lucken-Ardjomande and Martinou 2005; Hsu and Youle 1998; Vogel et al. 2012). Furthermore, activated BH3-only protein tBid is unable to interact with full-length Bax or Bcl-x_L in solution (Lovell et al. 2008). In cells, in particular in cultured cells, Bax is not only cytosolic, but also weakly attached to the mitochondria (Desagher et al. 1999). Therefore, Bax translocation either occurs irrespective of interactions with a potentially low affinity binding site or else this interaction is not necessarily related to apoptosis signaling. In the latter case, binding to the low affinity binding site might play a role in increasing Bax translocation to the mitochondria following an apoptotic stimulus. Notably, single amino acid substitutions within a low affinity binding site of Bax failed to interfere with Bax activation (Okamoto et al. 2013; Peng et al. 2013). However, constant translocation of Bax to the OMM includes the risk of Bax activation.

Constant retrotranslocation of mitochondrial Bax into the cytosol stabilizes the inactive monomeric Bax conformation and protects proliferating cells from Bax activity (Edlich et al. 2011). Bax retrotranslocation is conformation-dependent. In the inactive form, the Bax BH3 domain interacts with the hydrophobic groove of a pro-survival Bcl-2 protein (Bcl-2, Bcl-x_L or Mcl-1) and both Bcl-2 proteins co-retrotranslocate into the cytosol (Fig. 4.2, Edlich et al. 2011). The rate of retrotranslocation defines the size of the mitochondrial Bax pool, and as a consequence, the cell fate following induction of apoptosis (Todt et al. 2013). Bax translocation to the OMM and Bax activation, although often observed in concert, are two separate processes (Todt et al. 2013). Therefore, Bax can shift from the cytosol to the mitochondria and back during anoikis and reattachment without committing the cells to apoptosis (Owens et al. 2009; Valentijn et al. 2003). It remains to be elucidated whether this shift is caused by altered Bcl-2 protein-dependent retrotranslocation or by a Bcl-2 protein-independent mechanism (Schellenberg et al. 2013).

Interestingly, retrotranslocation also regulates predominantly mitochondrial Bak (Todt et al. 2015) and the differential localization pattern of Bax and Bak in proliferating cells results simply from a higher rate of Bax shuttling compared to Bak. As with Bax retrotranslocation, Bak shuttling depends on binding of its BH3 motif to the hydrophobic groove of pro-survival Bcl-2 proteins, although in contrast to Bax, only Bcl-x_L and Mcl-1 but not Bcl-2 can accelerate Bak shuttling (Todt et al. 2015). This observation corroborates earlier findings that Bcl-2 may not be involved in Bak regulation (Oltersdorf et al. 2005). Bax and Bak engineered to contain the same MA display similar shuttling rates and steady state localizations: the Bax MA shifts Bak into the cytosol, while the Bak MA localizes Bax to the OMM (Ferrer et al. 2012; Todt et al. 2015). However, the differential retrotranslocation rates of Bax and Bak appear to be determined by the hydrophobicity of the C-terminal MA rather than its specific amino acid sequence (Todt et al. 2015).

Strikingly, reduced Bax retrotranslocation rates result in Bax activity and commitment of the cell to apoptosis even in the absence of apoptotic stimulation (Todt et al. 2015). This observation indicates that Bax-specific activation occurs on the OMM and implies that Bax, unlike Bak, must be retrotranslocated quickly in order to prevent its activation. Mitochondrial Bax activation provides a rationale for differential shuttling of Bax and Bak and the resulting differential cellular localization pattern of both pro-apoptotic Bcl-2 proteins. Of note, the absence of the "activator" BH3-only proteins Bid, Bim, and Puma does not lead to the same severe phenotype seen in the Bax/Bak double knockout (Ren et al. 2010). This observation suggests the presence of an additional activation mechanism. Possibly, Bax and Bak could be activated by different BH3-only proteins or by a completely different and as yet unexplained mechanism (Moldoveanu et al. 2014). Perhaps a p53-induced, transcription-independent mechanism could account for BH3-only independent Bax activation (Chipuk et al. 2003). In this case, the cellular signals discriminating between cytosolic and nuclear activities of p53 and thus the contribution of this mechanism to Bax activation remain to be discovered (Schuler and Green 2005). Bax activation at the OMM, and the resulting inhibition of retrotranslocation, is possible if BH3-only proteins are available and can interact with Bax. This is not the case when pro-survival Bcl-2 proteins are overexpressed (Willis et al. 2005, 2007). Regardless of the nature of Bax-specific mitochondrial activation, the cell is required to ensure fast Bax retrotranslocation and stabilization of Bax in the cytosol to constantly prevent apoptosis.

4.3 Commitment to Apoptosis

The activation of Bax and Bak in the OMM is usually an irreversible process. In most situations, Bax activation is triggered by accumulation of the membrane integral form of Bax, which adopts a protein fold recognized by the monoclonal antibody 6A7 (Hsu and Youle 1998; Martinou et al. 1999). The experimental demonstration of Bak activation is more difficult due to the predominantly mitochondrial localization of the inactive form of Bak. Concomitant with activation, Bax oligomerizes to large clusters and cytochrome *C* is released from the IMS into the cytosol (Youle and Strasser 2008). The coincidence of these processes has led to the widely accepted view that Bax forms pores in the OMM.

The idea that Bax can form pores was initially based on similarities between the 3D structure of Bcl-x_L with some bacterial toxins, colicins, and the TM domain of diphtheria toxins, which are known to form pores (Muchmore et al. 1996). Based on these observations, the pore-forming ability of Bcl-2 family proteins, including Bcl-x_L, Bcl-2, and Bax, was tested in synthetic membranes and Bax was indeed shown to oligomerize and form pores (Antonsson et al. 1997; Antonsson et al. 2000; Minn et al. 1997; Korsmeyer et al. 2000; Minn et al. 1997). In fact, Bax activity seems to be capable of permeabilizing lipid membranes of very different shapes and composition (Annis et al. 2005; Bleicken et al. 2013). Investigation of Bax pore-forming activity in liposome systems has also produced a broad variety of potential active Bax states, ranging from small to large proteinaceous pores and lipidic pores (Annis et al. 2005; Kuwana et al. 2002). In addition to proteinaceous Bax or Bax/Bak pores and lipidic pores containing either Bax or Bak or both, pore complexes containing Bax and porins voltage-dependent anion channel 1-3 (VDAC1-3) and Bax-induced open or closed states of VDAC have been suggested as underlying processes in OMM permeabilization. Extensive efforts in structural biology in the last 15 years have failed to elucidate the structure of active Bax. However, several steps in Bax activation have been defined by biochemical investigations. The first stage of Bax activation appears to be the exposure of N-terminal residues (Griffiths et al. 2001; Hsu and Youle 1997), which was primarily observed in cultured cells using an antibody directed against the N-terminus of Bax (Desagher et al. 1999). The exposure of the N-terminus is followed by exposure of the C-terminal membrane anchor (Suzuki et al. 2000). Recent structural evidence indicates that in addition to the C-terminal helix, α helices 6–8, coined "the latch," are also exposed (Czabotar et al. 2013). Subsequently, transient exposure of the Bax BH3 domain is thought to occur (Dewson et al. 2008). It should be noted that these processes may be involved not only in Bax activation, but also be part of Bax shuttling between cytosol and mitochondria (Fig. 4.2, Edlich et al. 2011; Todt et al. 2015). Only if the transiently exposed BH3 motif binds to the hydrophobic cleft of another Bax molecule that oligomerization may ensue (Moldoveanu et al. 2013). However, it is not entirely clear how this oligomerization occurs, since the hydrophobic cleft which accommodates the BH3 domain is in part formed by the BH3 domain of the acceptor Bax molecule (Suzuki et al. 2000). Two competing models have been proposed to explain the nature of active Bax. On the one hand, Bax might insert into the OMM with the α helices 5 and 6 together with the C-terminal MA and form a "head-to-tail" chain (Bogner et al. 2010). This model is based on structural similarities between Bcl-2 proteins bacterial toxins and labeling studies using cysteine variants of Bax (Dlugosz et al. 2006; Muchmore et al. 1996). However, full Bax activity can occur in Bax variants lacking α helix 6 and recent labeling experiments of cysteine variants seem to disfavor OMM-insertion of α helix 6 (George et al. 2007; Westphal et al. 2014). The alternative model suggests the formation of a symmetric homodimer of Bax or Bak inserting the BH3 domain in the hydrophobic groove of the binding partner (Dewson et al. 2008). Structural insight using a Bax variant containing α helices 2, 3, 4, and a partial α helix 5 supports this view (Czabotar et al. 2013). On the other hand, interactions of full-length proteins in the presence of their physiological lipid environment in the OMM may differ from the structure formed by truncated proteins in vitro. According to this model, Bax or Bak oligomerization is mediated by α helix 1 – α helix 6 interactions (Dewson et al. 2009), although it remains to be clarified how such an OMM-associated structure can permeabilize the OMM. Perhaps hydrophobic patches on one side of the structure could displace lipids in the outer leaflet and thus destabilize the membrane (Moldoveanu et al. 2014). Recently, distance mapping of cysteine-containing Bax variants has led to the modelling of potential Bax pore structures in large unilamellar vesicles (Bleicken et al. 2014). The proposed model indicates a possible role of helices 2–5 in such Bax pores, supporting the previous analysis of minimal structural requirements of Bax activity in human cells (George et al. 2007), and X-ray cristallography (Czabotar et al. 2013). In addition, the C-terminal membrane anchor could represent an interface for dimer–dimer interactions during Bax activation (Gahl et al. 2014).

4.4 Influence of Bcl-2 Proteins on Mitochondrial Dynamics

Damage to mitochondria is reflected in changes in mitochondrial morphology (Mouli et al. 2009). Two opposing processes, fusion and fission, constantly shift the morphology of mitochondria from fragmented vesicles to interconnected networks of a tubular reticulum (Shutt and McBride 2013; Tatsuta and Langer 2008). Mitochondrial fusion can restore organelle function by combining the contents of damaged and healthy mitochondria (Chen et al. 2003). The asymmetric fission of mitochondria can lead to mitochondrial fragments that have forfeited their electrochemical membrane potential, resulting in fusion-incompetent mitochondrial remnants prone to degradation by mitophagy (Twig et al. 2008). The core components that govern mitochondrial morphology are the large dynamin-related proteins Drp1 and the mitofusins MFN1/2 on the OMM (Hales and Fuller 1997; Koshiba et al. 2004; Smirnova et al. 2001).

Interactions between the shape shifting machinery and Bcl-2 proteins have been largely associated with events around the commitment of cells to apoptosis. Bax translocates to the mitochondria following an apoptotic stimulus, concentrating to small focal regions on the OMM (Goping et al. 1998; Karbowski et al. 2002). Inactive Bak, although already present in the OMM, then migrates into these newly formed Bax-containing foci. The Bax/Bak clusters often appear at sites that later develop into mitochondrial fission sites, linking Bax activity to the process of mitochondrial fragmentation (Frank et al. 2001). Direct interactions between Bax and Drp1 seem to cause mitochondrial hyperfission following stimulation of apoptosis. In addition, sites of incomplete fusion or fission, hemifusion sites, can promote tBid-induced Bax oligomerization (Montessuit et al. 2010). Membrane remodeling was promoted by Drp1 in a cell-free system in a GTP-independent fashion. Thus, the role of hemifusion sites in Bax activation could explain the appearance of active Bax in focal structures associated with the tips and constriction sites of mitochondria (Montessuit et al. 2010). Interestingly, mitochondrial fragmentation also occurs upon Bax recruitment to the OMM induced by the viral protein vMIA, although under these conditions Bax is distributed more widely in the OMM rather than clustering into focal structures (Arnoult et al. 2004; Goldmacher et al. 1999). Another line of evidence suggesting an influence of Bax and Bak on mitochondrial dynamics beyond apoptosis induction is altered mitochondrial morphology found in proliferating Bax/Bak DKO cells (Karbowski et al. 2006). In the absence of Bax and Bak, cells appear to have a lower mitochondrial fusion rate. These effects are not limited to Bax and Bak, because ectopic expression of human Bcl-xL has also been shown to influence mitochondrial dynamics (Delivani et al. 2006). Whether Bcl-2 proteins recruit Drp1 directly or whether other partners are required remains unknown. To this end, a mechanism involving Bax/Bak-dependent Drp1 SUMOylation has been proposed (Wasiak et al. 2007). It should be noted that Drp1 inhibition delays but does not prevent mitochondrial fission (Ishihara et al. 2009). Therefore, Bcl-2 proteins might also induce mitochondrial fission via a Drp1-independent mechanism.

In this context, it is worth noting that Bax foci following apoptosis induction also contain MFN2 (Karbowski et al. 2002). Therefore, Bcl-2 proteins may influence mitochondrial shape changes not only via interactions with Drp1, but also by regulating the activities of MFN1 or MFN2. In a cell-free system of mitochondrial fusion, Bax was found to stimulate MFN2-mediated but not MFN1-mediated mitochondrial fusion (Hoppins et al. 2011). Furthermore, this activity of Bax is dependent on its protein conformation. Direct interactions between MFN1/2 and different Bcl-2 proteins have been previously shown to promote mitochondrial fusion (Brooks et al. 2007). Whether these interactions cause the stress-induced mitochondrial hyperfusion observed following a number of different stresses (Tondera et al. 2009) remains to be elucidated.

4.5 Conclusion

The mechanisms of activation of Bax and Bak remain incompletely elucidated. One of the central open questions is the nature of the active forms of Bax or Bak. Do Bax and Bak together or alone form a pore in response to mitochondrial apoptosis signaling? Regardless of the nature of active Bax and Bak, the solution of the structure of active pro-apoptotic Bcl-2 proteins and their associated components remains a main goal of apoptosis research.

References

Annis MG, Soucie EL, Dlugosz PJ, Cruz-Aguado JA, Penn LZ, Leber B, Andrews DW (2005) Bax forms multispanning monomers that oligomerize to permeabilize membranes during apoptosis. EMBO J 24:2096–2103

Antonsson B, Conti F, Ciavatta A, Montessuit S, Lewis S, Martinou I, Bernasconi L, Bernard A, Mermod JJ, Mazzei G et al (1997) Inhibition of Bax channel-forming activity by Bcl-2. Science 277:370–372

Antonsson B, Montessuit S, Lauper S, Eskes R, Martinou JC (2000) Bax oligomerization is required for channel-forming activity in liposomes and to trigger cytochrome c release from mitochondria. Biochem J 345(Pt 2):271–278

Arnoult D, Bartle LM, Skaletskaya A, Poncet D, Zamzami N, Park PU, Sharpe J, Youle RJ, Goldmacher VS (2004) Cytomegalovirus cell death suppressor vMIA blocks Bax- but not Bak-mediated apoptosis by binding and sequestering Bax at mitochondria. Proc Natl Acad Sci U S A 101:7988–7993

Billen LP, Shamas-Din A, Andrews DW (2008) Bid: a Bax-like BH3 protein. Oncogene 27(Suppl 1):S93–S104

Bleicken S, Wagner C, García-Sáez AJ (2013) Mechanistic differences in the membrane activity of Bax and Bcl-xL correlate with their opposing roles in apoptosis. Biophys J 104:421–431

Bleicken S, Jeschke G, Stegmueller C, Salvador-Gallego R, García-Sáez AJ, Bordignon E (2014) Structural model of active Bax at the membrane. Mol Cell 56:496–505

Bogner C, Leber B, Andrews DW (2010) Apoptosis: embedded in membranes. Curr Opin Cell Biol 22:845–851

Bratton SB, Cohen GM (2001) Caspase cascades in chemically-induced apoptosis. Adv Exp Med Biol 500:407–420

Brooks C, Wei Q, Feng L, Dong G, Tao Y, Mei L, Xie Z-J, Dong Z (2007) Bak regulates mitochondrial morphology and pathology during apoptosis by interacting with mitofusins. Proc Natl Acad Sci U S A 104:11649–11654

Colell A, Green DR, Ricci J-E (2009) Novel roles for GAPDH in cell death and carcinogenesis. Cell Death Differ 16:1573–1581

Chen H, Detmer SA, Ewald AJ, Griffin EE, Fraser SE, Chan DC (2003) Mitofusins Mfn1 and Mfn2 coordinately regulate mitochondrial fusion and are essential for embryonic development. J Cell Biol 160:189–200

Chipuk JE, Maurer U, Green DR, Schuler M (2003) Pharmacologic activation of p53 elicits Bax-dependent apoptosis in the absence of transcription. Cancer Cell 4:371–381

Chou JJ, Li H, Salvesen GS, Yuan J, Wagner G (1999) Solution structure of BID, an intracellular amplifier of apoptotic signaling. Cell 96:615–624

Clavería C, Giovinazzo G, Sierra R, Torres M (2013) Myc-driven endogenous cell competition in the early mammalian embryo. Nature 500:39–44

Czabotar PE, Westphal D, Dewson G, Ma S, Hockings C, Fairlie WD, Lee EF, Yao S, Robin AY, Smith BJ et al (2013) Bax crystal structures reveal how BH3 domains activate Bax and nucleate its oligomerization to induce apoptosis. Cell 152:519–531

Van Delft MF, Wei AH, Mason KD, Vandenberg CJ, Chen L, Czabotar PE, Willis SN, Scott CL, Day CL, Cory S et al (2006) The BH3 mimetic ABT-737 targets selective Bcl-2 proteins and efficiently induces apoptosis via Bak/Bax if Mcl-1 is neutralized. Cancer Cell 10:389–399

Delivani P, Adrain C, Taylor RC, Duriez PJ, Martin SJ (2006) Role for CED-9 and Egl-1 as regulators of mitochondrial fission and fusion dynamics. Mol Cell 21:761–773

Desagher S, Osen-Sand A, Nichols A, Eskes R, Montessuit S, Lauper S, Maundrell K, Antonsson B, Martinou JC (1999) Bid-induced conformational change of Bax is responsible for mitochondrial cytochrome c release during apoptosis. J Cell Biol 144:891–901

Dewson G, Kratina T, Sim HW, Puthalakath H, Adams JM, Colman PM, Kluck RM (2008) To trigger apoptosis, Bak exposes its BH3 domain and homodimerizes via BH3:groove interactions. Mol Cell 30:369–380

Dewson G, Kratina T, Czabotar P, Day CL, Adams JM, Kluck RM (2009) Bak activation for apoptosis involves oligomerization of dimers via their alpha6 helices. Mol Cell 36:696–703

Dlugosz PJ, Billen LP, Annis MG, Zhu W, Zhang Z, Lin J, Leber B, Andrews DW (2006) Bcl-2 changes conformation to inhibit Bax oligomerization. EMBO J 25:2287–2296

Echeverry N, Bachmann D, Ke F, Strasser A, Simon HU, Kaufmann T (2013) Intracellular localization of the BCL-2 family member BOK and functional implications. Cell Death Differ 20:785–799

Edlich F, Banerjee S, Suzuki M, Cleland MM, Arnoult D, Wang C, Neutzner A, Tjandra N, Youle RJ (2011) Bcl-x(L) retrotranslocates Bax from the mitochondria into the cytosol. Cell 145:104–116

Eskes R, Antonsson B, Osen-Sand A, Montessuit S, Richter C, Sadoul R, Mazzei G, Nichols A, Martinou JC (1998) Bax-induced cytochrome C release from mitochondria is independent of the permeability transition pore but highly dependent on Mg2+ ions. J Cell Biol 143:217–224

Ferrer PE, Frederick P, Gulbis JM, Dewson G, Kluck RM (2012) Translocation of a Bak C-terminus mutant from cytosol to mitochondria to mediate cytochrome C release: implications for Bak and Bax apoptotic function. PLoS One 7:e31510

Fletcher JI, Meusburger S, Hawkins CJ, Riglar DT, Lee EF, Fairlie WD, Huang DCS, Adams JM (2008) Apoptosis is triggered when prosurvival Bcl-2 proteins cannot restrain Bax. Proc Natl Acad Sci U S A 105:18081–18087

Frank S, Gaume B, Bergmann-Leitner ES, Leitner WW, Robert EG, Catez F, Smith CL, Youle RJ (2001) The role of dynamin-related protein 1, a mediator of mitochondrial fission, in apoptosis. Dev Cell 1:515–525

Gahl RF, He Y, Yu S, Tjandra N (2014) Conformational rearrangements in the pro-apoptotic protein, Bax, as it inserts into mitochondria: a cellular death switch. J Biol Chem 289(47):32871–32882

Gavathiotis E, Suzuki M, Davis ML, Pitter K, Bird GH, Katz SG, Tu H-C, Kim H, Cheng EH-Y, Tjandra N et al (2008) BAX activation is initiated at a novel interaction site. Nature 455:1076–1081

Geissler A, Haun F, Frank DO, Wieland K, Simon MM, Idzko M, Davis RJ, Maurer U, Borner C (2013) Apoptosis induced by the fungal pathogen gliotoxin requires a triple phosphorylation of Bim by JNK. Cell Death Differ 20:1317–1329

George NM, Evans JJD, Luo X (2007) A three-helix homo-oligomerization domain containing BH3 and BH1 is responsible for the apoptotic activity of Bax. Genes Dev 21:1937–1948

Goldmacher VS, Bartle LM, Skaletskaya A, Dionne CA, Kedersha NL, Vater CA, Han JW, Lutz RJ, Watanabe S, Cahir McFarland ED et al (1999) A cytomegalovirus-encoded mitochondria-localized inhibitor of apoptosis structurally unrelated to Bcl-2. Proc Natl Acad Sci U S A 96:12536–12541

Goping IS, Gross A, Lavoie JN, Nguyen M, Jemmerson R, Roth K, Korsmeyer SJ, Shore GC (1998) Regulated targeting of BAX to mitochondria. J Cell Biol 143:207–215

Green DR, Kroemer G (2004) The pathophysiology of mitochondrial cell death. Science 305:626–629

Griffiths GJ, Corfe BM, Savory P, Leech S, Esposti MD, Hickman JA, Dive C (2001) Cellular damage signals promote sequential changes at the N-terminus and BH-1 domain of the pro-apoptotic protein Bak. Oncogene 20:7668–7676

Hales KG, Fuller MT (1997) Developmentally regulated mitochondrial fusion mediated by a conserved, novel, predicted GTPase. Cell 90:121–129

Hinds MG, Lackmann M, Skea GL, Harrison PJ, Huang DCS, Day CL (2003) The structure of Bcl-w reveals a role for the C-terminal residues in modulating biological activity. EMBO J 22:1497–1507

Hinds MG, Smits C, Fredericks-Short R, Risk JM, Bailey M, Huang DCS, Day CL (2007) Bim, Bad and Bmf: intrinsically unstructured BH3-only proteins that undergo a localized conformational change upon binding to prosurvival Bcl-2 targets. Cell Death Differ 14:128–136

Hoppins S, Edlich F, Cleland MM, Banerjee S, McCaffery JM, Youle RJ, Nunnari J (2011) The soluble form of Bax regulates mitochondrial fusion via MFN2 homotypic complexes. Mol Cell 41:150–160

Hotchkiss RS, Strasser A, McDunn JE, Swanson PE (2009) Cell death. N Engl J Med 361:1570–1583

Hsu YT, Youle RJ (1997) Nonionic detergents induce dimerization among members of the Bcl-2 family. J Biol Chem 272:13829–13834

Hsu YT, Youle RJ (1998) Bax in murine thymus is a soluble monomeric protein that displays differential detergent-induced conformations. J Biol Chem 273:10777–10783

Ishihara N, Nomura M, Jofuku A, Kato H, Suzuki SO, Masuda K, Otera H, Nakanishi Y, Nonaka I, Goto Y-I et al (2009) Mitochondrial fission factor Drp1 is essential for embryonic development and synapse formation in mice. Nat Cell Biol 11:958–966

Jeong S-Y, Gaume B, Lee Y-J, Hsu Y-T, Ryu S-W, Yoon S-H, Youle RJ (2004) Bcl-x(L) sequesters its C-terminal membrane anchor in soluble, cytosolic homodimers. EMBO J 23:2146–2155

Karbowski M, Lee Y-J, Gaume B, Jeong S-Y, Frank S, Nechushtan A, Santel A, Fuller M, Smith CL, Youle RJ (2002) Spatial and temporal association of Bax with mitochondrial fission sites, Drp1, and Mfn2 during apoptosis. J Cell Biol 159:931–938

Karbowski M, Norris KL, Cleland MM, Jeong S-Y, Youle RJ (2006) Role of Bax and Bak in mitochondrial morphogenesis. Nature 443:658–662

Ke F, Bouillet P, Kaufmann T, Strasser A, Kerr J, Voss AK (2013) Consequences of the combined loss of BOK and BAK or BOK and BAX. Cell Death Dis 4:e650

Kerr JF, Wyllie AH, Currie AR (1972) Apoptosis: a basic biological phenomenon with wide-ranging implications in tissue kinetics. Br J Cancer 26:239–257

Kim H, Rafiuddin-Shah M, Tu H-C, Jeffers JR, Zambetti GP, Hsieh JJ-D, Cheng EH-Y (2006) Hierarchical regulation of mitochondrion-dependent apoptosis by BCL-2 subfamilies. Nat Cell Biol 8:1348–1358

Korsmeyer SJ, Wei MC, Saito M, Weiler S, Oh KJ, Schlesinger PH (2000) Pro-apoptotic cascade activates BID, which oligomerizes BAK or BAX into pores that result in the release of cytochrome c. Cell Death Differ 7:1166–1173

Koshiba T, Detmer SA, Kaiser JT, Chen H, McCaffery JM, Chan DC (2004) Structural basis of mitochondrial tethering by mitofusin complexes. Science 305:858–862

Kuwana T, Mackey MR, Perkins G, Ellisman MH, Latterich M, Schneiter R, Green DR, Newmeyer DD (2002) Bid, Bax, and lipids cooperate to form supramolecular openings in the outer mitochondrial membrane. Cell 111:331–342

Kuwana T, Bouchier-Hayes L, Chipuk JE, Bonzon C, Sullivan BA, Green DR, Newmeyer DD (2005) BH3 domains of BH3-only proteins differentially regulate Bax-mediated mitochondrial membrane permeabilization both directly and indirectly. Mol Cell 17:525–535

Kvansakul M, van Delft MF, Lee EF, Gulbis JM, Fairlie WD, Huang DCS, Colman PM (2007) A structural viral mimic of prosurvival Bcl-2: a pivotal role for sequestering proapoptotic Bax and Bak. Mol Cell 25:933–942

Leber B, Lin J, Andrews DW (2007) Embedded together: the life and death consequences of interaction of the Bcl-2 family with membranes. Apoptosis 12:897–911

Leshchiner ES, Braun CR, Bird GH, Walensky LD (2013) Direct activation of full-length proapoptotic BAK. Proc Natl Acad Sci U S A 110:E986–E995

Letai A, Bassik MC, Walensky LD, Sorcinelli MD, Weiler S, Korsmeyer SJ (2002) Distinct BH3 domains either sensitize or activate mitochondrial apoptosis, serving as prototype cancer therapeutics. Cancer Cell 2:183–192

Li H, Zhu H, Xu CJ, Yuan J (1998) Cleavage of BID by caspase 8 mediates the mitochondrial damage in the Fas pathway of apoptosis. Cell 94:491–501

Lindsten T, Ross AJ, King A, Zong WX, Rathmell JC, Shiels HA, Ulrich E, Waymire KG, Mahar P, Frauwirth K et al (2000) The combined functions of proapoptotic Bcl-2 family members bak and bax are essential for normal development of multiple tissues. Mol Cell 6:1389–1399

Llambi F, Moldoveanu T, Tait SWG, Bouchier-Hayes L, Temirov J, McCormick LL, Dillon CP, Green DR (2011) A unified model of mammalian BCL-2 protein family interactions at the mitochondria. Mol Cell 44:517–531

Lovell JF, Billen LP, Bindner S, Shamas-Din A, Fradin C, Leber B, Andrews DW (2008) Membrane binding by tBid initiates an ordered series of events culminating in membrane permeabilization by Bax. Cell 135:1074–1084

Lucken-Ardjomande S, Martinou JC (2005) Regulation of Bcl-2 proteins and of the permeability of the outer mitochondrial membrane. C R Biol 328:616–631

Martinou I, Desagher S, Eskes R, Antonsson B, André E, Fakan S, Martinou JC (1999) The release of cytochrome c from mitochondria during apoptosis of NGF-deprived sympathetic neurons is a reversible event. J Cell Biol 144:883–889

Minn AJ, Vélez P, Schendel SL, Liang H, Muchmore SW, Fesik SW, Fill M, Thompson CB (1997) Bcl-x(L) forms an ion channel in synthetic lipid membranes. Nature 385:353–357

Moldoveanu T, Grace CR, Llambi F, Nourse A, Fitzgerald P, Gehring K, Kriwacki RW, Green DR (2013) BID-induced structural changes in BAK promote apoptosis. Nat Struct Mol Biol 20:589–597

Moldoveanu T, Follis AV, Kriwacki RW, Green DR (2014) Many players in BCL-2 family affairs. Trends Biochem Sci 39:101–111

Montessuit S, Somasekharan SP, Terrones O, Lucken-Ardjomande S, Herzig S, Schwarzenbacher R, Manstein DJ, Bossy-Wetzel E, Basañez G, Meda P et al (2010) Membrane remodeling induced by the dynamin-related protein Drp1 stimulates Bax oligomerization. Cell 142:889–901

Mouli PK, Twig G, Shirihai OS (2009) Frequency and selectivity of mitochondrial fusion are key to its quality maintenance function. Biophys J 96:3509–3518

Muchmore SW, Sattler M, Liang H, Meadows RP, Harlan JE, Yoon HS, Nettesheim D, Chang BS, Thompson CB, Wong SL et al (1996) X-ray and NMR structure of human Bcl-xL, an inhibitor of programmed cell death. Nature 381:335–341

Okamoto T, Zobel K, Fedorova A, Quan C, Yang H, Fairbrother WJ, Huang DCS, Smith BJ, Deshayes K, Czabotar PE (2013) Stabilizing the pro-apoptotic BimBH3 helix (BimSAHB) does not necessarily enhance affinity or biological activity. ACS Chem Biol 8:297–302

Oltersdorf T, Elmore SW, Shoemaker AR, Armstrong RC, Augeri DJ, Belli BA, Bruncko M, Deckwerth TL, Dinges J, Hajduk PJ et al (2005) An inhibitor of Bcl-2 family proteins induces regression of solid tumours. Nature 435:677–681

Oltvai ZN, Milliman CL, Korsmeyer SJ (1993) Bcl-2 heterodimerizes in vivo with a conserved homolog, Bax, that accelerates programmed cell death. Cell 74:609–619

Owens TW, Valentijn AJ, Upton J-P, Keeble J, Zhang L, Lindsay J, Zouq NK, Gilmore AP (2009) Apoptosis commitment and activation of mitochondrial Bax during anoikis is regulated by p38MAPK. Cell Death Differ 16:1551–1562

Peng R, Tong J-S, Li H, Yue B, Zou F, Yu J, Zhang L (2013) Targeting Bax interaction sites reveals that only homo-oligomerization sites are essential for its activation. Cell Death Differ 20:744–754

Potts MB, Vaughn AE, McDonough H, Patterson C, Deshmukh M (2005) Reduced Apaf-1 levels in cardiomyocytes engage strict regulation of apoptosis by endogenous XIAP. J Cell Biol 171:925–930

Ren D, Tu H-C, Kim H, Wang GX, Bean GR, Takeuchi O, Jeffers JR, Zambetti GP, Hsieh JJ-D, Cheng EH-Y (2010) BID, BIM, and PUMA are essential for activation of the BAX- and BAK-dependent cell death program. Science 330:1390–1393

Roucou X, Montessuit S, Antonsson B, Martinou J-C (2002) Bax oligomerization in mitochondrial membranes requires tBid (caspase-8-cleaved Bid) and a mitochondrial protein. Biochem J 368:915–921

Sattler M, Liang H, Nettesheim D, Meadows RP, Harlan JE, Eberstadt M, Yoon HS, Shuker SB, Chang BS, Minn AJ et al (1997) Structure of Bcl-xL-Bak peptide complex: recognition between regulators of apoptosis. Science 275:983–986

Schellenberg B, Wang P, Keeble JA, Rodriguez-Enriquez R, Walker S, Owens TW, Foster F, Tanianis-Hughes J, Brennan K, Streuli CH et al (2013) Bax exists in a dynamic equilibrium between the cytosol and mitochondria to control apoptotic priming. Mol Cell 49:959–971

Schuler M, Green DR (2005) Transcription, apoptosis and p53: catch-22. Trends Genet 21:182–187

Sedlak TW, Oltvai ZN, Yang E, Wang K, Boise LH, Thompson CB, Korsmeyer SJ (1995) Multiple Bcl-2 family members demonstrate selective dimerizations with Bax. Proc Natl Acad Sci 92:7834–7838

Shimizu S, Narita M, Tsujimoto Y (1999) Bcl-2 family proteins regulate the release of apoptogenic cytochrome c by the mitochondrial channel VDAC. Nature 399:483–487

Shutt TE, McBride HM (2013) Staying cool in difficult times: mitochondrial dynamics, quality control and the stress response. Biochim Biophys Acta 1833:417–424

Smirnova E, Griparic L, Shurland DL, van der Bliek AM (2001) Dynamin-related protein Drp1 is required for mitochondrial division in mammalian cells. Mol Biol Cell 12:2245–2256

Sulston JE, Horvitz HR (1977) Post-embryonic cell lineages of the nematode, Caenorhabditis elegans. Dev Biol 56:110–156

Suzuki M, Youle RJ, Tjandra N (2000) Structure of Bax: coregulation of dimer formation and intracellular localization. Cell 103:645–654

Tatsuta T, Langer T (2008) Quality control of mitochondria: protection against neurodegeneration and ageing. EMBO J 27:306–314

Todt F, Cakir Z, Reichenbach F, Youle RJ, Edlich F (2013) The C-terminal helix of Bcl-x(L) mediates Bax retrotranslocation from the mitochondria. Cell Death Differ 20:333–342

Todt F, Cakir Z, Reichenbach F, Emschermann F, Lauterwasser J, Kaiser A, Ichim G, Tait SWG, Frank S, Langer HF, Edlich F (2015) Differential retrotranslocation of mitochondrial Bax and Bak. EMBO J 34:67–80

Tondera D, Grandemange S, Jourdain A, Karbowski M, Mattenberger Y, Herzig S, Da Cruz S, Clerc P, Raschke I, Merkwirth C et al (2009) SLP-2 is required for stress-induced mitochondrial hyperfusion. EMBO J 28:1589–1600

Tsujimoto Y, Cossman J, Jaffe E, Croce CM (1985) Involvement of the bcl-2 gene in human follicular lymphoma. Science 228:1440–1443

Twig G, Elorza A, Molina AJA, Mohamed H, Wikstrom JD, Walzer G, Stiles L, Haigh SE, Katz S, Las G et al (2008) Fission and selective fusion govern mitochondrial segregation and elimination by autophagy. EMBO J 27:433–446

Valentijn AJ, Metcalfe AD, Kott J, Streuli CH, Gilmore AP (2003) Spatial and temporal changes in Bax subcellular localization during anoikis. J Cell Biol 162:599–612

Vaux DL, Cory S, Adams JM (1988) Bcl-2 gene promotes haemopoietic cell survival and cooperates with c-myc to immortalize pre-B cells. Nature 335:440–442

Vogel S, Raulf N, Bregenhorn S, Biniossek ML, Maurer U, Czabotar P, Borner C (2012) Cytosolic Bax: does it require binding proteins to keep its pro-apoptotic activity in check? J Biol Chem 287:9112–9127

Wasiak S, Zunino R, McBride HM (2007) Bax/Bak promote sumoylation of DRP1 and its stable association with mitochondria during apoptotic cell death. J Cell Biol 177:439–450

Wei MC, Lindsten T, Mootha VK, Weiler S, Gross A, Ashiya M, Thompson CB, Korsmeyer SJ (2000) tBID, a membrane-targeted death ligand, oligomerizes BAK to release cytochrome c. Genes Dev 14:2060–2071

Westphal D, Dewson G, Menard M, Frederick P, Iyer S, Bartolo R, Gibson L, Czabotar PE, Smith BJ, Adams JM et al (2014) Apoptotic pore formation is associated with in-plane insertion of Bak or Bax central helices into the mitochondrial outer membrane. Proc Natl Acad Sci U S A 111(39):E4076–E4085

Willis SN, Chen L, Dewson G, Wei A, Naik E, Fletcher JI, Adams JM, Huang DCS (2005) Proapoptotic Bak is sequestered by Mcl-1 and Bcl-xL, but not Bcl-2, until displaced by BH3-only proteins. Genes Dev 19:1294–1305

Willis SN, Fletcher JI, Kaufmann T, van Delft MF, Chen L, Czabotar PE, Ierino H, Lee EF, Fairlie WD, Bouillet P et al (2007) Apoptosis initiated when BH3 ligands engage multiple Bcl-2 homologs, not Bax or Bak. Science 315:856–859

Yin XM, Wang K, Gross A, Zhao Y, Zinkel S, Klocke B, Roth KA, Korsmeyer SJ (1999) Bid-deficient mice are resistant to Fas-induced hepatocellular apoptosis. Nature 400:886–891

Youle RJ, Strasser A (2008) The BCL-2 protein family: opposing activities that mediate cell death. Nat Rev Mol Cell Biol 9:47–59

Zha J, Weiler S, Oh KJ, Wei MC, Korsmeyer SJ (2000) Posttranslational N-myristoylation of BID as a molecular switch for targeting mitochondria and apoptosis. Science 290:1761–1765

Zychlinsky A, Prevost MC, Sansonetti PJ (1992) Shigella flexneri induces apoptosis in infected macrophages. Nature 358:167–169

Chapter 5
Remodeling of Mitochondria in Apoptosis

Guy A. Perkins and Mark H. Ellisman

5.1 Introduction

Apoptosis was first defined by morphological approaches in the 1972 in a seminal article by Kerr and colleagues (Kerr et al. 1972). Characteristic cellular changes were described to include:

1. Cell shrinkage and rounding (because of caspase digestion of the cytoskeleton).
2. Chromatin condensation with dense patches adjacent to the nuclear envelope.
3. Disruption of the nuclear envelope followed by breakdown of the nucleus into discrete bodies.
4. Blebbing of the plasma membrane.
5. Finally, break-up of the cell into apoptotic bodies, which are then digested by phagocytes.

Following the early work describing apoptosis in *C. elegans* (Metzstein et al. 1998; Sulston and Brenner 1974), it wasn't until years later that a central role for mitochondria in apoptosis was discovered. Vaux, Hockenbery, and colleagues found

G.A. Perkins (✉)
Center for Research in Biological Systems, National Center for Microscopy and Imaging Research, University of California at San Diego, La Jolla, CA 92093-0608, USA
e-mail: gperkins@ucsd.edu

M.H. Ellisman
Center for Research in Biological Systems, National Center for Microscopy and Imaging Research, University of California at San Diego, La Jolla, CA 92093-0608, USA

Department of Neurosciences, University of California at San Diego, La Jolla, CA 92093-0608, USA

Department of Bioengineering, University of California at San Diego, La Jolla, CA 92093-0608, USA
e-mail: mark@ncmir.ucsd.edu

D.M. Hockenbery (ed.), *Mitochondria and Cell Death*, Cell Death in Biology and Diseases, DOI 10.1007/978-1-4939-3612-0_5

X.-M. Yin and Z. Dong (Series eds.), *Cell Death in Biology and Diseases*

that BCL2 family members associate with mitochondria and regulate apoptosis (Hockenbery et al. 1990; Vaux et al. 1988). Since the early days of that portion of apoptosis research centered on mitochondria, the molecular mechanisms behind this type of programmed cell death have been of great interest. As more molecular players that influence the mitochondrial convergence of apoptosis have been uncovered, researchers have endeavored to understand the sequence of actions, which molecules interact directly with each other, whether translocation to mitochondria or escape from this organelle is necessary, which mitochondrial membranes or compartments house these players, and why, when, and how do they remodel mitochondrial structure. The goal of each section of this chapter is to describe the current knowledge of what ways mitochondria are remodeled, which molecules play a role in this remodeling, what are their binding or regulatory partners, and where and when are they located on or inside mitochondria.

The principal pathway of mitochondrially related apoptosis is the BAX/BAK-dependent pathway. In this pathway, the release of cytochrome *c* and related pro-apoptotic proteins from mitochondria is a consequence of two carefully coordinated remodeling events: opening of crista junctions triggered by OPA1 oligomer disassembly and formation of outer membrane pores, facilitated by BAX and BAK. Both steps are necessary for the complete release of pro-apoptotic proteins. The remodeling of mitochondrial structure is intrinsic to this pathway, including mitochondrial fission, outer membrane remodeling, and cristae and crista junction alterations.

5.2 Mitochondrial Fission Can Be an Early Remodeling Event During Apoptosis

The term "remodeling" when applied to mitochondria is used in the literature to describe both fission/fusion and cristae/crista junction alterations with references to fission/fusion greatly predominating over the latter. Upon apoptotic stimulation in many situations, mitochondria undergo fission (Fig. 5.1), and this is often an early event and caspase-independent (Cleland and Youle 2011; Karbowski and Youle 2003; Lee et al. 2004). Related to the etiology of apoptosis, fissioning of mitochondria occurs when the mitochondrial inner membrane potential has been altered (Benard et al. 2007), via pathogenic nitric oxide signaling (Cho et al. 2009; Yuan et al. 2007) or in response to impairment of oxidative phosphorylation (Ishihara et al. 2006). However, the fissioning of mitochondria per se does not evoke apoptosis (Barsoum et al. 2006; Bereiter-Hahn et al. 2008; Sheridan et al. 2008; Sheridan and Martin 2010; Yuan et al. 2007). It is worthwhile to note that many studies of mitochondrial fission relied on conventional fluorescence microscopy. Yet, because the resolution of this microscopy (about 200 nm) is not always sufficient to distinguish individual mitochondrion when multiple mitochondria are close to each other, the determination of fission may be inaccurate. The higher resolution afforded by electron microscopy is preferred for the most accurate characterization of fission. With the advent of super-resolution light microscopy (Jakobs and Wurm 2014), an accurate quantitative

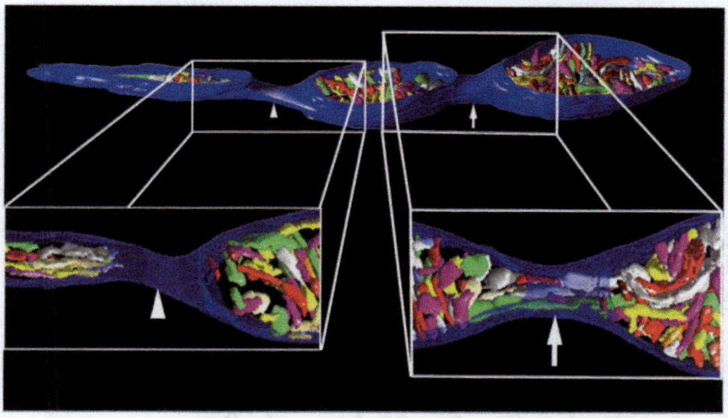

Fig. 5.1 Mitochondrion undergoing fission. Mitochondrial fission is often an early event of apoptosis. Electron tomography of a brain mitochondrion caught it fissioning into three parts. The 3D segmentation of the volume shows the mitochondrial outer membrane in blue and the many cristae in various colors. The two constriction sites are indicated by *arrowhead* (*left*) and *arrow* (*right*). The *boxed regions* are shown expanded and rotated to provide a top-view in the foreground to highlight that the segregation of cristae between left and middle parts has been completed, whereas cristae still are in the larger right-hand constriction, indicating that the progression to complete division has not advanced as far as that on the left. The outer membrane in the foreground was made translucent to see the cristae

evaluation of mitochondrial fission no longer requires electron microscopy and can be placed in the realm of fluorescence microscopy, albeit a more advanced technique of light microscopy than used for the majority of fission studies to date.

Since the first studies to describe the relationship between mitochondrial fission and cell death (Frank et al. 2001; Karbowski et al. 2002), much of the precise role fission plays in apoptosis has been deciphered. A key player in the apoptotic fission of mitochondria is DRP1, a member of the dynamin family of large GTPases. Mouse embryonic fibroblasts from *Drp1*-KO mice displayed a retardation in cytochrome *c* release, caspase activation, and nuclear DNA fragmentation (Ishihara et al. 2009). DRP1 is a mechanoenzyme that self-assembles at foci on the mitochondrial outer membrane to form a helical rope that wraps around mitochondria and pinches off the membrane constriction between two forming daughter mitochondria (Ingerman et al. 2005). DRP1 also appears to promote apoptosis independent of its mitochondria-severing activity (Montessuit et al. 2010). Because with *Drp1*-KO, the mitochondrial network can remain intact even after the release of cytochrome c with network fragmentation occurring only in the advanced stage of apoptosis, it is likely that DRP1-independent mitochondrial fragmentation occurs late downstream from the release of cytochrome *c* (Ishihara et al. 2009). Until the induction of apoptosis, DRP1 resides in the cytoplasm. Recently, DRP1's receptor and effector partner, mitochondrial fission factor (MFF) (Otera et al. 2010; Otera and Mihara 2011), was found to accumulate at sites of endoplasmic reticulum (ER)-mitochondrial contact even in the absence of DRP1, providing a scaffold for DRP1 recruitment to mitochondrial constriction sites (Friedman et al. 2011). MFF depletion by RNAi produces mitochondrial elongation, delayed cytochrome *c* release, and slowing of apoptosis.

Other DRP1-binding proteins are outer membrane-bound FIS1 and MIEF1, which may have opposite regulatory roles (Dikov and Reichert 2011). MIEF1 mutually exclusively binds to either DRP1 or FIS1. Upon binding to DRP1, MIEF1 appears to inhibit DRP1-induced mitochondrial fission. Interestingly, FIS1 may indirectly promote mitochondrial fission by its ability to sequester MIEF1, thus preventing it from inhibiting fission (Zhao et al. 2011). These findings provide a compelling rationale to further explore upstream regulators and proteins that interact with DRP1 so as to identify novel targets for pharmacotherapy of diseases whose etiology depends on mitochondrial fission.

5.3 How the Fission/Fusion Balance Influences Mitochondrial Inner Membrane Remodeling in Apoptosis

During apoptosis, mitochondrial inner-membrane topology is influenced by a balance between fusion and fission (Knott and Bossy-Wetzel 2008; Knott et al. 2008; Mannella 2006, 2008; Martinou and Youle 2011; Scorrano 2013; Youle and Karbowski 2005). Both pro-apoptotic and anti-apoptotic BCL2 family member proteins play roles in regulating mitochondrial remodeling. The pro-apoptotic BCL2 family members BAX and BAK colocalize with mitochondrial fission and fusion GTPases (Karbowski et al. 2002). tBid is usually the trigger for BAX/BAK activation, perhaps assisted by mitochondrial fragmentation (Youle and Karbowski 2005). However, mitochondria remodeling events and cytochrome *c* mobilization are also triggered by BIK (Germain et al. 2005). During apoptosis, cytosolic BAX inserts into the mitochondrial outer membrane, oligomerizes, and also associates with DRP1 and mitofusin2 at sites where mitochondrial fission subsequently occurs. BAK appears to be the facilitator of this fissioning (Brooks and Dong 2007). BAK, which initially localizes uniformly on the mitochondrial outer membrane, clusters at discrete foci at mitochondrial fission sites during apoptosis. Through interaction with mitofusins on the mitochondrial outer membrane, BAK blocks mitochondrial fusion and so induces fragmentation (Cleland et al. 2011). Overexpression of mitofusin2 diminishes or delays mitochondrial fragmentation, BAK activation, cytochrome *c* release, and apoptosis (Jahani-Asl et al. 2007; Sugioka et al. 2004). Not surprisingly, loss of mitofusin2 has profound effects on mitochondrial structure and function, including matrix swelling, loss of mitochondrial DNA nucleoids, and cristae vesiculation or disorganization (Chen et al. 2007). On the other hand, BAK switches from interacting with mitofusin2 to mitofusin1 during apoptosis (Brooks et al. 2007). In a non-apoptotic mitochondrion, BAK is sequestered by the anti-apoptotic proteins MCL1 and BCLXL. Only when BAK is released from both MCL1 and BCLXL, it can induce apoptosis. 2-deoxyglucose (2DG) can prime highly glycolytic cells by dissociating the BAK-MCL1 complex (Yamaguchi and Perkins 2012a, b). The BCL2 antagonist ABT-263/737 can bind to BCLXL, dissociating the BAK-BCLXL complex. Freed BAK induces apoptosis through its association with BAX and the activity of the resulting complex. Neurons challenged with toxic levels of nitric oxide display BAX foci on mitochondria undergoing fission (Barsoum et al. 2006; Goyal et al. 2007; Yuan et al. 2007). Inhibiting DRP1 delays fission and BAX foci formation.

It has been proposed that other pro-apoptotic members of the BCL2 family that are thought to be active on the mitochondrial outer membrane, namely BIK, BNIP3, BIM, and PUMA, remodel cristae in response to a variety of apoptotic stimuli, not unlike the proposed mechanism of tBID, thus linking the outer membrane to the inner membrane (Germain et al. 2005; Kuwana et al. 2005; Landes et al. 2010b; Ren et al. 2010; Yamaguchi et al. 2008). This remodeling appears to be essential for the complete release of cytochrome *c* (Frezza et al. 2006; Scorrano 2009) and may involve widening of both cristae and crista junction. In this model, inhibition of DRP1 impedes the widening of cristae by BIK overexpression (Germain et al. 2005). When discussing remodeling, it is important to have an up-to-date picture of the mitochondrion, specifically that cristae are not the membrane baffles too often still drawn in cartoons and found in textbooks and articles. Instead, they are now identified as a separate compartment, resembling pleomorphic bags, which are consistently connected (and throughout nature) by a relatively small and uniformly sized neck called the crista junction to the inner boundary membrane (Fig. 5.2) (Perkins et al. 1997). The cristae remodeling may otherwise be subtle, only slightly affecting the crista junction opening size (Yamaguchi et al. 2008) or occur during late stages of apoptosis (Sun et al. 2007) characterized by elongated crista junctions and conversion of cristae to vesiculated structures. The trigger for this vesiculation appears to be loss of mitochondrial membrane potential initiated by etoposide-induced sensitivity

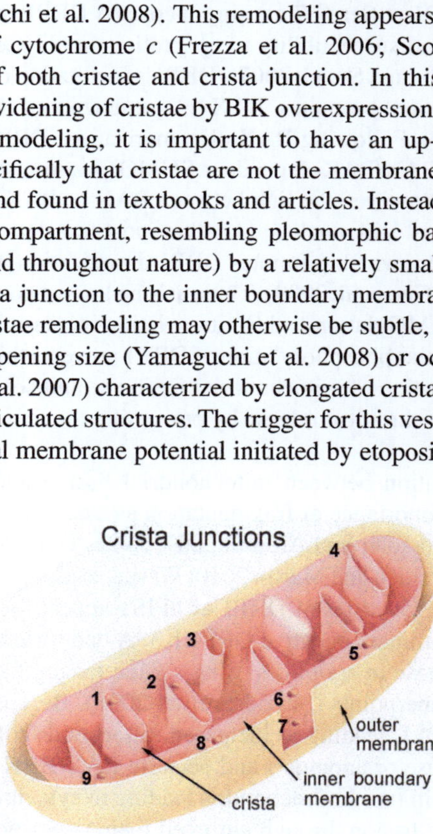

Fig. 5.2 Model of the crista junction architecture. The crista junction is a ubiquitous feature of mitochondria and is consistently seen to be a narrow, tubular opening connecting cristae to the inner boundary membrane. Because of the narrowness of its opening, it is thought to be a diffusion barrier to the passage of macromolecules into or out of the cristae. Because of the high curvature of its neck and the large complexes thought to reside there, the crista junction is also thought to be a diffusion barrier to the movement of integral membrane proteins between the crista membrane and inner boundary membrane. Opening of the crista junction is essential for release of cytochrome *c* during apoptosis. Nine crista junctions are depicted, including a pair (6 and 7) on the same side of the crista, which is sometimes observed

to tBID. The mechanism for vesiculation is likely increased elongation and fusion of crista junctions around the perimeter of cristae until a complete opening and separation of membranes occurs. This remodeling achieves fragmentation of the matrix without compromising the integrity of the membranes surrounding it. The observation that siRNA suppression of OPA1 expression produces a similar cristae remodeling after dissipation of the mitochondrial membrane potential (Griparic et al. 2004; Olichon et al. 2003) suggests that this conversion may represent the mechanism for inner-membrane fission. BIK regulates cristae remodeling through association with DRP1. The signal for DRP1 to be translocated to mitochondria appears to be via calcium. When the calcium-dependent phosphatase calcineurin is activated, it dephosphorylates DRP1 at Ser 637 causing increased DRP1 translocation to mitochondria, with increased susceptibility to apoptotic stimuli (Cereghetti et al. 2008; Costa et al. 2010; Cribbs and Strack 2007). DRP1's facilitation of curved membranes as a precursor to fission promotes BAX oligomerization (Brooks et al. 2011; Montessuit et al. 2010). Following BAX oligomerization, DRP1 clings to the outer membrane due to its BAX/BAK-dependent SUMO post-translational modification (Wasiak et al. 2007). However, the link between BAX or BAK oligomerization and DR{1 SUMOylation is still unclear. Another post-translational modification of DRP1 is S-nitrosylation, which promotes DRP1 dimerization, GTPase activity, and mitochondrial fission (Cho et al. 2009). Thus, among known post-translational modifications of DRP1, SUMOylation and S-nitrosylation increase DRP1-induced mitochondrial fission, whereas phosphorylation of DRP1 prevents this fission (see below). Pharmacological inhibition of mitochondrial fission would seem to be an attractive approach to retard cytochrome c release by preventing the remodeling of mitochondrial cristae.

Despite the correlation between mitochondrial fission and cell death, studies have questioned the importance of fragmentation as a causative agent for apoptosis (Alirol et al. 2006; Estaquier and Arnoult 2007; James et al. 2003; Lee et al. 2004; Parone et al. 2006; Sheridan and Martin 2010; Wasiak et al. 2007). The pro-apoptotic and fission-promoting functions of DRP1 or hFIS1 appear to be distinct (Cassidy-Stone et al. 2008). Dominant negative DRP1 K38A had little effect on cytochrome c release and no decrease in apoptosis using a wide range of apoptotic stimuli and monitoring various time points (Sheridan et al. 2008). Because overexpression of MFN1, MFN2, or OPA1, leading to increased mitochondrial fusion, had no effect on cytochrome c release or apoptosis and because BAX/BAK-dependent pore formation can still occur in fused mitochondria leading to cytochrome *c* loss (Sheridan et al. 2008), inhibiting fission through enforced fusion does not dampen apoptosis. Also, BAX- or BAK-induced mitochondrial fission can be uncoupled from cytochrome *c* release (Barsoum et al. 2006; Sheridan et al. 2008; Yuan et al. 2007). More telling, large-scale fragmentation of mitochondria has been observed in healthy cells, providing evidence that mitochondrial fission alone is not a sufficient event for cell death (Karbowski et al. 2006; Sheridan et al. 2008; Szabadkai et al. 2004). Further, in HeLa cells, cytochrome *c* release preceded mitochondrial fragmentation by at least 10 min (Arnoult et al. 2005; Gao et al. 2001). The major portion of mitochondrial fragmentation may be instigated by caspases in the later stages of apoptosis (Ricci et al. 2004; Sun et al. 2007). All told, mitochondrial fission likely attends cytochrome c release as opposed to orchestrating it. Thus, mitochondrial fission per se does not cause apoptosis.

5.4 Signaling Pathways Convergent on Mitochondria That Sensitize to Apoptotic Remodeling

The best characterized regulatory mechanism of mitochondrial fission is (de)phosphorylation of DRP1 through opposing kinase and phosphatase activity. Multiple signals converge at the mitochondrial outer membrane where a signalosome complex resides (Fig. 5.3) whose anchor and chief recruiter is *A k*inase *a*nchoring *p*rotein *1* (AKAP1). A pivotal phosphorylation site was identified in DRP1, conserved in all metazoans, which acts as an on/off switch for mitochondrial fission (Merrill et al. 2011). DRP1 amino acid Ser 656 (in the conserved domain of rat and equivalent to Ser 637 in the variable domain

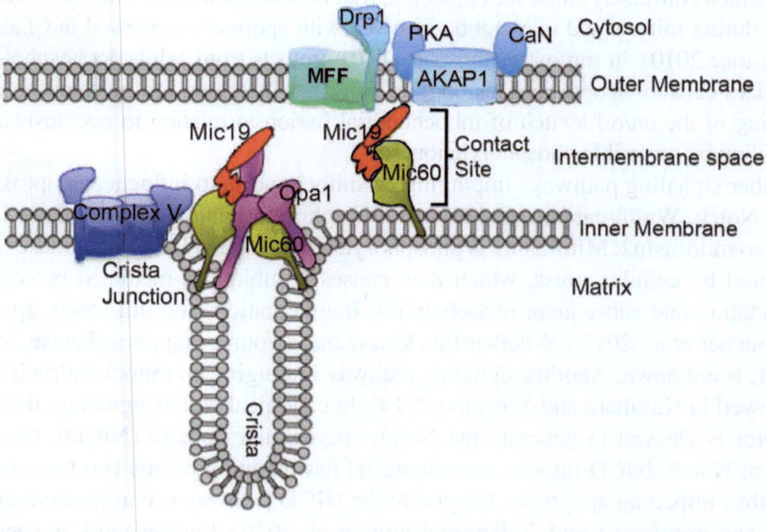

Fig. 5.3 *Cartoon* depicting the signalosome complex residing on the mitochondrial outer membrane that governs mitochondrial fission and whose anchor and chief recruiter is AKAP1. The key mitochondrial fission protein, DRP1, is phosphorylated by PKA bound to AKAP1 to inhibit mitochondrial fission. In the same signalosome complex is CaN, which dephosphorylates DRP1 to promote mitochondrial fission. The recruitment of DRP1 to the mitochondrial outer membrane requires the phosphorylation of its receptor protein MFF. It is thought that the signalosome communicates to the inner membrane at contact sites, interacting with MIC19 in the intermembrane space, which associates with MIC60 anchored to the inner membrane. Interestingly, both MIC19 and MIC60 are found at the crista junction, along with OPA1 and complex V of the electron transport chain, and MIC19's known association with OPA1 may influence the sequestration and release of cytochrome *c* during apoptosis

of humans) is phosphorylated by protein kinase A (PKA) bound to AKAP1 to inhibit mitochondrial fission (reviewed by Knott et al. 2008). Opposite PKA and in the same signalosome complex is the calcium-dependent protein phosphatase calcineurin (CaN), which dephosphorylates Ser 656 to promote mitochondrial fission (Cereghetti et al. 2008; Cribbs and Strack 2007; Wang et al. 2011). Consistent with CaN's role opposite PKA, a PKA binding-deficient AKAP1 mutant, ΔPKA (L313,319P), fragments mitochondria almost as potently as knockdown of endogenous AKAP1 (Dickey and Strack 2011). Further evidence for the central role of PKA in fission is that the PKA inhibitor H89 induces mitochondrial fragmentation, which depends on DRP1 dephosphorylation at Ser 656 (Merrill et al. 2011). Moreover, in cells expressing AKAP1 ΔCaN, a CaN-binding deficient AKAP1 mutant, H89-induced fission, is abolished, indicating that AKAP1 must recruit CaN to the OMM for DRP1 to be activated. Thus, scaffolded CaN and PKA engage in inhibitory crosstalk. Recently, another phosphorylation site on DRP1 was identified, Ser 693, which is mediated by GSK3β and induces the elongation of mitochondria and down-regulation of apoptosis (Chou et al. 2012). Even the recruitment of DRP1 to the mitochondrial outer membrane requires the phosphorylation of its receptor protein MFF. An exception to the rule that phospho-DRP1 inhibits fission is the phosphorylation of Ser 616 of DRP1 by the cyclin-dependant kinase 1/cyclin B complex, which conversely stimulates mitochondrial fission, but this has only been shown to occur during mitosis and may not be involved with apoptosis (reviewed in Chang and Blackstone 2010). In summary, phospho-DRP1 protects from, while dephospho-DRP1 sensitizes cells to, apoptosis. This body of work has increased our mechanistic understanding of the on/off switch of mitochondrial fission in relation to apoptosis and its regulation by reversible phosphorylation.

Other signaling pathways impinging on mitochondria to influence apoptosis are: JNK, Notch, Wnt/β-catenin, and NF-κB. Phosphorylation not only regulates DRP1, but also mitofusin2. Mitofusin2 is phosphorylated by c-Jun N terminal kinase (JNK), triggered by cellular stress, which then causes its ubiquitin-mediated proteasomal degradation and subsequent mitochondrial fragmentation, and ultimately apoptosis (Leboucher et al. 2012). Whether this kinase has a countering phosphatase, as does DRP1, is unknown. Another signaling pathway impinging on mitochondria is Notch (reviewed in Kasahara and Scorrano 2014). In canonical Notch signaling, the Notch receptor is cleaved to generate the Notch intercellular domain (NICD), the active form of Notch. NICD inhibits mitochondrial fission and prevents Bax oligomerization, thus impeding apoptosis. Integral to the NICD pathway is complementation by AKT and mitofusin1 and 2 (Perumalsamy et al. 2010). Further work is needed to extend this link between mitochondrial remodeling proteins and Notch signaling to understand the molecular mechanism governing the NICD–AKT cascade as it relates to mitofusin-mediated inhibition of apoptosis. Yet another recently discovered signaling pathway that converges on mitochondria to influence apoptosis is Wnt/β-catenin activation (Ming et al. 2012). The activation of β-catenin leads to loss of mitochondrial membrane potential and subsequent caspase 3 and 9 cleavage, which cascades into apoptosis. Tests of general applicability of the Wnt/β-catenin-apoptosis axis are needed as it has only been shown in hematopoietic progenitor cells to date. Finally, even OPA1 appears to be the target of a multi-functional signaling pathway.

The NF-κB essential modulator (NEMO) is ubiquitinated by Parkin, which increases OPA1 expression, thus inhibiting apoptosis (Muller-Rischart et al. 2013).

5.5 Mitochondrial Remodeling by ER Constriction During Apoptosis

It has been known for some time that, at the contact sites between mitochondria and the endoplasmic reticulum (ER), their membranes do not fuse but instead are tethered together, suggesting a molecular bridge structure (Fig. 5.4) (Perkins et al. 1997). The term now used for this physical tethering is mitochondria-associated ER membranes (MAM) and its molecular components are now being teased out. Mitofusin2 localizes in the MAM, where it tethers the ER to mitochondria by a homotypic interaction with its counterpart mitofusin2 on the mitochondrial outer membrane, or by a heterotypic interaction with mitofusin1, also on the mitochondrial outer membrane (de Brito and Scorrano 2008). Mitofusin2 depletion causes ER fragmentation and calcium overload, which can initiate apoptosis. The mitofusin2 tether is modulated by MITOL, a mitochondrial ubiquitin ligase that ubiquinates mitochondrial mitofusin2, but not ER-associated mitofusin2 and is required for GTP-dependent mitofusin2 oligomerization, activation, and ER tethering (Sugiura et al. 2013). Trichoplein/mitostatin acts on mitofusin2 contrary to MITOL to disrupt the tethering (Cerqua et al. 2010); the disrupted tethering by trichoplein/mitostatin inhibits Ca^{2+}-dependent apoptosis that relies on ER–mitochondria juxtaposition.

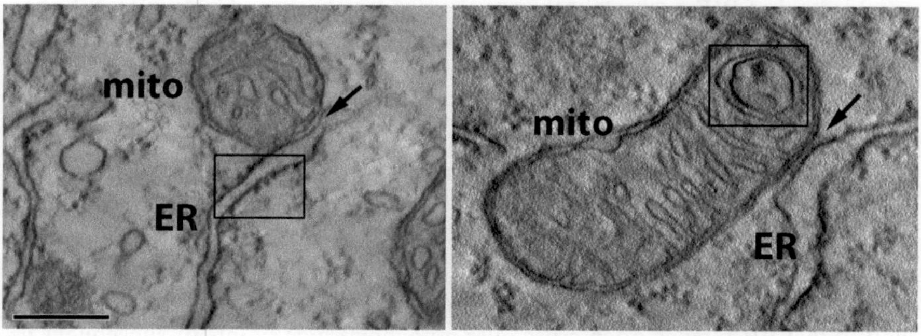

Fig. 5.4 Electron tomography showing the fine details of mitochondria-ER contact. Mitochondria and ER are often seen tethered together. Their membranes do not fuse but instead are tethered by molecular bridge structures (*arrows*). Sometimes arrays of ribosomes decorate the ER (*box on the left*), but never are seen at the mitochondrial-ER contact regions. An abnormal ring-like crista encompassing a dark inclusion body is seen in the mitochondrion on the right (*box on the right*). Intriguingly, ER contact started to form nearby, suggesting that a pinch-off of the abnormal region will proceed at this site

Several other proteins of the MAM have been identified that affect apoptosis-induced ER-mitochondria remodeling, including Fis1 (Yoon et al. 2003), B cell receptor-associated protein 31 (BAP31) (Iwasawa et al. 2011), and PACS2 (Simmen et al. 2005). FIS1 recruits DRP1 to mitochondria, although it is apparently not necessary for DRP1-dependent mitochondrial fission in mammals (Otera et al. 2010). However, overexpression of human FIS1 induces mitochondrial fission concomitant with BAX/BAK-dependent release of cytochrome c from mitochondria (James et al. 2003). Interestingly, human FIS1 does not directly activate BAX and BAK, but instead acts on the MAM to instigate calcium-dependent apoptosis (Alirol et al. 2006). This MAM interaction goes through BAP31, which forms a platform for the recruitment of the initiator procaspase 8 (Iwasawa et al. 2011). Further, apoptotic signals induce cleavage of BAP31 by caspases into the pro-apoptotic fragment p20BAP31, which causes the rapid opening of inositol triphosphate receptors at the MAM, allowing calcium influx into mitochondria that then stimulates DRP1-dependent mitochondrial fission and cytochrome *c* release. In mammalian cells, it is likely that calcium transfers directly between the inositol triphosphate receptor on the ER part of the MAM to the mitochondrial porin VDAC (voltage-dependent anion channel) (reviewed by Helle et al. 2013). The glue between receptor and porin appears to be glucose-regulated-protein 75 (GRP75) as knockdown of GRP75 decreases ER–mitochondria calcium exchange. Another MAM complex consists of the ER protein VAPB and the mitochondrial protein PTPIP51. Knocking-down either VAPB or PTPIP51 causes about 10 % decrease in ER–mitochondrial Ca2+ upload, which may affect calcium-induced apoptosis (reviewed by Helle et al. 2013). Further investigation is needed, though, to determine what role, if any, the VAPB/PTPIP51 complex has on apoptosis. PACS2 is tied into the FIS1/BAP31 signalosome because it was found that depletion of PACS2 causes BAP31-dependent mitochondria fragmentation and uncoupling from the MAM (Simmen et al. 2005). PACS2's relationship to apoptotic mitochondrial remodeling is that it causes the translocation of BID to the mitochondrial outer membrane, which initiates the cascade of BID truncation to tBID, the release of cytochrome *c*, and the activation of caspase-3. Another possible player in MAM-related apoptosis is the small GTP-binding protein RAB32, which modulates ER calcium handling and recruits PKA for phosphorylation and inactivation of DRP1 (Bui et al. 2010). However, the target of RAB32 on the ER is not yet known. Further work is needed to determine how FIS1, BAP31, RAB32, and PACS2 link tBID-induced apoptosis to MAM-calcium-induced apoptosis.

Recent microscopy work has shed light on the physical basis for how mitochondrial fission actually occurs. High-resolution light microscopy on living cells showed that ER tubules often cross over mitochondria at points of subsequent fission or they wrap around mitochondria at those points (Friedman et al. 2011). It was also shown that ER and mitochondrial contacts were still made and mitochondrial constriction still occurred even when the recruitment of DRP1 to mitochondria was inhibited by siRNA knockdown of DRP1 or MFF. These results suggest that ER has a role early in fission, before DRP1 further constricts and severs the mitochondrion (Fig. 5.5). ER tubule constriction of mitochondria might aid DRP1 coiling and

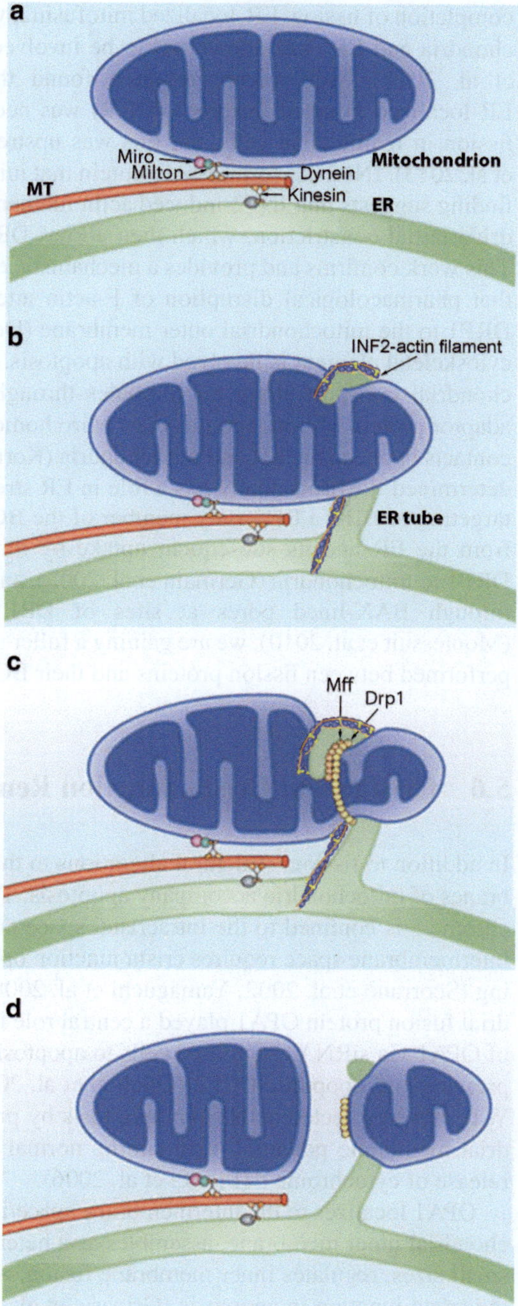

Fig. 5.5 Mechanism of ER-induced mitochondrial fission. (**a**) Miro and its adaptor protein Milton are involved with mitochondrial transport along microtubules (MT) through interactions with dynein, whereas kinesin is the motor that tethers ER to MTs. MTs position mitochondria and ER in close proximity. (**b**) ER tubes and INF2-induced actin filament ropes wrap around mitochondria as an early step in ER-induced fission. (**c**) ER tubule/actin filament wrapping allows MFF to recruit DRP1 to the fission site. Constriction of mitochondria subsequently allows DRP1 coiling and completion of fission (**d**)

completion of fission. ER-localized mitofusin2, which is involved in tethering mitochondria and ER, was shown not to be involved with this constriction (Friedman et al. 2011). More recent research found that actin polymerization through ER-localized inverted formin 2 (INF2) was necessary for efficient mitochondrial fission in mammalian cells and this was upstream of Drp1 activation (Korobova et al. 2013). INF2 is a formin-like protein that initiates actin filament assembly. This finding suggests that INF2-induced actin filament roping around mitochondria may drive initial constriction, which then allows DRP1-driven secondary constriction. This work confirms and provides a mechanistic explanation for the previous finding that pharmacological disruption of F-actin attenuates fission and recruitment of DRP1 to the mitochondrial outer membrane (De Vos et al. 2005). Perhaps another cytoskeletal element is involved with apoptosis. In mammals, Miro regulates mitochondrial transport along microtubules through interactions with dynein and its adaptor protein Milton, and the yeast Miro homolog Gem1p was shown to regulate contacts between the ER and mitochondria (Kornmann et al. 2011). It remains to be determined if Miro actually has a role in ER stress-induced apoptosis. Because ER targeting of BIK, a BH3-only member of the BCL2 family, causes calcium release from the ER and its subsequent uptake by mitochondria, which in turn recruits DRP1 to mitochondria (Germain et al. 2005), and because cytochrome *c* is released through BAX-lined pores at sites of DRP1-mediated mitochondrial fission (Montessuit et al. 2010), we are gaining a fuller understanding of the intricate dance performed between fission proteins and their BCL2 family partners.

5.6 Cristae and Crista Junction Remodeling in Apoptosis

In addition to fission, structural alterations to the internal compartments and membranes of mitochondria accompany apoptosis. This is because the majority of cytochrome *c* is confined to the intracristal space and its complete mobilization to the intermembrane space requires crista junction opening and perhaps cristae remodeling (Scorrano et al. 2002; Yamaguchi et al. 2008). The first hint that the mitochondrial fusion protein OPA1 played a central role in apoptosis was the report that loss of OPA1 via siRNA committed cells to apoptosis, which was prevented by overexpressing anti-apoptotic BCL2 (Olichon et al. 2003). In contrast, overexpression of WT OPA1 protected cells from apoptosis by preventing the loss of the mitochondrial membrane potential, maintaining normal cristae structure, and delaying the release of cytochrome *c* (Frezza et al. 2006).

OPA1 localizes to the intermembrane space/intracristal space leaflet of the mitochondrial inner membrane, assembles as a hetero-oligomeric complex of large and small sizes, regulates inner membrane fusion, and is necessary for maintenance of the crista junction in apoptosis (Griparic et al. 2004; Ishihara et al. 2006; Olichon et al. 2007). *OPA1* is a nuclear-encoded gene, and in humans, there are eight splice variants of *OPA1* (Griparic et al. 2004; Song et al. 2007). In human and mouse cells, six isoforms of OPA1 are detectable by western blot: two long membrane-bound isoforms and four short membrane-free isoforms. Loss of membrane-bound isoforms causes aberrant cristae morphogenesis and impaired cellular proliferation (Merkwirth et al. 2008; Song et al. 2007). It appears that OPA1 can block apoptosis independently from its role in mitochondrial fusion because knock-down of the OPA1 splice variants containing exon 4 shows that it is important for fusion, whereas knock-down of those variants containing either exon 4b or 5b shows that they regulate cytochrome *c* release (Olichon et al. 2007). Furthermore, expression of an oligomer disassembly resistant mutant of OPA1 (Q297A) blocked the full release of cytochrome *c* and apoptosis (Yamaguchi et al. 2008; Yamaguchi and Perkins 2009). There are two conditions under which the complete conversion of all long isoforms of OPA1 to short isoforms takes place: one, during tBID-induced apoptosis, and two, during the treatment of cells with the mitochondrial uncoupler, CCCP. When isolated mitochondria are incubated with tBID, all the long isoforms of OPA1 are quickly converted to short isoforms and it is only the short isoforms that are released from mitochondria when OPA1 oligomers are disassembled (Yamaguchi et al. 2008). It was found that this disassembly required the presence of either BAX or BAK. Remarkably, the BAX/BAK-dependent events at the inner membrane (OPA1 complex disassembly and crista remodeling) could be uncoupled from BAX/BAK-dependent events at the outer membrane (oligomerization and outer membrane permeabilization). However, it is worth noting that the release of the short isoforms is not completed during the 30 min in which 100 % of cytochrome *c* and OMI molecules are released. This might be because the membrane-free short isoforms are still attached to a processing protease. Thus, disassembly of the OPA1 oligomer is a key early event in apoptosis and represents a potential point of therapeutic intervention to modulate apoptosis. Expression of a disassembly resistant mutant Opa1 blocked both the full release of cytochrome *c* and apoptosis.

Disruption of the OPA1 hetero-oligomer and release of OPA1 from inside mitochondria during early apoptosis precipitates structural alterations, including mitochondrial fragmentation (Arnoult et al. 2005). Different forms of sparse and shortened cristae, and crista vesiculation, indicative of unbalanced membrane fission, are observed in an *OPA1* mutant (Frezza et al. 2006) and after etoposide treatment (Sun et al. 2007). Further, OPA1 appears to be a chaperone for subunit *e* of ATP synthase and mutation in *OPA1* or its yeast homologue, *MGM1*, inhibits the formation of ATP synthase dimers and abolishes normal tubular crista junctions (Amutha et al. 2004). OPA1 depletion by RNAi leads to fragmented mitochondria, disrupted cristae, and an increase in sensitivity to apoptotic stimuli (Frezza et al. 2006; Lee et al. 2004; Olichon et al. 2003). Interestingly, OPA1/DRP1 and OPA1/MFF double-RNAi depletion produce elongated mitochondria, yet with disrupted cristae, and exhibit a significant delay in the release of cytochrome *c* in response to apoptotic stimuli, but this did not affect BAX translocation to mitochondria or the release of SMAC/DIABLO (Otera et al. 2010). Genetic manipulation of several proteins has generated mitochondria with altered cristae structure that are unable to release all of their cytochrome *c* (Arnoult et al. 2005; Cipolat et al. 2006; Frezza et al. 2006; Griparic et al. 2004; John et al. 2005; Lee et al. 2004), showing that cytochrome *c* can be trapped inside cristae.

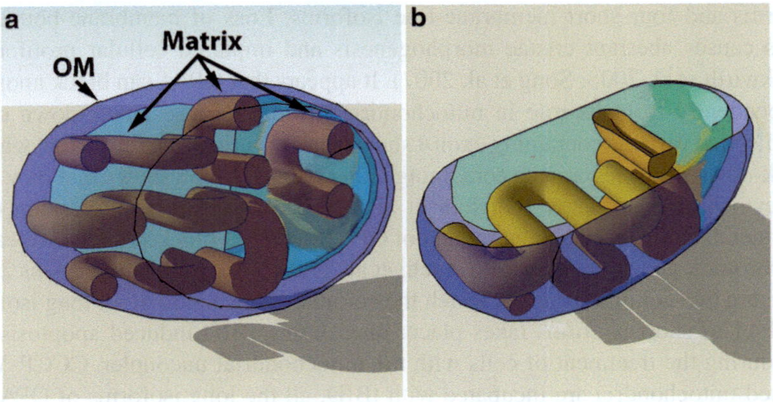

Fig. 5.6 Ultracondensed mitochondria have been observed with tBID-mediated cytochrome *c* release. Later during apoptosis, there is a reversal of the membrane curvature of cristae to form tubes that enclose the matrix instead of the intracristal space. These tubes look like sausages that snake through the mitochondrial interior; these mitochondria are termed ultracondensed. (**a**) Model of an ultracondensed mitochondrion. The mitochondrial membrane, shown in *blue*, was made translucent to better visualize the matrix sausages (*red*). (**b**) Another view of the model with the top portion cut away

The structural remodeling of cristae and crista junctions during apoptosis has received attention from several groups. One group found that with tBID-mediated cytochrome *c* release, there was a fivefold widening of crista junctions and a reversal of the membrane curvature of cristae to form tubes that enclose the matrix instead of the intracristal space, a feature seen in ultracondensed mitochondria (Fig. 5.6) (Frezza et al. 2006; Scorrano et al. 2002). Here, the OPA1 complexes disappeared rapidly, leaving only the monomeric form. Another group reported that remodeling to the condensed state with dilated crista junctions occurs during apoptosis induced by growth factor withdrawal and a subsequent decrease in the inner membrane potential (Gottlieb et al. 2003). Once in the intermembrane space, cytochrome *c* can escape to the cytosol when pores are formed on the outer membrane (see section below). However, the work of two other groups questions whether this cristae/crista junction remodeling indeed is required for rapid and complete cytochrome *c* release or is simply a consequence of the released cytochrome *c*. One group reported that during tBID-induced apoptosis, a narrowing instead of a widening of crista junction openings was observed (Yamaguchi et al. 2008). The average diameter of crista junctions was about 9 nm, half the diameter of crista junctions in untreated mitochondria (average 16 nm), and no cristae remodeling was found. Remarkably, in the presence of MG132 plus tBID or tBID alone, crista junctions narrowed while the accessibility of cytochrome *c* to the outer membrane was increased. Thus, only a subtle modification to mitochondrial structure occurred, but was sufficient for complete cytochrome *c* mobilization and release. Still, another report did not note a change in crista junction diameters in cells treated with etoposide and z-VAD until after cytochrome *c* release (Sun et al. 2007). Thus, it appears that OPA1 complex assembly and disassembly regulate cytochrome *c* sequestration inside cristae, perhaps like a dam. Under non-apoptotic conditions, this dam would block the crista junction opening. It is envisioned that the dam would block soluble proteins, but not membrane proteins because of evidence that no physical barrier blocks the lateral movement of small inner membrane proteins across crista junctions (Vogel et al. 2006). Oligomerized OPA1 complexes formed at this junction may exceed several hundred kDaltons and so may be large enough to act as a dam (Landes et al. 2010a). During apoptosis, though, the OPA1 dam is demolished (Fig. 5.7) by a process independent of caspases, leaving holes initially about 9 nm across (and then growing much larger from cristae remodeling), large enough to let pass 60–100 kDa proteins (Kinnally and Antonsson 2007); hence it would allow passage of a cytochrome *c* molecule that has a hydrodynamic diameter of 3.4 nm (Mirkin et al. 2008). In summary, the pro-apoptotic BH3-only proteins cause the

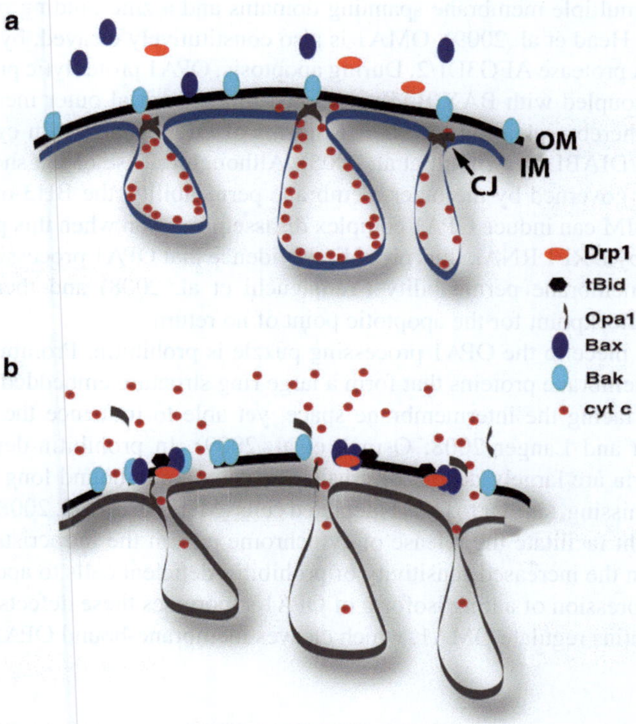

Fig. 5.7 A large OPA1 complex at the crista junction regulates cytochrome *c* sequestration inside cristae. (**a**) Under non-apoptotic conditions, the OPA1 complex blocks the crista junction (CJ) opening. These complexes may exceed several hundred kilo Dalton and so would be large enough to act as a dam. (**b**) During apoptosis, though, the OPA1 dam is demolished by tBID activation leaving holes large enough to allow passage of a cytochrome *c* molecule from the cristae into the intermembrane space. BAX and BAK activation, perhaps mediated by DRP1, forms holes in the mitochondrial outer membrane releasing cytochrome *c*, OPA1, and other pro-apoptotic effectors into the cytosol

OPA1 complexes at the crista junction to rapidly dissociate, leading to mobilization of cytochrome *c* from the intracristal space to the intermembrane space and as described below the subsequent release of cytochrome *c* and other pro-apoptotic effectors through the mitochondrial outer membrane into the cytosol.

What is the protease that cleaves OPA1 and why is this important for apoptotic remodeling of mitochondria? As shall be seen, regulating the apoptosis-induced proteolytic cleavage of OPA1 offers an appealing strategy to either promote or prevent apoptosis. Early efforts to identify this protease focused on the rhomboid protease PARL (Presenilin-associated rhomboid-like) (Herlan et al. 2003), Paraplegin (Ishihara et al. 2006), and the i-AAA protease YME1 (Griparic et al. 2007). However, the current consensus is that YME1 mediates apoptosis-driven processing only of a subset of OPA1 splice variants, whereas PARL and Paraplegin have minor, if any, roles in OPA1 processing. More recently, two groups identified that OPA1-inducible cleavage is governed by the mitochondrial zinc metalloprotease OMA1, which has multiple membrane spanning domains and a zinc-binding motif (Ehses et al. 2009; Head et al. 2009). OMA1 is also constitutively cleaved, by subunits of the m-AAA protease AFG3L1/2. During apoptosis, OPA1 proteolytic processing by OMA1 is coupled with BAX/BAK-promoted mitochondrial outer membrane permeability, thereby releasing the short isoforms of OPA1 along with cytochrome *c* and SMAC/DIABLO (Arnoult et al. 2005). Although release of the short isoforms of OPA1 is governed by the outer membrane permeability, the BH3-only proteins tBID and BIM can induce OPA1 complex disassembly even when this permeability is blocked by DRP1 RNAi, thus providing evidence that OPA1 processing precedes the outer membrane permeability (Yamaguchi et al. 2008) and therefore is an appealing checkpoint for the apoptotic point of no return.

Another piece to the OPA1 processing puzzle is prohibitin. Prohibitins 1 and 2 are inner membrane proteins that form a large ring structure embedded in the inner membrane facing the intermembrane space, yet able to influence the matrix side (Merkwirth and Langer 2008; Osman et al. 2009). In prohibitin-depleted cells, mitochondria are largely devoid of cristae and membrane-bound long isoforms of OPA1 are missing, similar to OPA1-depleted cells (Merkwirth et al. 2008). Disturbed cristae might facilitate the release of cytochrome *c* from the intracristal space and thus explain the increased sensitivity of prohibitin-deficient cells to apoptotic stimuli. The expression of a long isoform of OPA1 suppresses these defects, suggesting that prohibitins regulate OMA1, which cleaves membrane-bound OPA1 isoforms.

5.7 The Role of Cardiolipin in the Full Release of Cytochrome c During Apoptosis

It appears that the binding of cytochrome *c* to cardiolipin causes it to be retained inside the cristae even after the opening of the crista junction and permeabilization of the mitochondrial outer membrane (Schug and Gottlieb 2009). Consistent with this concept, permeabilization of the mitochondrial outer membrane with digitonin was not sufficient for complete cytochrome *c* release (Yamaguchi et al. 2008). Interestingly, depletion of cardiolipin releases cytochrome *c* from the mitochondrial inner membrane and accelerates apoptosis (Choi et al. 2007). There is an electrostatic force between cytochrome *c* and cardiolipin, and unless this weak interaction is broken, cytochrome *c* can remain inside mitochondria (Rytomaa and Kinnunen 1995; Rytomaa et al. 1992; Tuominen et al. 2002). Cardiolipin is a charged lipid found almost exclusively in the mitochondrial inner membrane where it constitutes about 20 % of the total lipid content (de Kroon et al. 1997). It was reported that cardiolipin must be oxidized to obtain the complete release of cytochrome *c* (Ott et al. 2002). However, the influx of calcium observed in certain pathways of apoptosis does not appear to be required for the disruption of the cardiolipin–cytochrome *c* association (reviewed by Perkins et al. 2009). Might loss or modification of cardiolipin disrupt cardiolipin–cytochrome *c* association during apoptosis? A 50 % reduction in cardiolipin content was observed in mitochondria treated with tBID-G94E as well as mitochondria treated with tBID plus BCLXL (Yamaguchi et al. 2008). It may be that cardiolipin modification by reactive oxygen species (ROS) affects its binding to cytochrome *c* (Iverson and Orrenius 2004). In STS-induced apoptosis, cardiolipin underwent oxidation as measured by mass spectroscopy (Tyurin et al. 2008). Furthermore, in actinomycin-D-induced apoptosis, oxidation of cardiolipin (6 h) preceded cytochrome *c* release (8 h), caspase 3 and 7 activation (8 h), annexin V positivity (9 h), and decrease in the transmembrane potential (12–14 h) (Kagan et al. 2006), suggesting that peroxidation of this lipid is not the consequence of cytochrome *c* release nor a drop in the transmembrane potential. Interestingly, *OPA1* mutation showed enhanced sensitivity to elevated ROS (Kanazawa et al. 2008) and abnormal cristae.

5.8 Remodeling of the Mitochondrial Outer Membrane During Apoptosis

Two mitochondrial remodeling steps lead to cytochrome *c* release: mobilization from cristae and translocation through the mitochondrial outer membrane. Even though not as well-studied, several other proteins have also been shown to be released from mitochondria following mitochondrial outer membrane permeability during apoptosis including SMAC/DIABLO, adenylate kinase, OMI/HTRA2, DDP, ENDOG, and cleaved OPA1 (Wang and Youle 2009). Upon triggering apoptosis, the BH3-only proteins, such as BID, BIK, and BIM, induce the mitochondrial localization of BAX and its accumulation at sites on the mitochondrial outer membrane (Kuwana et al. 2002; Montessuit et al. 2010). After BAX is at the mitochondrial outer membrane, Drp1 stimulates its oligomerization and subsequent remodeling of the outer membrane. Reports of the past 20 years provide evidence that BAX/BAK-dependent mitochondrial outer membrane permeability is the key event leading to the execution phase of the mitochondrial pathway of apoptosis. Supporting this model are the observations that in the absence of BAX or BAK oligomerization, there is no cytochrome *c* release (Kuwana et al. 2002; Scorrano et al. 2003; Wei et al. 2001; Yamaguchi et al. 2008). BAX and/or BAK form the pore on the outer membrane through which cytochrome *c* leaves the mitochondrion, while BCL2, BCLXL, or MCL1 inhibit its formation (Yamaguchi and Perkins 2009). Because blocking mitochondrial outer membrane permeability did not block OPA1 processing or crista junction opening, but did block cytochrome *c* release and BAK oligomerization (Yamaguchi et al. 2008), mitochondrial outer membrane permeability and crista junction opening are separate events.

Not only does DRP1 self-assemble into helical structures that wrap around mitochondria and coordinate the division of the outer and inner membranes (Ingerman et al. 2005), but it also impinges specifically on mitochondrial outer membrane permeabilization. The precise mechanism by which DRP1 and other mitochondrial shaping proteins influence outer membrane permeabilization during apoptosis is still a mystery. However, under apoptotic conditions, DRP1 is found in foci with BAX and mitofusin2 on the mitochondrial outer membrane (Cleland and Youle 2011; Karbowski et al. 2002). Yet, mitochondrial fission is not a key factor in mitochondrial outer membrane permeabilization during apoptosis, indicating that the role of DRP1 in outer membrane permeabilization is independent of its role in mitochondrial division. Martinou and coworkers recently proposed a model wherein DRP1 promotes the formation of a nonbilayer hemifission intermediate that facilitates oligomerized BAX formation of a hole, central to the concept of mitochondrial outer membrane permeability (Montessuit et al. 2010). Hoppins and Nunnari (2012) propose that the specialized ER tubules that wrap around mitochondria and mark mitochondrial division sites (see section above) create a geometric "hot spot" for an ER-mitochondria microdomain facilitating BAX insertion and oligomerization. Evidence for such microdomains is that the DRP1 effector, MFF, accumulates at sites of ER-mitochondrial contact in the absence of DRP1 that forms a spatial mark for DRP1 recruitment to constriction sites (Friedman et al. 2011). More work is needed to ascertain the dynamic nature of the assembly, organization, and number of ER-mitochondrial hot spots. Contacts between mitochondria and ER are also involved in regulating the levels of sphingolipid metabolites required for BAX/BAK activation (Chipuk et al. 2012).

The sphingolipid, ceramide, is involved with mitochondrial outer membrane permeabilization through the channels it forms in the outer membrane. Anti-apoptotic BCL2 family proteins destabilize these channels, whereas the pro-apoptotic proteins act synergistically with ceramide to form and stabilize the channels (reviewed in Colombini 2013). For example, BAX was found to act synergistically with ceramide to increase the permeabilization of the mitochondrial outer membrane (Ganesan et al. 2010). However, BAX does not need ceramide to form channels and vice versa. It was speculated that activated BAX acts as a molecular scaffold that governs the specific radius of curvature of the ceramide channel. The action of BAX is dependent on access to ceramide's amide nitrogen. This interaction is abrogated when the nitrogen is methylated. Ceramide molecules are proposed to form columns that span the hydrophobic inner core of the bilayer membrane. Each column is proposed to consist of six ceramides and the number of columns, arranged as staves of a barrel, determines the opening size of the channel. The relatively large size of the ceramide channel allows individual channels to be imaged by negative stain electron microscopy.

The most common pore opening size was measured to be about 10 nm in diameter, intriguingly similar in size to the 9 nm crista junction opening size measured upon disassembly of the OPA1 oligomers (Yamaguchi et al. 2008) which allows cytochrome *c* to exit the intracristal space into the intermembrane space.

Interestingly, ceramide and its transport protein (CERT) are key molecules signaling the accelerated incidence of apoptosis in oocytes of aged female mice (Kujjo et al. 2013). With increasing age, changes in levels of both ceramide and its transport protein contribute to the deterioration of oocyte mitochondrial structure and function. These mitochondria possessed fewer cristae and degenerated membranes similar to the phenotype observed in oocytes when loss of mitochondrial activity triggered apoptosis (Perez et al. 2007).

5.9 The Role of Cardiolipin in the Remodeling of the Mitochondrial Outer Membrane

Several reports show that tBID associates with cardiolipin (Gonzalvez et al. 2005; Kim et al. 2004; Kuwana et al. 2002; Lutter et al. 2000). Cardiolipin is present at contact sites between the outer and the inner boundary membrane of mitochondria (Ardail et al. 1990). It has been suggested that cardiolipin even draws tBID to the mitochondrial inner membrane (Epand et al. 2002; Kim et al. 2004). Interestingly, tBID produced ultracondensed mitochondria showing a reversal of the curvature of cristae membranes (Frezza et al. 2006; Scorrano et al. 2002), indicating its influence on this lipid bilayer. Consistent with this report, a cardiolipin deficiency disease also produces reversed curvature of the cristae membranes (Acehan et al. 2007). It has even been proposed that cardiolipin anchors the apoptosis-initiating caspase, caspase 8, at contact sites between inner and outer mitochondrial membranes, where this caspase oligomerizes facilitating its self-activation (Gonzalvez et al. 2008). tBID also localized to contact sites (Lutter et al. 2001). Kuwana and coworkers (2002) first reported that cardiolipin is required for the formation of tBID-induced BAX oligomers and of pores more than large enough for cytochrome *c* release, which was later confirmed by two other groups (Lucken-Ardjomande et al. 2008; Terrones et al. 2004). In summary, it appears that cardiolipin is needed for concentrating tBID into the contact site microdomain that would help orient it for proper presentation of its BH3 domain to BAX or BAK [15] and confer prime positioning for cristae remodeling (Schug and Gottlieb 2009; Zinkel et al. 2013).

5.10 Conclusion

Here we focused on how mitochondrial architecture and its remodeling regulate apoptotic cell death. The execution phase or "point of no return" of apoptosis requires the activation of specific signaling pathways and the arrival and interplay of DRP1, tBID, BAX, and BAK at the mitochondrial outer membrane. The key remodeling step to prime the release of cytochrome *c* to the cytoplasm, thus initiating the cellular apoptotic cascade of enzyme action, is the opening of crista junctions triggered by disassembly of OPA1 oligomers. What follows next is the key remodeling step for the execution phase of apoptosis: the formation of mitochondrial outer membrane pores for the rapid and complete release of cytochrome *c*. Along the way, it is likely that ER-actin filament–mitochondria association is important for mitochondrial fission, which has a role in apoptosis. Although whether fission is obligatory for mitochondrial-induced apoptosis is still controversial. It is also likely that cytochrome *c* needs to dissociate from cardiolipin to be released from the intracristal space, but further work needs to confirm this. All these events represent checkpoints for strategies that would enhance or impede apoptotic progression. These checkpoints are important to understand how cancer evades the body's mechanisms to kill cells that threaten our well-being. As cancer progresses, cancer cells accumulate more mutations, abrogating checkpoint functions and apoptosis-inducing programs, making them resistant to genotoxin-based treatments. Another approach that holds promise would be to directly activate apoptosis at the most critical juncture in the apoptotic pathway—the execution phase. Further work needs to be done on combination therapies that promote the execution phase (Yamaguchi and Perkins 2012a, b).

References

Acehan D, Xu Y, Stokes DL, Schlame M (2007) Comparison of lymphoblast mitochondria from normal subjects and patients with Barth syndrome using electron microscopic tomography. Lab Invest 87:40–48

Alirol E, James D, Huber D et al (2006) The mitochondrial fission protein hFis1 requires the endoplasmic reticulum gateway to induce apoptosis. Mol Biol Cell 17:4593–4605

Amutha B, Gordon DM, Gu Y, Pain D (2004) A novel role of Mgm1p, a dynamin related GTPase, in ATP synthase assembly and cristae formation/maintenance. Biochem J 381:19–23

Ardail D, Privat JP, Egret-Charlier M, Levrat C, Lerme F, Louisot P (1990) Mitochondrial contact sites. Lipid composition and dynamics. J Biol Chem 265:18797–18802

Arnoult D, Grodet A, Lee YJ, Estaquier J, Blackstone C (2005) Release of Opa1 during apoptosis participates in the rapid and complete release of cytochrome c and subsequent mitochondrial fragmentation. J Biol Chem 280:35742–35750

Barsoum MJ, Yuan H, Gerencser AA, Liot G, Kushnareva Y, Gräber S, Kovacs I, Lee WD, Waggoner J, Cui J, White AD, Bossy B, Martinou JC, Youle RJ, Lipton SA, Ellisman MH, Perkins GA, Bossy-Wetzel E (2006) Nitric oxide-induced mitochondrial fission is regulated by dynamin-related GTPases in neurons. EMBO J 25:3900–3911

Benard G, Bellance N, James D, Parrone P, Fernandez H, Letellier T, Rossignol R (2007) Mitochondrial bioenergetics and structural network organization. J Cell Sci 120:838–848

Bereiter-Hahn J, Vöth M, Mai S, Jendrach M (2008) Structural implications of mitochondrial dynamics. Biotechnol J 3:765–780

Brooks C, Dong Z (2007) Regulation of mitochondrial morphological dynamics during apoptosis by Bcl-2 family proteins: a key in Bak? Cell Cycle 6:3043–3047

Brooks C, Wei Q, Feng L, Dong G, Tao Y, Mei L, Xie Z-J, Dong Z (2007) Bak regulates mitochondrial morphology and pathology during apoptosis by interacting with mitofusins. Proc Natl Acad Sci U S A 104:11649–11654

Brooks C, Cho SG, Wang CY, Yang T, Dong Z (2011) Fragmented mitochondria are sensitized to Bax insertion and activation during apoptosis. Am J Physiol Cell Physiol 300:C447–C455

Bui M, Gilady SY, Fitzsimmons RE, Benson MD, Lynes EM, Gesson K, Alto NM, Strack S, Scott JD, Simmen T (2010) Rab32 modulates apoptosis onset and mitochondria-associated membrane (MAM) properties. J Biol Chem 285:31590–31602

Cassidy-Stone A, Chipuk JE, Ingerman E, Song C, Yoo C, Kuwana T, Kurth MJ, Shaw JT, Hinshaw JE, Green DR, Nunnari J (2008) Chemical inhibition of the mitochondrial division dynamin reveals its role in Bax/Bak-dependent mitochondrial outer membrane permeabilization. Dev Cell 14:193–204

Cereghetti GM, Stangherlin A, Martins de Brito O, Chang CR, Blackstone C, Bernardi P, Scorrano L (2008) Dephosphorylation by calcineurin regulates translocation of Drp1 to mitochondria. Proc Natl Acad Sci U S A 105:15803–15808

Cerqua C et al (2010) Trichoplein/mitostatin regulates endoplasmic reticulum-mitochondria juxtaposition. EMBO Rep 11:854–860

Chang CR, Blackstone C (2010) Dynamic regulation of mitochondrial fission through modification of the dynamin-related protein Drp1. Ann N Y Acad Sci 1201:34–39

Chen H, McCaffery JM, Chan DC (2007) Mitochondrial fusion protects against neurodegeneration in the cerebellum. Cell 130:548–562

Chipuk JE, McStay GP, Bharti A et al (2012) Sphingolipid metabolism cooperates with BAK and BAX to promote the mitochondrial pathway of apoptosis. Cell 148:988–1000

Cho D-H, Nakamura T, Fang J, Cieplak P, Godzik A, Gu Z, Lipton SA (2009) S-nitrosylation of Drp1 mediates beta-amyloid-related mitochondrial fission and neuronal injury. Science 324:102–105

Choi SY, Gonzalvez F, Jenkins GM, Slomianny C, Chretien D, Arnoult D, Petit PX, Frohman MA (2007) Cardiolipin deficiency releases cytochrome c from the inner mitochondrial membrane and accelerates stimuli-elicited apoptosis. Cell Death Differ 14:597–606

Chou CH, Lin CC, Yang MC, Wei CC, Liao HD, Lin RC, Tu WY, Kao TC, Hsu CM, Cheng JT, Chou AK, Lee CI, Loh JK, Howng SL, Hong YR (2012) GSK3beta-mediated Drp1 phosphorylation induced elongated mitochondrial morphology against oxidative stress. PLoS One 7(11):e49112

Cipolat S, Rudka T, Hartmann D, Costa V, Serneels L, Craessaerts K, Metzger K, Frezza C, Annaert W, D'Adamio L et al (2006) Mitochondrial rhomboid PARL regulates cytochrome c release during apoptosis via Opa1-dependent cristae remodeling. Cell 126:163–175

Cleland MM, Youle RJ (2011) Chapter 4: Mitochondrial dynamics and apoptosis. In: Lu B (ed) Mitochondrial dynamics and neurodegeneration. Springer Science, Berlin

Cleland MM, Norris KL, Karbowski M, Wang C, Suen DF, Jiao S, George NM, Luo X, Li Z, Youle RJ (2011) Bcl-2 family interaction with the mitochondrial morphogenesis machinery. Cell Death Differ 18:235–247

Colombini M (2013) Membrane channels formed by ceramide. Handbook Exp Pharmacol 215:109–126

Costa V, Giacomello M, Hudec R, Lopreiato R, Ermak G, Lim D et al (2010) Mitochondrial fission and cristae disruption increase the response of cell models of Huntington's disease to apoptotic stimuli. EMBO Mol Med 2:490–503

Cribbs JT, Strack S (2007) Reversible phosphorylation of Drp1 by cyclic AMP-dependent protein kinase and calcineurin regulates mitochondrial fission and cell death. EMBO Rep 8:939–944

de Brito OM, Scorrano L (2008) Mitofusin 2 tethers endoplasmic reticulum to mitochondria. Nature 456:605–610

de Kroon AI, Dolis D, Mayer A, Lill R, de Kruijff B (1997) Phospholipid composition of highly purified mitochondrial outer membranes of rat liver and Neurospora crassa. Is cardiolipin present in the mitochondrial outer membrane. Biochim Biophys Acta 1325:108–116

De Vos KJ, Allan VJ, Grierson AJ, Sheetz MP (2005) Mitochondrial function and actin regulate dynamin-related protein 1-dependent mitochondrial fission. Curr Biol 15:678–683

Dickey AS, Strack S (2011) PKA/AKAP1 and PP2A/Bβ2 regulate neuronal morphogenesis via Drp1 phosphorylation and mitochondrial bioenergetics. J Neurosci 31:15716–15726

Dikov D, Reichert AS (2011) How to split up: lessons from mitochondria. EMBO J 30:2751–2753
Ehses S, Raschke I, Mancuso G, Bernacchia A, Geimer S, Tondera D, Martinou J-C, Westermann B, Rugarli EI, Langer T (2009) Regulation of OPA1 processing and mitochondrial fusion by m-AAA protease isoenzymes and OMA1. J Cell Biol 187:1023–1036
Epand RF, Martinou JC, Fornallaz-Mulhauser M, Hughes DW, Epand RM (2002) The apoptotic protein tBid promotes leakage by altering membrane curvature. J Biol Chem 277:32632–32639
Estaquier J, Arnoult D (2007) Inhibiting Drp1-mediated mitochondrial fission selectively prevents the release of cytochrome c during apoptosis. Cell Death Differ 14:1086–1094
Frank S, Gaume B, Bergmann-Leitner ES, Leitner WW, Robert EG, Catez F, Smith CL, Youle RJ (2001) The role of dynamin-related protein 1, a mediator of mitochondrial fission, in apoptosis. Dev Cell 1:515–525
Frezza C, Cipolat S, Martins de Brito O et al (2006) Opa1 controls apoptotic cristae remodeling independently from mitochondrial fusion. Cell 126:177–189
Friedman JR, Lackner LL, West M, Dibenedetto JR, Nunnari J, Voeltz GK (2011) ER tubules mark sites of mitochondrial division. Science 334:358–362
Ganesan V, Perera MN, Colombini D, Datskovskiy D, Chadha K, Colombini M (2010) Ceramide and activated Bax act synergistically to permeabilize the mitochondrial outer membrane. Apoptosis 15:553–562
Gao W, Pu Y, Luo KQ, Chang DC (2001) Temporal relationship between cytochrome c release and mitochondrial swelling during UV-induced apoptosis in living HeLa cells. J Cell Sci 114:2855–2862
Germain M, Mathai JP, McBride HM, Shore GC (2005) Endoplasmic reticulum BIK initiates DRP1-regulated remodeling of mitochondrial cristae during apoptosis. EMBO J 24:1546–1556
Gonzalvez F, Pariselli F, Dupaigne P, Budihardjo I, Lutter M, Antonsson B, Diolez P, Manon S, Martinou JC, Goubern M, Wang X, Bernard S, Petit PX (2005) tBid interaction with cardiolipin primarily orchestrates mitochondrial dysfunctions and subsequently activates Bax and Bak. Cell Death Differ 12:614–626
Gonzalvez F, Schug ZT, Houtkooper RH, MacKenzie ED, Brooks DG, Wanders RJ, Petit PX, Vaz FM, Gottlieb E (2008) Cardiolipin provides an essential activating platform for caspase-8 on mitochondria. J Cell Biol 183:681–696
Gottlieb E, Armour SM, Harris MH, Thompson CB (2003) Mitochondrial membrane potential regulates matrix configuration and cytochrome c release during apoptosis. Cell Death Differ 10:709–717
Goyal G, Fell B, Sarin A, Youle RJ, Sriram V (2007) Role of mitochondrial remodeling in programmed cell death in Drosophila melanogaster. Dev Cell 12:807–816
Griparic L, van der Wel NN, Orozco IJ, Peters PJ, van der Bliek AM (2004) Loss of the intermembrane space protein Mgm1/Opa1 induces swelling and localized constrictions along the lengths of mitochondria. J Biol Chem 279:18792–18798
Griparic L, Kanazawa T, Van Der Bliek AM (2007) Regulation of the mitochondrial dynamin-like protein Opa1 by proteolytic cleavage. J Cell Biol 178:757–764
Head B, Griparic L, Amiri M, Gandre-Babbe S, Van Der Bliek AM (2009) Inducible proteolytic inactivation of OPA1 mediated by the OMA1 protease in mammalian cells. J Cell Biol 187:959–966
Helle SCJ, Kanfer G, Kolar K, Lang A, Agnès H, Kornmann MB (2013) Organization and function of membrane contact sites. Biochim Biophys Acta 1833:2526–2541
Herlan M, Vogel F, Bornhovd C, Neupert W, Reichert AS (2003) Processing of Mgm1 by the rhomboid-type protease Pcp1 is required for maintenance of mitochondrial morphology and of mitochondrial DNA. J Biol Chem 278:27781–27788
Hockenbery D, Nuñez G, Milliman C, Schreiber RD, Korsmeyer SJ (1990) Bcl-2 is an inner mitochondrial membrane protein that blocks programmed cell death. Nature 348:334–336
Hoppins S, Nunnari J (2012) Mitochondrial dynamics and apoptosis--the ER connection. Science 337:1052–1054

Ingerman E, Perkins EM, Marino M, Mears JA, McCaffery JM, Hinshaw JE, Nunnari J (2005) Dnm1 forms spirals that are structurally tailored to fit mitochondria. J Cell Biol 170:1021–1027
Ishihara N, Fujita Y, Oka T, Mihara K (2006) Regulation of mitochondrial morphology through proteolytic cleavage of OPA1. EMBO J 25:2966–2977
Ishihara N, Nomura M, Jofuku A et al (2009) Mitochondrial fission factor Drp1 is essential for embryonic development and synapse formation in mice. Nat Cell Biol 11:958–966
Iverson SL, Orrenius S (2004) The cardiolipin–cytochrome c interaction and the mitochondrial regulation of apoptosis. Arch Biochem Biophys 423:37–46
Iwasawa R, Mahul-Mellier A-L, Datler C, Pazarentzos E, Grimm S (2011) Fis1 and Bap31 bridge the mitochondria–ER interface to establish a platform for apoptosis induction. EMBO J 30:556–568
Jahani-Asl A, Cheung EC, Neuspiel M, Maclaurin JG, Fortin A, Park DS, Mcbride HM, Slack RS (2007) Mitofusin 2 protects cerebellar granule neurons against injury induced cell death. J Biol Chem 282:23788–23798
Jakobs S, Wurm CA (2014) Super-resolution microscopy of mitochondria. Curr Opin Chem Biol 20C:9–15
James DI, Parone PA, Mattenberger Y, Martinou JC (2003) hFis1, a novel component of the mammalian mitochondrial fission machinery. J Biol Chem 278:36373–36379
John GB, Shang Y, Li L, Renken C, Mannella CA, Selker JM, Rangell L, Bennett MJ, Zha J (2005) The mitochondrial inner membrane protein mitofilin controls cristae morphology. Mol Biol Cell 16:1543–1554
Kagan VE, Tyurina YY, Bayir H, Chu CT, Kapralov AA, Vlasova II, Belikova NA, Tyurin VA, Amoscato A, Epperly M, Greenberger J, Dekosky S, Shvedova AA, Jiang J (2006) The "pro-apoptotic genies" get out of mitochondria: oxidative lipidomics and redox activity of cytochrome c/cardiolipin complexes. Chem Biol Interact 163:15–28
Kanazawa T, Zappaterra MD, Hasegawa A, Wright AP, Newman-Smith ED, Buttle KF, McDonald K, Mannella CA, van der Bliek AM (2008) The C. elegans Opa1 homologue EAT-3 is essential for resistance to free radicals. PLoS Genet 4(2):e1000022
Karbowski M, Lee YJ, Gaume B, Jeong SY, Frank S, Nechushtan A, Santel A, Fuller M, Smith CL, Youle RJ (2002) Spatial and temporal association of Bax with mitochondrial fission sites, Drp1, and Mfn2 during apoptosis. J Cell Biol 159:931–938
Karbowski M, Youle RJ (2003) Dynamics of mitochondrial morphology in healthy cells and during apoptosis. Cell Death Differ 10:870–880
Karbowski M, Norris KL, Cleland MM, Jeong SY, Youle RJ (2006) Role of Bax and Bak in mitochondrial morphogenesis. Nature 443:658–662
Kasahara A, Scorrano L (2014) Mitochondria: from cell death executioners to regulators of cell differentiation. Trends Cell Biol 24:761–770
Kerr JF, Wyllie AH, Currie AR (1972) Apoptosis: a basic biological phenomenon with wide-ranging implications in tissue kinetics. Br J Cancer 26:239–257
Kim TH, Zhao Y, Ding WX, Shin JN, He X, Seo YW, Chen J, Rabinowich H, Amoscato AA, Yin XM (2004) Bid-cardiolipin interaction at mitochondrial contact site contributes to mitochondrial cristae reorganization and cytochrome C release. Mol Biol Cell 15:3061–3072
Kinnally KW, Antonsson B (2007) A tale of two mitochondrial channels, MAC and PTP, in apoptosis. Apoptosis 12:857–868
Knott AB, Bossy-Wetzel E (2008) Impairing the mitochondrial fission and fusion balance: a new mechanism of neurodegeneration. Ann N Y Acad Sci 1147:283–292
Knott AB, Perkins G, Schwarzenbacher R, Bossy-Wetzel E (2008) Mitochondrial fragmentation in neurodegeneration. Nat Rev Neurosci 9:505–518
Kornmann B, Osman C, Walter P (2011) The conserved GTPase Gem1 regulates endoplasmic reticulum-mitochondria connections. Proc Natl Acad Sci 108:14151–14156
Korobova F, Ramabhadran V, Higgs HN (2013) An actin-dependent step in mitochondrial fission mediated by the ER-associated formin INF2. Science 339:464–467
Kujjo LL, Acton BM, Perkins GA, Ellisman MH, Giscard D'Estaing S, Casper RF, Jurisicova A, Perez GI (2013) Ceramide and its transport protein (CERT) contribute to deterioration of mitochondrial structure and function in aging oocytes. Mech Ageing Dev 134:43–52

Kuwana T, Mackey MR, Perkins G, Ellisman MH, Latterich M, Schneiter R, Green DR, Newmeyer DD (2002) Bid, Bax, and lipids cooperate to form supramolecular openings in the outer mitochondrial membrane. Cell 111:331–342

Kuwana T, Bouchier-Hayes L, Chipuk JE, Bonzon C, Sullivan BA, Green DR, Newmeyer DD (2005) BH3 domains of BH3-only proteins differentially regulate Bax-mediated mitochondrial membrane permeabilization both directly and indirectly. Mol Cell 17:525–535

Landes T et al (2010a) OPA1 (dys)functions. Semin Cell Dev Biol 21:593–598

Landes T, Emorine LJ, Courilleau D, Rojo M, Belenguer P, Arnaune-Pelloquin L (2010b) The BH3-only Bnip3 binds to the dynamin Opa1 to promote mitochondrial fragmentation and apoptosis by distinct mechanisms. EMBO Rep 11:459–465

Leboucher GP et al (2012) Stress-induced phosphorylation and proteasomal degradation of mitofusin 2 facilitates mitochondrial fragmentation and apoptosis. Mol Cell 47:547–557

Lee YJ, Jeong SY, Karbowski M, Smith CL, Youle RJ (2004) Roles of the mammalian mitochondrial fission and fusion mediators Fis1, Drp1, and Opa1 in apoptosis. Mol Biol Cell 15:5001–5011

Lucken-Ardjomande S, Montessuit S, Martinou JC (2008) Bax activation and stress-induced apoptosis delayed by the accumulation of cholesterol in mitochondrial membranes. Cell Death Differ 15:484–493

Lutter M, Fang M, Luo X, Nishijima M, Xie X, Wang X (2000) Cardiolipin provides specificity for targeting of tBid to mitochondria. Nat Cell Biol 2:754–761

Lutter M, Perkins GA, Wang X (2001) The pro-apoptotic Bcl-2 family member tBid localizes to mitochondrial contact sites. BMC Cell Biol 2:22

Mannella CA (2006) The relevance of mitochondrial membrane topology to mitochondrial function. Biochim Biophys Acta 1762:140–147

Mannella CA (2008) Structural diversity of mitochondria: functional implications. Ann N Y Acad Sci 1147:171–179

Martinou JC, Youle RJ (2011) Mitochondria in apoptosis: Bcl-2 family members and mitochondrial dynamics. Dev Cell 21:92–101

Merkwirth C, Langer T (2008) Prohibitin function within mitochondria: essential roles for cell proliferation and cristae morphogenesis. Biochim Biophys Acta 1793:27–32

Merkwirth C, Dargazanli S, Tatsuta T, Geimer S, Lower B, Wunderlich FT, von Kleist-Retzow JC, Waisman A, Westermann B, Langer T (2008) Prohibitins control cell proliferation and apoptosis by regulating OPA1-dependent cristae morphogenesis in mitochondria. Genes Dev 22:476–488

Merrill RA, Dagda RK, Dickey AS, Cribbs JT, Green SH, Usachev YM, Strack S (2011) Mechanism of neuroprotective mitochondrial remodeling by PKA/AKAP1. PLoS Biol 9:e1000612

Metzstein MM, Stanfield GM, Horvitz HR (1998) Genetics of programmed cell death in C. elegans: past, present and future. Trends Genet 14:410–416

Ming M et al (2012) Activation of Wnt/beta-catenin protein signaling induces mitochondria-mediated apoptosis in hematopoietic progenitor cells. J Biol Chem 287:22683–22690

Mirkin N, Jaconcic J, Stojanoff V, Moreno A (2008) High resolution X-ray crystallographic structure of bovine heart cytochrome c and its application to the design of an electron transfer biosensor. Proteins 70:83–92

Montessuit S, Somasekharan SP, Terrones O, Lucken-Ardjomande S, Herzig S, Schwarzenbacher R, Manstein DJ, Bossy-Wetzel E, Basanez G, Meda P, Martinou JC (2010) Membrane remodeling induced by the dynamin-related protein Drp1 stimulates Bax oligomerization. Cell 142:889–901

Muller-Rischart AK et al (2013) The E3 ligase parkin maintains mitochondrial integrity by increasing linear ubiquitination of NEMO. Mol Cell 49:908–921

Olichon A, Baricault L, Gas N et al (2003) Loss of OPA1 perturbates the mitochondrial inner membrane structure and integrity, leading to cytochrome c release and apoptosis. J Biol Chem 278:7743–7746

Olichon A, Landes T, Arnaune-Pelloquin L, Emorine LJ, Mils V, Guichet A, Delettre C, Hamel C, Amati-Bonneau P, Bonneau D, Reynier P, Lenaers G, Belenguer P (2007) Effects of OPA1 mutations on mitochondrial morphology and apoptosis: relevance to ADOA pathogenesis. J Cell Physiol 211:423–430

Osman C, Merkwirth C, Langer T (2009) Prohibitins and the functional compartmentalization of mitochondrial membranes. J Cell Sci 122:3823–3830

Otera H, Mihara K (2011) Discovery of the membrane receptor for mitochondrial fission GTPase Drp1. Small GTPases 2:167–172

Otera H, Wang C, Cleland MM, Setoguchi K, Yokota S, Youle RJ, Mihara K (2010) Mff is an essential factor for mitochondrial recruitment of Drp1 during mitochondrial fission in mammalian cells. J Cell Biol 191:1141–1158

Ott M, Robertson JD, Gogvadze V, Zhivotovsky B, Orrenius S (2002) Cytochrome c release from mitochondria proceeds by a two-step process. Proc Natl Acad Sci U S A 99:1259–1263

Parone PA, James DI, Da Cruz S, Mattenberger Y, Donzé O, Barja F, Martinou JC (2006) Inhibiting the mitochondrial fission machinery does not prevent Bax/Bak-dependent apoptosis. Mol Cell Biol 26:7397–7408

Perez GI, Acton BM, Jurisicova A, Perkins GA, White A, Brown J, Trbovich AM, Kim M-R, Fissore R, Xu J, Ahmady A, D'Estaing H, Kagawa W, Kurumizaka H, Yokoyama S, Okada Mak TW, Ellisman MH, Casper RF, Tilly JL (2007) Genetic variance modifies apoptosis susceptibility in mature oocytes via alterations in DNA repair capacity and mitochondrial ultrastructure. Cell Death Differ 14:524–533

Perkins G, Bossy-Wetzel E, Ellisman MH (2009) New insights into mitochondrial structure during cell death. Exp Neurol 218:183–192

Perkins G, Renken C, Martone M, Young S, Ellisman MH, Frey T (1997) Electron tomography of neuronal mitochondria: 3-D structure and organization of cristae and membrane contacts. J Struct Biol 119:260–272

Perumalsamy LR et al (2010) Notch-activated signaling cascade interacts with mitochondrial remodeling proteins to regulate cell survival. Proc Natl Acad Sci U S A 107:6882–6887

Ren D, Tu HC, Kim H, Wang GX, Bean GR, Takeuchi O, Jeffers JR, Zambetti GP, Hsieh JJ, Cheng EH (2010) BID, BIM, and PUMA are essential for activation of the BAX- and BAK-dependent cell death program. Science 330:1390–1393

Ricci JE, Munoz-Pinedo C, Fitzgerald P, Bailly-Maitre B, Perkins GA, Yadava N, Scheffler IE, Ellisman MH, Green DR (2004) Disruption of mitochondrial function during apoptosis is mediated by caspase cleavage of the p75 subunit of complex I of the electron transport chain. Cell 117:773–786

Rytomaa M, Kinnunen PK (1995) Reversibility of the binding of cytochrome c to liposomes. Implications for lipid–protein interactions. J Biol Chem 270:3197–3202

Rytomaa M, Mustonen P, Kinnunen PK (1992) Reversible, nonionic, and pH dependent association of cytochrome c with cardiolipin–phosphatidylcholine liposomes. J Biol Chem 267:22243–22248

Schug ZT, Gottlieb E (2009) Cardiolipin acts as a mitochondrial signalling platform to launch apoptosis. Biochim Biophys Acta 1788:2022–2031

Scorrano L (2009) Opening the doors to cytochrome c: changes in mitochondrial shape and apoptosis. Int J Biochem Cell Biol 41:1875–1883

Scorrano L (2013) Keeping mitochondria in shape: a matter of life and death. Eur J Clin Investig 43:886–893

Scorrano L, Ashiya M, Buttle K et al (2002) A distinct pathway remodels mitochondrial cristae and mobilizes cytochrome c during apoptosis. Dev Cell 2:55–67

Scorrano L, Oakes SA, Opferman JT, Cheng EH, Sorcinelli MD, Pozzan T, Korsmeyer SJ (2003) BAX and BAK regulation of endoplasmic reticulum Ca2+: a control point for apoptosis. Science 300:135–139

Sheridan C, Martin SJ (2010) Mitochondrial fission/fusion dynamics and apoptosis. Mitochondrion 10:640–648

Sheridan C, Delivani P, Cullen SP, Martin SJ (2008) Bax- or Bak-induced mitochondrial fission can be uncoupled from cytochrome C release. Mol Cell 31:570–585

Simmen T et al (2005) PACS-2 controls endoplasmic reticulum-mitochondria communication and Bid-mediated apoptosis. EMBO J 24:717–729

Song Z, Chen H, Fiket M, Alexander C, Chan DC (2007) OPA1 processing controls mitochondrial fusion and is regulated by mRNA splicing, membrane potential, and Yme1L. J Cell Biol 178:749–755

Sugioka R, Shimizu S, Tsujimoto Y (2004) Fzo1, a protein involved in mitochondrial fusion, inhibits apoptosis. J Biol Chem 279:52726–52734

Sugiura A et al (2013) MITOL regulates endoplasmic reticulum-mitochondria contacts via Mitofusin2. Mol Cell 51:20–34

Sulston JE, Brenner S (1974) The DNA of Caenorhabditis elegans. Genetics 77:95–104

Sun MG, Williams J, Munoz-Pinedo C, Perkins GA, Brown JM, Ellisman MH, Green DR, Frey TG (2007) Correlated three-dimensional light and electron microscopy reveals transformation of mitochondria during apoptosis. Nat Cell Biol 9:1057–1065

Szabadkai G, Simoni AM, Chami M, Wieckowski MR, Youle RJ, Rizzuto R (2004) Drp-1-dependent division of the mitochondrial network blocks intraorganellar Ca2+ waves and protects against Ca2+-mediated apoptosis. Mol Cell 16:59–68

Terrones O, Antonsson B, Yamaguchi H, Wang HG, Liu Y, Lee RM, Herrmann A, Basanez G (2004) Lipidic pore formation by the concerted action of pro-apoptotic BAX and tBID. J Biol Chem 279:30081–30091

Tuominen EK, Wallace CJ, Kinnunen PK (2002) Phospholipid–cytochrome c interaction: evidence for the extended lipid anchorage. J Biol Chem 277:8822–8826

Tyurin VA, Tyurina YY, Feng W, Mnuskin A, Jiang J, Tang M, Zhang X, Zhao Q, Kochanek PM, Clark RS, Bayir H, Kagan VE (2008) Mass-spectrometric characterization of phospholipids and their primary peroxidation products in rat cortical neurons during staurosporine-induced apoptosis. J Neurochem 107:1614–1633

Vaux DL, Cory S, Adams JM (1988) Bcl-2 gene promotes haemopoietic cell survival and cooperates with c-myc to immortalize pre-B cells. Nature 335:440–442

Vogel F, Bornhovd C, Neupert W, Reichert AS (2006) Dynamic subcompartmentalization of the mitochondrial inner membrane. J Cell Biol 175:237–247

Wang JX, Jiao JQ, Li Q, Long B, Wang K, Liu JP, Li YR, Li PF (2011) miR-499 regulates mitochondrial dynamics by targeting calcineurin and dynamin-related protein-1. Nat Med 17:71–78

Wang C, Youle R (2009) The role of mitochondria in apoptosis. Annu Rev Genet 43:95–118

Wasiak S, Zunino R, McBride HM (2007) Bax/Bak promote sumoylation of DRP1 and its stable association with mitochondria during apoptotic cell death. J Cell Biol 177:439–450

Wei MC, ZongWX CEH, Lindsten T, Panoutsakopoulou V, Ross AJ, Roth KA, MacGregor GR, Thompson CB, Korsmeyer SJ (2001) Proapoptotic BAX and BAK: a requisite gateway to mitochondrial dysfunction and death. Science 292:727–730

Yamaguchi R, Perkins G (2009) Dynamics of mitochondrial structure during apoptosis and the enigma of Opa1. Biochim Biophys Acta 1787:963–972

Yamaguchi R, Perkins G (2012a) Challenges in targeting cancer metabolism for cancer therapy. EMBO Rep 13:1034–1035

Yamaguchi R, Perkins G (2012b) Finding a panacea among combination cancer therapies. Cancer Res 72:18–23

Yamaguchi R, Lartigue L, Perkins G, Scott RT, Dixit A, Ellisman MH, Kuwana T, Newmeyer DD (2008) Opa1-mediated cristae opening Is Bax/Bak and BH3 dependent, required for apoptosis, and independent of Bak oligomerization. Mol Cell 31:557–569

Yoon Y, Krueger EW, Oswald BJ, McNiven MA (2003) The mitochondrial protein hFis1 regulates mitochondrial fission in mammalian cells through an interaction with the dynamin like protein DLP1. Mol Cell Biol 23:5409–5420

Youle RJ, Karbowski M (2005) Mitochondrial fission in apoptosis. Nat Rev Mol Cell Biol 6:657–663

Yuan H, Gerencser AA, Liot G, Lipton SA, Ellisman MH, Perkins G, Bossy-Wetzel E (2007) Mitochondrial fission is an upstream and required event for bax foci formation in response to nitric oxide in cortical neurons. Cell Death Differ 14:462–471

Zhao J, Liu T, Jin S, Wang X, Qu M, Uhlén P, Tomilin N, Shupliakov O, Lendahl U, Nistér M (2011) Human MIEF1 recruits Drp1 to mitochondrial outer membranes and promotes mitochondrial fusion rather than fission. EMBO J 30:2762–2778

Zinkel SS, Yin XM, Gross A (2013) Rejuvenating Bi(d)ology. Oncogene 32:3213–3219

Chapter 6
Physiologic and Pathologic Functions of Mitochondrial ROS

Bryan G. Allen and Douglas R. Spitz

6.1 Introduction

By virtue of providing the reducing equivalents and energy necessary for living systems, metabolic oxidation–reduction (redox) reactions occurring in mitochondria and other organelles govern the flow of electrons through complex higher order biological structures that play pivotal roles in all normal cellular functions including energy production, biosynthesis, transcription, translation, differentiation, and proliferation as well as cell death pathways (Fig. 6.1). In normal healthy mammalian cells, there is delicate nonequilibrium steady-state relationship between the relatively low fluxes of metabolic prooxidant production (such as superoxide and hydrogen peroxide) that is balanced by a much greater reductive capacity (through antioxidant pathways) that maintain the cell in a highly reducing environment that oscillates in a nonequilibrium steady-state between approximately −200 mV and −240 mV (Schafer and Buettner 2001). This oscillation under steady-state conditions is believed to allow for changes in the flow and flux of electrons through redox sensitive signaling circuits controlling gene expression pathways coordinating the availability of metabolic capacity with most if not all of normal cellular functions

B.G. Allen, M.D., Ph.D. (✉)
Division of Free Radical and Radiation Biology, Department of Radiation Oncology, University of Iowa Hospitals and Clinics,
200 Hawkins Drive, Iowa City, IA 52242-1009, USA

The University of Iowa Carver College of Medicine,
B180 Medical Laboratories, Iowa City, IA 52242, USA
e-mail: bryan-allen@uiowa.edu

D.R. Spitz
Division of Free Radical and Radiation Biology, Department of Radiation Oncology, University of Iowa Hospitals and Clinics, 200 Hawkins Drive, Iowa City, IA 52242-1009, USA

D.M. Hockenbery (ed.), *Mitochondria and Cell Death*, Cell Death in Biology and Diseases, DOI 10.1007/978-1-4939-3612-0_6

X.-M. Yin and Z. Dong (Series eds.), *Cell Death in Biology and Diseases*

(Fig. 6.1) (Schafer and Buettner 2001; Spitz et al. 2000; Zhou et al. 2014). When this redox balance is disrupted from its normal steady-state oscillations due to increased fluxes of prooxidants and/or decreased antioxidant capacity, oxidative stress can occur (Sies 1991). Initially, during a condition of potential oxidative stress, redox sensitive signaling pathways respond rapidly to redirect metabolism by shifting to a more reductive state through decreasing the utilization of prooxidant metabolic pathways, increasing antioxidant capacity, and enhancing repair pathways for critical biomolecules that may be accumulating oxidative damage (Fig. 6.1). This initial cellular response to oxidative stress can lead to transient alterations in redox sensitive signaling/gene expression pathways that govern cell growth, development, and differentiated cellular functions. If oxidative stress becomes a chronic condition and cannot be compensated for by shifting metabolism to a more reductive state then redox sensitive signaling becomes chronically dysregulated leading to the accumulation oxidative damage to proteins, DNA, and lipids as well as tissue injury/dysfunction, inflammation, transformation, senescence, dysregulated proliferation/differentiation, and eventually cell death. Thus oxidative

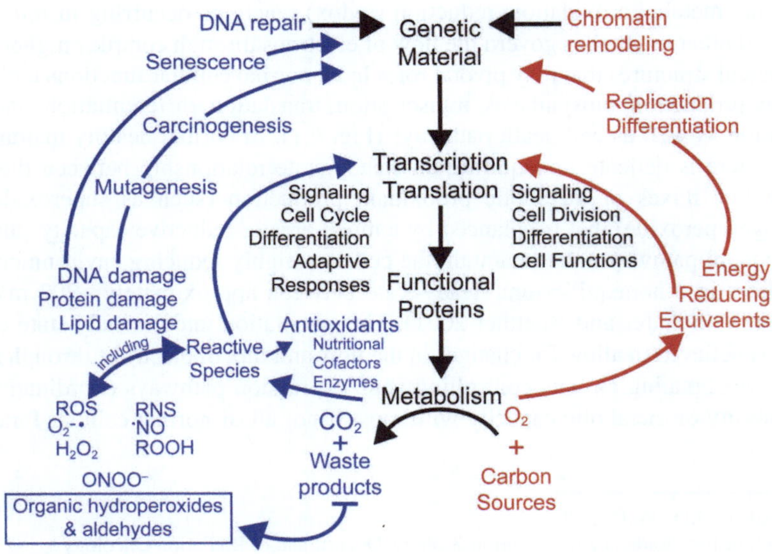

Fig. 6.1 "Follow the Electrons": an integrated approach to understanding the relationship between metabolism, signal transduction, and gene expression in mammalian systems. Genetic material (DNA) undergoes transcription and translation to form functional proteins that execute metabolic processes responsible for extracting, storing, and moving electrons to derive the energy and reducing equivalents necessary to accomplish biosynthesis and maintain the living system. Reactive oxygen species are produced as normal by-products of oxidative metabolism and allow the signaling and gene expression pathways to "sip the flux" of electrons through metabolic processes to coordinately regulate cell division, differentiation, and adaptive responses necessary to maintain homeostasis in mammalian organisms (*Red*). The slow accumulation of damage during the process of aging caused by reactive oxygen species may result in persistent DNA, protein, and lipid damage leading to inefficiencies in metabolism that then accelerate the processes of carcinogenesis, senescence, and degenerative diseases associated with aging (*Blue*)

stress is believed to contribute to a wide range of chronic pathologic conditions including atherosclerosis, cancer, neurological diseases, and a host of degenerative diseases associated with aging (Fig. 6.1) (Schafer and Buettner 2001; Spitz et al. 2000, 2004; Zhou et al. 2014; Sies 1991; Finkel et al. 2007; Fridovich 1978; Fridovich 2004; Oberley et al. 1980, 1981; Park et al. 2011).

In this regard, the relationship between oxidative metabolism and gene expression can be viewed as a "cart and horse" (Fig. 6.2). The "horse" of metabolism pulls the "cart" of gene expression via the flow of electrons from metabolic processes to signal transduction and gene expression pathways allowing the coordinate regulation of the life cycles of living mammalian systems (Fig. 6.2). As the "waste products" of mitochondrial metabolism [one electron leakage from electron transport chains (ETCs) and oxidative damage to critical biomolecules] accumulate and begin to damage the "cart" of gene expression; then the "horse" of metabolism becomes more inefficient producing more reactive species and damage until the "horse" inevitably takes the "cart" to "death." This death may occur by senescence where normal cells become gradually unable to perform normal differentiated function leading to degenerative diseases of aging or cancer via malignant cell growth (Fig. 6.2).

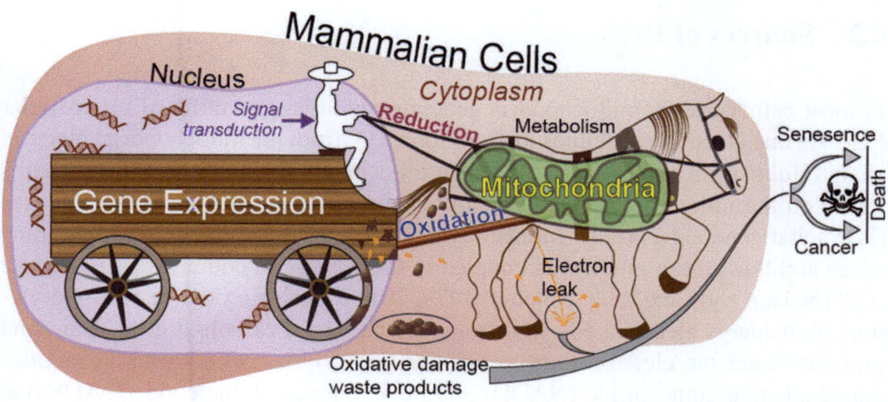

Fig. 6.2 The "Horse and the Cart" relationship between metabolism, signal transduction, and gene expression in mammalian systems biology. Mitochondrial oxidative metabolism (the horse) and gene expression (the cart) are tightly coupled by redox signaling pathways that allow for the flow of electrons from metabolic reactions governing energy production and biosynthesis to gene expression pathways governing growth and development. In this way oxidation/reduction reactions provide the energy and reducing equivalents necessary for cell survival, biosynthesis, and all cellular functions. Unfortunately, oxidation reactions are not entirely efficient and waste products are formed via one electron reductions of O_2 (electron leak) in electron transport chains leading to the formation of reaction oxygen species and oxidative damage. This inefficiency in metabolism slowly erodes the genetic integrity of the cart governing the efficiency of the intracellular communication as well as coordination of metabolism and gene expression necessary for the maintenance of normal cellular functions. Reactive oxygen species are therefore necessary for life as redox signaling molecules as well as leading to the accumulation of oxidative damage to the genome causing increased metabolic inefficiency as a function of aging which accelerates the processes associated with senescence as well as onset of degenerative diseases associated with aging and cancer

Reactive oxygen species (ROS) and reactive nitrogen species (RNS) represent a group of chemically diverse molecules derived from the formation of superoxide ($O_2^{\bullet -}$) and nitric oxide ($^{\bullet}NO$) (Sullivan and Chandel 2014; Thomas et al. 2015). ROS contain oxygen and their metabolic formation occurs predominantly during mitochondrial respiration, by the action of oxidoreductase enzymes, and during peroxisomal/microsomal metabolism. Examples of ROS include oxygen containing free radicals such as superoxide ($O_2^{\bullet -}$) and the hydroxyl radical ($OH^{\bullet}$) as well as non-radical oxidants such as hydrogen peroxide (H_2O_2). While a disturbance in the oxidant to antioxidant balance with an increase in ROS may lead to oxidative stress, lower steady-state levels of ROS formed during redox metabolism have important roles in intracellular signaling (Fig. 6.1). For example, ROS acting as signaling moieties may serve as potential messengers triggering cell growth (Bae et al. 1997; Sundaresan et al. 1995; Sarsour et al. 2014). At resting steady-state levels of metabolism in most normal mammalian cells, mitochondria represent the largest source of O_2 consumption and therefore ROS production. This chapter focuses on mitochondrial ROS production, targets of mitochondrial ROS, cellular approaches to control ROS, ROS and cancer, as well as ROS and aging.

6.2 Sources of ROS

In most cell types, the mitochondria generate the largest quantity of intracellular ROS via one electron reductions of O_2 to form $O_2^{\bullet -}$ from the mitochondrial electron chain (Holmstrom and Finkel 2014). The mitochondrial electron transport chains reside on the inner mitochondrial membrane and include four protein complexes (I–IV) that generate a proton gradient by pumping protons into the intermembrane space and this proton gradient is coupled to the ATPase in complex V to generate ATP (Nelson and Cox 2000; Mitchell 1979, 2011). During oxidative phosphorylation, high energy electrons (derived from metabolism of carbohydrates, lipids, and proteins) enter the electron transport chains (ETCs) via carriers such as nicotinamide adenine dinucleotide (NADH) and flavin adenine dinucleotide ($FADH_2$) at complexes I and II. As the electrons flow down the ETCs moving from higher energy to lower energy, protons (H^+) are pumped from the mitochondrial matrix to the intermembrane space generating a proton gradient that reequilibrates through complex V (aka ATPase) resulting in the formation of ATP from ADP + P_i. Once the electrons reach complex IV (cytochrome oxidase), they flow onto O_2 through 4 e^- reductions to generate 2 molecules of H_2O. However, during respiration associated with oxidative phosphorylation, 0.1–2 % of electrons may result in 1 e^- reductions of O_2 to form $O_2^{\bullet -}$ at sites up stream of complex IV (complexes I–III) that are accessible to molecular oxygen (Boveris 1977; Boveris and Cadenas 1982; Forman and Kennedy 1975; Forman and Kennedy 1974; Orrenius et al. 2007; Rigoulet et al. 2011). Complex I and II produce ROS in the mitochondrial matrix while complex III may form superoxide in either the matrix or intermembrane space (Muller et al. 2004). Superoxide formed in the intermembrane space may then pass through voltage dependent anion channels to enter into the cellular cytosol and potentially participate in cellular signaling. Blocking or slowing electron movement from complexes I–III to complex IV increases the probability of one electron reductions of O_2 to form $O_2^{\bullet-}$ and H_2O_2. This is because a less efficient flow of electrons to complex IV where 4 e^- reductions of O_2 to H_2O occur increases the residence time of electrons on sites accessible to O_2 (upstream of complex IV) where $O_2^{\bullet-}$ can form. Additional potential sources of cellular superoxide include enzymes capable of doing 1 e^- reduction of O_2 including uncoupled nitric oxide synthases, cytochrome p450 enzyme isoforms, alpha-ketoglutarate dehydrogenase, glycerol-3-phosphate dehydrogenase, and NADPH-dependent oxidases (Handy and Loscalzo 2012).

ROS production in each cell is determined by the availability of electron donors and carriers, local concentrations of molecular oxygen, and the second-order rate constants for reactions between electron carriers and molecular oxygen (Murphy 2009). Thus, ROS production depends upon cells current respiratory rate, the mitochondrial membrane potential, and inherent damage to the electron transport chain carriers.

Once formed, superoxide may react with nitric oxide to form peroxynitrite (Szabo et al. 2007), superoxide may be protonated to form the very reactive hydroperoxyl radical, or superoxide may be reduced spontaneously or by superoxide dismutases located within the mitochondria, extracellular environment, and cytosol to form hydrogen peroxide. Hydrogen peroxide may also be formed by xanthine oxidase (Kelley et al. 2010), NOX enzymes (Dikalov et al. 2008), peroxisomal enzymes, or through the oxidation of vitamin C (Levine et al. 2011). Hydrogen peroxide may react with metals capable of redox cycling through Fenton Chemistry to form the highly reactive hydroxyl radical capable of oxidizing any biomolecule. Despite the presence of mitochondrial antioxidants including superoxide dismutase (MnSOD), glutathione peroxidases, thioredoxin and thioredoxin reductase, peroxiredoxins, in addition to water and fat soluble antioxidants (vitamins C and E), the mitochondria is still believed to be the major source of ROS during normal metabolism with steady-state concentrations estimated to be five to tenfold higher than in the cytosol or nucleus (Cadenas and Davies 2000).

Mitochondria are the major hub of ROS generation and signaling during normal metabolism that both produce ROS and oxidative damage as well as participate in redox signaling to coordinate metabolism and gene expression governing most if not all cellular functions (Fig. 6.2). For example, cells exposed to hypoxia lead to increased mitochondrial ROS formation which causes upregulation and stabilization of hypoxia-inducible factors resulting in the expression of adaptive response genes including erythropoietin, vascular endothelial growth factor and glycolytic enzymes (Bell et al. 2007; Chandel et al. 2000). Mitochondrial ROS also activate AMP-activated protein kinase in hypoxic conditions resulting in Na/K-ATPase inhibition perhaps as a mechanism to conserve energy (Emerling et al. 2009; Gusarova et al. 2009). Other examples of mitochondrial ROS signaling include insulin signaling and insulin resistance (Loh et al. 2009; McClung et al. 2004) as well as TNFalpha-induced cell death (Kamata et al. 2005).

6.3 Targets of ROS

ROS formed in the mitochondria or in other parts of the cell may react with proteins, lipids or nucleic acids. Because of its charge, superoxide tends to be limited to the water soluble intracellular compartments in which it was made (Handy and Loscalzo 2012). Alternatively, hydrogen peroxide is uncharged and can readily diffuse across membranes (Winterbourn and Hampton 2008). Hydrogen peroxide can also interact with proteins leading to oxidation of sulfhydryl groups, reactions to form aldehydes, protein–protein cross-linking, and protein fragmentation (Stadtman and Oliver 1991; Starke-Reed and Oliver 1989). Additionally, some amino acids, such as lysine, arginine, proline, and threonine, may be directly oxidized forming protein carbonyls. Protein carbonyls can change the protein tertiary structure which may dramatically alter protein function. Phospholipids containing polyunsaturated fatty acids are very sensitive to oxidation with the hydroxyl radical being one of the most potent stimulators of lipid peroxidation (Esterbauer et al. 1991). Oxidatively damaged DNA bases, such as 8-oxoguanine, as well as damage to the DNA backbone, can lead to DNA strand breaks or mutations. These DNA modifications may be mutagenic resulting in cancer induction, cancer progression, neurodegenerative diseases, or aging (Bohr 2002).

6.4 Antioxidants

Since excessive ROS may lead to DNA, protein, or lipid damage, cells have evolved a variety of antioxidant enzymes and systems. The superoxide dismutase (SOD) family of enzymes converts superoxide in hydrogen peroxide. SOD1 or copper/zinc SOD is located in the cytoplasm, the intermembrane space in the mitochondria, and in the nucleus of the cell. Mutations in SOD1 may result in the formation of amyotrophic lateral sclerosis (ALS) (Barber et al. 2006). SOD2 or manganese-dependent SOD (MnSOD), is expressed in mitochondria. Mouse SOD2 knockout results in neurodegeneration, cardiac injury, and perinatal death (Lebovitz et al. 1996). SOD3 or EcSOD is a secreted glycoprotein that reacts with extracellular superoxide forming hydrogen peroxide; in this way, SOD3 potentially limits the oxidative damage that may be caused by peroxynitrite, extracellularly. Additionally, angiotensin-II induced hypertension increases vascular ecSOD as a potential compensatory mechanism to hypertension (Gongora et al. 2006).

Catalase reduces hydrogen peroxide into water and molecular oxygen. Even though it is expressed predominantly in peroxisomes many cells, including erythrocytes and rat heart mitochondria, depend on catalase to assist with redox homeostasis (Gaetani et al. 1989; Radi et al. 1991).

Intracellularly, glutathione predominantly exists in the reduced state (GSH) with a GSH/GSSG ratio over 100:1 in many normal tissues (Schafer and Buettner 2001). Sources of GSH include cellular biosynthesis, uptake of exogenous glutathione and the reduction of oxidized glutathione (GSSG) back into its reduced state (GSH) by glutathione reductase (Meister and Tate 1976). The majority of intracellular GSH is found in the cytosol (1–11 mM) (Smith et al. 1996). In contrast, the endoplasmic reticulum is more oxidizing than the cytosol to allow for protein folding with GSH/GSSG ratios ranging from 1:1 to 3:1 (Hwang et al. 1992). The mitochondria (1–14 mM) and nucleus have similar GSH concentrations to that in the cytosol (Smith et al. 1996; Griffith and Meister 1985). Deficiency of mitochondrial glutathione results in mitochondrial damage and apoptosis due at least in part from loss of glutathione peroxidase-mediated protection against oxidants (Meister 1995). Extracellular reduced glutathione concentrations are dramatically reduced relative to intracellular levels with concentrations ranging from 0.1 to 0.001 mM (Jones et al. 1998).

Glutathione peroxidases are a family of at least 4 enzymes that a contain selenocysteine at their active site and are involved in the reduction of hydrogen peroxide and lipid hydroperoxides into less reactive alcohols (Arthur 2000). Some members of the family utilize reduced glutathione (GSH) as an electron donor in this process. Oxidized glutathione (GSSG) may then be reduced by glutathione reductase and NADPH (Rotruck et al. 1973). Glutathione peroxidase 1 (GPX1) is expressed in both the cytoplasm and mitochondria and is a key antioxidant enzyme in the endothelium. Mice with a heterozygous GPX1 deficiency have abnormal endothelial and cardiac function and structure (Forgione et al. 2002). Reduced GPX1 activity in humans is associated with an increased risk of cardiovascular events (Blankenberg et al. 2003). When cells are exposed to oxidative stress, there is an increase in the oxidized form of glutathione (GSSG) and decrease in the GSH/GSSG ratio that is related to GPX1, as GPX1 knockout cells exposed to a prooxidant have less intracellular changes in GSSG (Fu et al. 2001). GPX2 is similar to GPX1 in function but is predominantly expressed in epithelial cells of the GI tract and may play a gastrointestinal cancer prevention (Bartel et al. 2007). GPX3 is a secreted glycoprotein with deficiency being associated with endothelial dysfunction, platelet activation, and thrombus formation (Jin et al. 2011). GPX4 reduces phospholipid hydroperoxides in situ in lipid bilayers and knockouts are lethal in mice; heterozygous knockouts have increased sensitivity to gamma radiation and hydrogen peroxide (Yant et al. 2003). One GPX4 isoform protects the mitochondria against membrane lipid peroxidation and regulates the release of apoptogenic proteins (Liang et al. 2009). GPX6 is predominantly expressed in adult olfactory epithelium and less is known about its role in biology (Reeves and Hoffmann 2009).

Glutaredoxins are small proteins that utilize glutathione as a cofactor that catalyze the reduction of proteins containing oxidized disulfide bonds either by a dithiol or monothiol mechanism (Bushweller et al. 1992; Holmgren and Aslund 1995). Three mammalian glutaredoxin isoforms (GRX) have been described, while in yeast five isoforms have been described. The GRX1 isoform is located predominantly in the cytoplasm while alternative splicing of GRX2 results in isoforms that reside in the nucleus and mitochondrial matrix (Lundberg et al. 2001). GRX1 is involved in multiple cellular processes including disulfide reduction (Holmgren 1989), dehydroascorbate (DHA) reduction (Wells et al. 1990), DNA transcription regulation (Bandyopadhyay et al. 1998; Hirota et al. 2000), and apoptosis (Chrestensen et al. 2000;

Daily et al. 2001). Dependent upon the GSH/GSSG ratio, GRX2 mediates the glutathiolation of proteins including complex I in the mitochondrial electron transport chain (Beer et al. 2004). Interestingly, oxidative stress which reduces the GSH/GSSG ratio reduces complex I activity (Beer et al. 2004). Cells depleted of mitochondrial GRX2 exposed to oxidative stress undergo apoptosis (Lillig et al. 2004), while the overexpression of GRX2 resulted in the protection against oxidative stress-induced apoptosis (Enoksson et al. 2005). Similar to GRX2, GRX5 is predominantly found in mitochondria and protects against oxidative stress-induced apoptosis (Linares et al. 2009). Peroxiredoxins are a large family of antioxidant proteins that reduce hydroperoxides and peroxynitrite using reducing agents as cofactors, such as thioredoxin. Peroxiredoxins are involved in a wide array of cellular functions including cellular proliferation and differentiation (Nemoto et al. 1990; Prosperi et al. 1994), enhancing the immune system (Shau et al. 1994), heme metabolism (Iwahara et al. 1995), and intracellular signaling (Wen and Van Etten 1997). The active site contains a conserved cysteine residue that cycles between peroxide-dependent oxidation and thiol-dependent reduction (Rhee et al. 2005). Six mammalian peroxiredoxin isoforms (PRX I–VI) have been identified which have been further classified into three subgroups based on the number and position of the cysteine residues (Rhee et al. 2005). PRX I–IV are typical 2-Cys isoforms, PRX V is an atypical 2-Cys isoform, and PRX VI is a 1-Cys isoform. Mice lacking PRX I have shortened life spans and develop an oxidative stress-induced hemolytic anemia and a variety of malignancies including lymphomas, sarcomas, and carcinomas (Neumann et al. 2003). PRX II knockout mice also develop an oxidative stress-induced hemolytic anemia (Lee et al. 2003). PRX III is localized to the mitochondria and knockdown increases intracellular hydrogen peroxide levels, cytochrome c release, and caspase activation leading towards apoptosis (Chang et al. 2004). PRX VI-deficient mice hearts were found to be more susceptible to reperfusion injury following ischemia and were also subjected to a higher amount of oxidative stress than control littermates (Nagy et al. 2006). Peroxiredoxins also play critical roles in hydrogen peroxide signaling (Wood et al. 2003). Hydrogen peroxide acts as a second messenger in the NF-κB and mitogen-activated protein (MAP) kinase pathways by reacting with thiols (Claiborne et al. 1999). Peroxiredoxins, in combination with catalase and glutathione peroxidases, may act as a switch by which the oxidative inactivation of phosphatases allows for hydrogen peroxide signaling (Wood et al. 2003).

Thioredoxin is a small protein (approximately 12 kDa) that serves as the electron donor to reduce oxidized disulfides, including peroxiredoxins. Thioredoxins have many cellular roles, including being involved in transcription, cell division, and immune function (Matsui et al. 1996). Two major forms of thioredoxin have been identified, TRX 1 located in the cytosol and TRX 2 located in the mitochondria. TRX 1 and TRX 2 have two active cysteine residues that may be oxidized to form a disulfide bond. Oxidized thioredoxin is then reduced by NADPH-dependent thioredoxin reductases. TRX 1 and TRX 2 homozygous mutant mice die in utero shortly after implantation with TRX 2 mutant embryonic lethality coinciding with mitochondrial maturation (Matsui et al. 1996; Nonn et al. 2003). Similarly, knockout of thioredoxin reductase 1 (TRXR 1) or 2 (TRXR 2), expressed in the cytosol and mitochondria respectively, results in embryonic lethality. Inactivation of TRXR 1 leads to early embryonic lethality with embryos displaying growth retardation (Jakupoglu et al. 2005) while inactivation of TRXR 2 results growth retardation, anemia, increased liver apoptosis, and heart defects (Conrad et al. 2004).

NADPH has an essential role in maintaining the activity of cellular antioxidants including catalase, glutathione, and thioredoxin-dependent metabolism. NADPH is produced in the pentose phosphate shunt by reducing NADP+. Glucose-6-phosphate dehydrogenase (G6PD) is a cytosolic enzyme that is ubiquitously expressed and serves as the entry into as well as the rate limiting step of the pentose phosphate shunt. In humans, G6PD deficiency protects against malaria because of the accumulation of oxidative damage (Cappadoro et al. 1998). However, G6PD deficiency may also make individuals more susceptible to vascular diseases such as hypertension (Leopold and Loscalzo 2005). Complete deficiency of G6PD is incompatible with life and results in decreased cell proliferation and migration, while G6PD overexpression promotes cell proliferation and migration. Other potential sources of NADPH regeneration include isocitrate dehydrogenases, malate dehydrogenase, and glutamate dehydrogenase.

6.5 ROS and Cancer

Mitochondrial function was initially proposed to be fundamentally related to the process of carcinogenesis and cancer progression more than 50 years ago when Otto Warburg proposed that tumor cells demonstrated increased aerobic glycolysis compared to normal cells (Dai et al. 2014; Warburg 1956). Warburg's original proposal posited that tumor cells metabolize more glucose to generate ATP in order to compensate for "damage" to their respiratory mechanism leading to a deficit in ATP production. The need for cancer cells to use increased glucose metabolism to generate ATP, reducing equivalents, and ribose sugars was also thought to be required to support high levels of proliferation associated with uncontrolled growth (Warburg 1956; Vander Heiden et al. 2009). However, tumor cells were found to have an adequate capacity to generate ATP from mitochondrial metabolism (Maity and Tuttle 2006). Furthermore, normal cells that proliferate more rapidly than tumor cells as determined by FLT-PET imaging (such as bone marrow cells) were found to take up and metabolize less glucose than cancer cells (Salskov et al. 2007; Shields et al. 1998). These data clearly support the hypothesis that proliferation and increasing biomass was not the only prerequisite driving increased glucose metabolism in cancer vs. normal cells.

The discovery that glucose deprivation selectively induced metabolic oxidative stress and cell killing in cancer vs. normal cells, coupled with the fact that glucose deprivation caused redox sensitive activation of signal transduction and gene expression pathways associated with the maintenance of the transformed phenotype led to a new theoretical construct for explaining the increased consumption of glucose in tumor vs. normal cells (Spitz et al. 2000; Ahmad et al. 2005; Aykin-Burns et al. 2009; Blackburn et al. 1999; Graham et al. 2012; Lee et al. 1998). In this model, tumor cells are posited to have an accumulation of damage to the ~3000 genes necessary for the assembly and efficient function of electron transport proteins that result in stoichiometric mismatches leading to increased one electron reductions of O_2 to form $O_2^{\bullet-}$ and H_2O_2, relative to normal cells. This defect in tumor cell respiration is compensated for by increased consumption of glucose to generate pyruvate (from glycolysis) as well as NADPH (from the pentose cycle) to detoxify increased fluxes of prooxidants derived from this respiratory defect. thereby allowing the transformed cells to escape oxidative stress-induced cell death (Ahmad et al. 2005; Aykin-Burns et al. 2009; Blackburn et al. 1999; Graham et al. 2012; Lee et al. 1998; Allen et al. 2014).

In support of this hypothesis, numerous studies have shown that tumors accumulate defective mitochondria (Bize et al. 1980; Chatterjee et al. 2011; Galluzzi et al. 2010; Ralph et al. 2010; Springer 1980). Mitochondrial DNA mutations are associated with both tumor growth and metastasis (Ishikawa et al. 2008; Petros et al. 2005). Unlike the nucleus which houses two copies of the genome, mitochondria may have up to ten copies of the mitochondrial genome (He et al. 2010). The human mitochondrial genome is 16.6 kb long and encodes for 13 polypeptides involved in the electron transport chain, 22 tRNAs, and 2 rRNAs (Attardi and Schatz 1988). The remaining components of the electron transport chain are encoded in the nuclear DNA. Cancers often have somatic mitochondrial DNA mutations that are believed to be caused from the continuous exposure to ROS (Petros et al. 2005; Polyak et al. 1998; Yu 2012). In addition, mitochondrial DNA has up to a tenfold higher accumulation rate of mutations than nuclear DNA and since mitochondrial DNA lacks large introns, many of the mutations will have a biological effect (Chatterjee et al. 2011; Carew and Huang 2002). For example, certain mitochondrial polymorphisms that may affect electron transport chain efficiency and thereby ROS production are associated with increased risk of developing breast cancer (Bai et al. 2007). NADH dehydrogenase, complex I in the mitochondrial electron transport chain, is located in the inner mitochondrial space and accepts electrons from NADH and transfers electrons to Co-enzyme Q. Complex I consists of 46 protein subunits, 7 of which are encoded in the mitochondrial genome. A mutation in the first subunit (*NDI*) of the mitochondria-encoded complex I is associated with renal cell carcinoma (Horton et al. 1996). Similarly, a mutation in mitochondrial cytochrome b, which is part of complex III, promotes bladder tumor growth (Dasgupta et al. 2008). To repair mitochondrial DNA mutations, mitochondria have endogenous DNA repair systems include base excision repair (Liu et al. 2008; Szczesny et al. 2008). In addition to mitochondrial mutations, reduced electron transport chain efficiency and increased ROS production results in a chronic oxidative stress environment associated with increased cell proliferation, cellular motility, and invasive behaviors (Pelicano et al. 2009). Normally, cells combat ROS through antioxidant systems and ROS scavengers including SOD, catalases, glutathione peroxidases; however, some cancer cells are deficient in these antioxidant systems (Chatterjee et al. 2011; Baker et al. 1997; Bostwick et al. 2000). As a biological example, biopsies from and adjacent to precancerous lesions in ulcerative colitis were shown to have progressively reduced cytochrome c oxidase the closer the proximity to the dysplasia (Ussakli et al. 2013).

6.6 ROS and Aging

In 1956, it was proposed that aging and associated degenerative diseases may be attributed to free radical-induced cellular damage (Harman 1956). Aging is associated with metabolic and structural changes in the mitochondrial electron transport chains and lipid composition of the inner mitochondrial membrane, resulting in increased ROS generation (Szeto et al. 2014). Activity of electron transport chain complexes I, III, and IV have all been reported to decrease with age (Lenaz et al. 1997; Lesnefsky et al. 2001; Moghaddas et al. 2003; Paradies et al. 1997; Petrosillo et al. 2008). In support of the free radical theory of aging, yeast and *Drosophila* SOD1 knockouts have shorter life spans (Phillips et al. 1989; Wawryn et al. 1999). SOD1-deficient mice similarly have shorter life spans and develop hepatocellular carcinoma (Elchuri et al. 2005). Similarly, SOD2 knockouts also show reduced life span in yeast (Longo et al. 1999) and a SOD2 null mutation reduced life span in *Drosophila* (Duttaroy et al. 2003). Mice SOD2 knockouts experienced neonatal death (Li et al. 1995). Loss of murine SOD3 demonstrated no aging effects until stressed by exposure to >99 % oxygen when mice developed lung edema and had reduced life spans (Carlsson et al. 1995). In addition, SOD1 and SOD2 knockdowns or SOD mutations in *C. elegans* had a significant effect on longevity (Van Raamsdonk and Hekimi 2009; Yang et al. 2007). In contrast, mouse transgenic overexpression of SOD1 and SOD2 failed to extend life span (Jang et al. 2009; Perez et al. 2009). Many human clinical trials have also looked at the effects of antioxidant nutritional supplements on mortality. A meta-analysis of 68 randomized trials comparing beta carotene, vitamin A, vitamin C, vitamin E, and selenium either alone or in combination versus placebo showed no effect on mortality (Bjelakovic et al. 2007). Therefore, the theory of free radical-induced aging continues to be highly controversial.

One possible explanation for the lack of efficacy of overexpressing antioxidants in murine models and nutritional antioxidant supplements in clinical trials is the lack of antioxidant distribution to the mitochondria (Dai et al. 2014). Mice with transgenic mitochondrial-targeted catalase had significantly increased life span by an average of 5 months while there was not a significant effect on aging with nuclear-targeted catalase (Schriner and Linford 2006; Schriner et al. 2005). Similarly, mice with homozygous mutations in the mitochondrial proof reading subunit (PolgA) have increased mitochondrial point mutations and deletions, premature onset of age related diseases including weight loss, alopecia, kyphosis, osteoporosis, anemia, reduced fertility, and heart enlargement, as well as reduced life span (Trifunovic et al. 2004).

The discovery of mitochondria-specific oxidative stress associated with H_2O_2 as a possible source of aging has led to the development of agents that specifically target mitochondrial ROS. Examples include SkQ compounds that contain a plastoquinone, a cation, and a decane or pentane linkage and have been shown to prolong life span in *P. anserine, C. affinis, Drosophila*, and mice (Skulachev et al. 2009). Another strategy, Szeto-Schiller (SS) peptides, targets cardiolipin on the inner mitochondrial membrane improving mitochondrial bioenergetics and reducing mitochondria ROS generation (Birk et al. 2013). The more specific targeting of mitochondrial oxidative metabolism holds great promise for the future of both cancer therapy as well as agents to retard the process of aging on an organismal level.

Acknowledgements The authors would like to thank Gareth Smith for assistance with manuscript preparation and graphics. Funding to B.G.A. was provided by the American Society of Therapeutic Radiation Oncology (JF2014-1). Funding to D.R.S. was provided by the National Institutes of Health (R01 CA182804).

References

Ahmad IM, Aykin-Burns N, Sim JE, Walsh SA, Higashikubo R, Buettner GR, Venkataraman S, Mackey MA, Flanagan SW, Oberley LW, Spitz DR (2005) Mitochondrial $O_2^{\bullet-}$ and H_2O_2 mediate glucose deprivation-induced stress in human cancer cells. J Biol Chem 280(6):4254–4263. doi:10.1074/jbc.M411662200

Allen BG, Bhatia SK, Anderson CM, Eichenberger-Gilmore JM, Sibenaller ZA, Mapuskar KA, Schoenfeld JD, Buatti JM, Spitz DR, Fath MA (2014) Ketogenic diets as an adjuvant cancer therapy: History and potential mechanism. Redox Biol 2C:963–970. doi:10.1016/j.redox.2014.08.002

Arthur JR (2000) The glutathione peroxidases. Cell Mol Life Sci 57(13-14):1825–1835. doi:10.1007/PL00000664

Attardi G, Schatz G (1988) Biogenesis of mitochondria. Annu Rev Cell Biol 4:289–333. doi:10.1146/annurev.cb.04.110188.001445

Aykin-Burns N, Ahmad IM, Zhu Y, Oberley LW, Spitz DR (2009) Increased levels of superoxide and H_2O_2 mediate the differential susceptibility of cancer cells versus normal cells to glucose deprivation. Biochem J 418(1):29–37. doi:10.1042/bj20081258

Bae YS, Kang SW, Seo MS, Baines IC, Tekle E, Chock PB, Rhee SG (1997) Epidermal growth factor (EGF)-induced generation of hydrogen peroxide. Role in EGF receptor-mediated tyrosine phosphorylation. J Biol Chem 272(1):217–221. doi:10.1074/jbc.272.1.217

Bai RK, Leal SM, Covarrubias D, Liu A, Wong LJ (2007) Mitochondrial genetic background modifies breast cancer risk. Cancer Res 67(10):4687–4694. doi:10.1158/0008-5472.CAN-06-3554

Baker AM, Oberley LW, Cohen MB (1997) Expression of antioxidant enzymes in human prostatic adenocarcinoma. Prostate 32(4):229–233. doi:10.1002/(SICI)1097-0045(19970901)32:4<229::AID-PROS1>3.0.CO;2-E

Bandyopadhyay S, Starke DW, Mieyal JJ, Gronostajski RM (1998) Thioltransferase (glutaredoxin) reactivates the DNA-binding activity of oxidation-inactivated nuclear factor I. J Biol Chem 273(1):392–397. doi:10.1074/jbc.273.1.392

Barber SC, Mead RJ, Shaw PJ (2006) Oxidative stress in ALS: a mechanism of neurodegeneration and a therapeutic target. Biochim Biophys Acta 1762(11-12):1051–1067. doi:10.1016/j.bbadis.2006.03.008

Bartel J, Bartz T, Wolf C, Charkiewicz E, Kuhbacher M, Pohl T, Kyriakopoulos A (2007) Activity of the glutathione peroxidase-2. Differences in the selenium-dependent expression between colon and small intestine. Cancer Genomics Proteomics 4(5):369–372

Beer SM, Taylor ER, Brown SE, Dahm CC, Costa NJ, Runswick MJ, Murphy MP (2004) Glutaredoxin 2 catalyzes the reversible oxidation and glutathionylation of mitochondrial membrane thiol proteins: implications for mitochondrial redox regulation and antioxidant DEFENSE. J Biol Chem 279(46):47939–47951. doi:10.1074/jbc.M408011200

Bell EL, Klimova TA, Eisenbart J, Schumacker PT, Chandel NS (2007) Mitochondrial reactive oxygen species trigger hypoxia-inducible factor-dependent extension of the replicative life span during hypoxia. Mol Cell Biol 27(16):5737–5745. doi:10.1128/MCB.02265-06

Birk AV, Liu S, Soong Y, Mills W, Singh P, Warren JD, Seshan SV, Pardee JD, Szeto HH (2013) The mitochondrial-targeted compound SS-31 re-energizes ischemic mitochondria by interacting with cardiolipin. J Am Soc Nephrol 24(8):1250–1261. doi:10.1681/ASN.2012121216

Bize IB, Oberley LW, Morris HP (1980) Superoxide dismutase and superoxide radical in Morris hepatomas. Cancer Res 40(10):3686–3693

Bjelakovic G, Nikolova D, Gluud LL, Simonetti RG, Gluud C (2007) Mortality in randomized trials of antioxidant supplements for primary and secondary prevention: systematic review and meta-analysis. JAMA 297(8):842–857. doi:10.1001/jama.297.8.842

Blackburn RV, Spitz DR, Liu X, Galoforo SS, Sim JE, Ridnour LA, Chen JC, Davis BH, Corry PM, Lee YJ (1999) Metabolic oxidative stress activates signal transduction and gene expression during glucose deprivation in human tumor cells. Free Radic Biol Med 26(3-4):419–430. doi:10.1016/S0891-5849(98)00217-2

Blankenberg S, Rupprecht HJ, Bickel C, Torzewski M, Hafner G, Tiret L, Smieja M, Cambien F, Meyer J, Lackner KJ, AtheroGene I (2003) Glutathione peroxidase 1 activity and cardiovascular events in patients with coronary artery disease. N Engl J Med 349(17):1605–1613. doi:10.1056/NEJMoa030535

Bohr VA (2002) Repair of oxidative DNA damage in nuclear and mitochondrial DNA, and some changes with aging in mammalian cells. Free Radic Biol Med 32(9):804–812. doi:10.1016/S0891-5849(02)00787-6

Bostwick DG, Alexander EE, Singh R, Shan A, Qian J, Santella RM, Oberley LW, Yan T, Zhong W, Jiang X, Oberley TD (2000) Antioxidant enzyme expression and reactive oxygen species damage in prostatic intraepithelial neoplasia and cancer. Cancer 89(1):123–134. doi:10.1002/1097-0142(20000701)89:1<123::AID-CNCR17>3.0.CO;2-9

Boveris A (1977) Mitochondrial production of superoxide radical and hydrogen peroxide. Adv Exp Med Biol 78:67–82. doi:10.1007/978-1-4615-9035-4_5

Boveris A, Cadenas E (1982) Production of Superoxide Radicals and Hydrogen Peroxide in Mitochondria. In: Oberley LW (ed) Superoxide dismutase, vol II. CRC Press, Boca Raton, Florida

Bushweller JH, Aslund F, Wuthrich K, Holmgren A (1992) Structural and functional characterization of the mutant Escherichia coli glutaredoxin (C14----S) and its mixed disulfide with glutathione. Biochemistry 31(38):9288–9293. doi:10.1021/bi00153a023

Cadenas E, Davies KJ (2000) Mitochondrial free radical generation, oxidative stress, and aging. Free Radic Biol Med 29(3-4):222–230. doi:10.1016/S0891-5849(00)00317-8

Cappadoro M, Giribaldi G, O'Brien E, Turrini F, Mannu F, Ulliers D, Simula G, Luzzatto L, Arese P (1998) Early phagocytosis of glucose-6-phosphate dehydrogenase (G6PD)-deficient erythrocytes parasitized by Plasmodium falciparum may explain malaria protection in G6PD deficiency. Blood 92(7):2527–2534

Carew JS, Huang P (2002) Mitochondrial defects in cancer. Mol Cancer 1:9. doi:10.1186/1476-4598-1-9

Carlsson LM, Jonsson J, Edlund T, Marklund SL (1995) Mice lacking extracellular superoxide dismutase are more sensitive to hyperoxia. Proc Natl Acad Sci U S A 92(14):6264–6268

Chandel NS, McClintock DS, Feliciano CE, Wood TM, Melendez JA, Rodriguez AM, Schumacker PT (2000) Reactive oxygen species generated at mitochondrial complex III stabilize hypoxia-inducible factor-1alpha during hypoxia: a mechanism of O2 sensing. J Biol Chem 275(33):25130–25138. doi:10.1074/jbc.M001914200

Chang TS, Cho CS, Park S, Yu S, Kang SW, Rhee SG (2004) Peroxiredoxin III, a mitochondrion-specific peroxidase, regulates apoptotic signaling by mitochondria. J Biol Chem 279(40):41975–41984. doi:10.1074/jbc.M407707200

Chatterjee A, Dasgupta S, Sidransky D (2011) Mitochondrial subversion in cancer. Cancer Prev Res (Phila) 4(5):638–654. doi:10.1158/1940-6207.CAPR-10-0326

Chrestensen CA, Starke DW, Mieyal JJ (2000) Acute cadmium exposure inactivates thioltransferase (Glutaredoxin), inhibits intracellular reduction of protein-glutathionyl-mixed disulfides, and initiates apoptosis. J Biol Chem 275(34):26556–26565. doi:10.1074/jbc.M004097200

Claiborne A, Yeh JI, Mallett TC, Luba J, Crane EJ 3rd, Charrier V, Parsonage D (1999) Protein-sulfenic acids: diverse roles for an unlikely player in enzyme catalysis and redox regulation. Biochemistry 38(47):15407–15416. doi:10.1021/bi992025k

Conrad M, Jakupoglu C, Moreno SG, Lippl S, Banjac A, Schneider M, Beck H, Hatzopoulos AK, Just U, Sinowatz F, Schmahl W, Chien KR, Wurst W, Bornkamm GW, Brielmeier M (2004) Essential role for mitochondrial thioredoxin reductase in hematopoiesis, heart development, and heart function. Mol Cell Biol 24(21):9414–9423. doi:10.1128/MCB.24.21.9414-9423.2004

Dai DF, Chiao YA, Marcinek DJ, Szeto HH, Rabinovitch PS (2014) Mitochondrial oxidative stress in aging and healthspan. Longev Healthspan 3:6. doi:10.1186/2046-2395-3-6

Daily D, Vlamis-Gardikas A, Offen D, Mittelman L, Melamed E, Holmgren A, Barzilai A (2001) Glutaredoxin protects cerebellar granule neurons from dopamine-induced apoptosis by activating NF-kappa B via Ref-1. J Biol Chem 276(2):1335–1344. doi:10.1074/jbc.M008121200

Dasgupta S, Hoque MO, Upadhyay S, Sidransky D (2008) Mitochondrial cytochrome B gene mutation promotes tumor growth in bladder cancer. Cancer Res 68(3):700–706. doi:10.1158/0008-5472.CAN-07-5532

Dikalov SI, Dikalova AE, Bikineyeva AT, Schmidt HH, Harrison DG, Griendling KK (2008) Distinct roles of Nox1 and Nox4 in basal and angiotensin II-stimulated superoxide and hydrogen peroxide production. Free Radic Biol Med 45(9):1340–1351. doi:10.1016/j.freeradbiomed.2008.08.013

Duttaroy A, Paul A, Kundu M, Belton A (2003) A Sod2 null mutation confers severely reduced adult life span in Drosophila. Genetics 165(4):2295–2299

Elchuri S, Oberley TD, Qi W, Eisenstein RS, Jackson Roberts L, Van Remmen H, Epstein CJ, Huang TT (2005) CuZnSOD deficiency leads to persistent and widespread oxidative damage and hepatocarcinogenesis later in life. Oncogene 24(3):367–380. doi:10.1038/sj.onc.1208207

Emerling BM, Weinberg F, Snyder C, Burgess Z, Mutlu GM, Viollet B, Budinger GR, Chandel NS (2009) Hypoxic activation of AMPK is dependent on mitochondrial ROS but independent of an increase in AMP/ATP ratio. Free Radic Biol Med 46(10):1386–1391. doi:10.1016/j.freeradbiomed.2009.02.019

Enoksson M, Fernandes AP, Prast S, Lillig CH, Holmgren A, Orrenius S (2005) Overexpression of glutaredoxin 2 attenuates apoptosis by preventing cytochrome c release. Biochem Biophys Res Commun 327(3):774–779. doi:10.1016/j.bbrc.2004.12.067

Esterbauer H, Schaur RJ, Zollner H (1991) Chemistry and biochemistry of 4-hydroxynonenal, malonaldehyde and related aldehydes. Free Radic Biol Med 11(1):81–128. doi:10.1016/0891-5849(91)90192-6

Finkel T, Serrano M, Blasco MA (2007) The common biology of cancer and ageing. Nature 448(7155):767–774. doi:10.1038/nature05985

Forgione MA, Cap A, Liao R, Moldovan NI, Eberhardt RT, Lim CC, Jones J, Goldschmidt-Clermont PJ, Loscalzo J (2002) Heterozygous cellular glutathione peroxidase deficiency in the mouse: abnormalities in vascular and cardiac function and structure. Circulation 106(9):1154–1158. doi:10.1161/01.CIR.0000026820.87824.6A

Forman HJ, Kennedy JA (1974) Role of superoxide radical in mitochondrial dehydrogenase reactions. Biochem Biophys Res Commun 60(3):1044–1050. doi:10.1016/0006-291X(74)90418-5

Forman HJ, Kennedy J (1975) Superoxide production and electron transport in mitochondrial oxidation of dihydroorotic acid. J Biol Chem 250(11):4322–4326

Fridovich I (1978) The biology of oxygen radicals. Science 201(4359):875–880

Fridovich I (2004) Mitochondria: are they the seat of senescence? Aging Cell 3(1):13–16. doi:10.1046/j.1474-9728.2003.00075.x

Fu Y, Sies H, Lei XG (2001) Opposite roles of selenium-dependent glutathione peroxidase-1 in superoxide generator diquat- and peroxynitrite-induced apoptosis and signaling. J Biol Chem 276(46):43004–43009. doi:10.1074/jbc.M106946200

Gaetani GF, Galiano S, Canepa L, Ferraris AM, Kirkman HN (1989) Catalase and glutathione peroxidase are equally active in detoxification of hydrogen peroxide in human erythrocytes. Blood 73(1):334–339

Galluzzi L, Morselli E, Kepp O, Vitale I, Rigoni A, Vacchelli E, Michaud M, Zischka H, Castedo M, Kroemer G (2010) Mitochondrial gateways to cancer. Mol Aspects Med 31(1):1–20. doi:10.1016/j.mam.2009.08.002

Gongora MC, Qin Z, Laude K, Kim HW, McCann L, Folz JR, Dikalov S, Fukai T, Harrison DG (2006) Role of extracellular superoxide dismutase in hypertension. Hypertension 48(3):473–481. doi:10.1161/01.HYP.0000235682.47673.ab

Graham NA, Tahmasian M, Kohli B, Komisopoulou E, Zhu M, Vivanco I, Teitell MA, Wu H, Ribas A, Lo RS, Mellinghoff IK, Mischel PS, Graeber TG (2012) Glucose deprivation activates a metabolic and signaling amplification loop leading to cell death. Mol Syst Biol 8:589. doi:10.1038/msb.2012.20

Griffith OW, Meister A (1985) Origin and turnover of mitochondrial glutathione. Proc Natl Acad Sci U S A 82(14):4668–4672

Gusarova GA, Dada LA, Kelly AM, Brodie C, Witters LA, Chandel NS, Sznajder JI (2009) Alpha1-AMP-activated protein kinase regulates hypoxia-induced Na, K-ATPase endocytosis via direct phosphorylation of protein kinase C zeta. Mol Cell Biol 29(13):3455–3464. doi:10.1128/MCB.00054-09

Handy DE, Loscalzo J (2012) Redox regulation of mitochondrial function. Antioxid Redox Signal 16(11):1323–1367. doi:10.1089/ars.2011.4123

Harman D (1956) Aging: a theory based on free radical and radiation chemistry. J Gerontol 11(3):298–300

He Y, Wu J, Dressman DC, Iacobuzio-Donahue C, Markowitz SD, Velculescu VE, Diaz LA Jr, Kinzler KW, Vogelstein B, Papadopoulos N (2010) Heteroplasmic mitochondrial DNA mutations in normal and tumour cells. Nature 464(7288):610–614. doi:10.1038/nature08802

Hirota K, Matsui M, Murata M, Takashima Y, Cheng FS, Itoh T, Fukuda K, Yodoi J (2000) Nucleoredoxin, glutaredoxin, and thioredoxin differentially regulate NF-kappaB, AP-1, and CREB activation in HEK293 cells. Biochem Biophys Res Commun 274(1):177–182. doi:10.1006/bbrc.2000.3106

Holmgren A (1989) Thioredoxin and glutaredoxin systems. J Biol Chem 264(24):13963–13966

Holmgren A, Aslund F (1995) Glutaredoxin. Methods Enzymol 252:283–292. doi:10.1016/0076-6879(95)52031-7

Holmstrom KM, Finkel T (2014) Cellular mechanisms and physiological consequences of redox-dependent signalling. Nat Rev Mol Cell Biol 15(6):411–421. doi:10.1038/nrm3801

Horton TM, Petros JA, Heddi A, Shoffner J, Kaufman AE, Graham SD Jr, Gramlich T, Wallace DC (1996) Novel mitochondrial DNA deletion found in a renal cell carcinoma. Genes Chromosomes Cancer 15(2):95–101. doi:10.1002/(SICI)1098-2264(199602)15:2<95::AID-GCC3>3.0.CO;2-Z

Hwang C, Sinskey AJ, Lodish HF (1992) Oxidized redox state of glutathione in the endoplasmic reticulum. Science 257(5076):1496–1502. doi:10.1126/science.1523409

Ishikawa K, Takenaga K, Akimoto M, Koshikawa N, Yamaguchi A, Imanishi H, Nakada K, Honma Y, Hayashi J (2008) ROS-generating mitochondrial DNA mutations can regulate tumor cell metastasis. Science 320(5876):661–664. doi:10.1126/science.1156906

Iwahara S, Satoh H, Song DX, Webb J, Burlingame AL, Nagae Y, Muller-Eberhard U (1995) Purification, characterization, and cloning of a heme-binding protein (23 kDa) in rat liver cytosol. Biochemistry 34(41):13398–13406. doi:10.1021/bi00041a017

Jakupoglu C, Przemeck GK, Schneider M, Moreno SG, Mayr N, Hatzopoulos AK, de Angelis MH, Wurst W, Bornkamm GW, Brielmeier M, Conrad M (2005) Cytoplasmic thioredoxin reductase is essential for embryogenesis but dispensable for cardiac development. Mol Cell Biol 25(5):1980–1988. doi:10.1128/MCB.25.5.1980-1988.2005

Jang YC, Perez VI, Song W, Lustgarten MS, Salmon AB, Mele J, Qi W, Liu Y, Liang H, Chaudhuri A, Ikeno Y, Epstein CJ, Van Remmen H, Richardson A (2009) Overexpression of Mn superoxide dismutase does not increase life span in mice. J Gerontol A Biol Sci Med Sci 64(11):1114–1125. doi:10.1093/gerona/glp100

Jin RC, Mahoney CE, Coleman Anderson L, Ottaviano F, Croce K, Leopold JA, Zhang YY, Tang SS, Handy DE, Loscalzo J (2011) Glutathione peroxidase-3 deficiency promotes platelet-dependent thrombosis in vivo. Circulation 123(18):1963–1973. doi:10.1161/CIRCULATIONAHA.110.000034

Jones DP, Carlson JL, Samiec PS, Sternberg P Jr, Mody VC Jr, Reed RL, Brown LA (1998) Glutathione measurement in human plasma. Evaluation of sample collection, storage and derivatization conditions for analysis of dansyl derivatives by HPLC. Clin Chim Acta 275(2):175–184. doi:10.1016/S0009-8981(98)00089-8

Kamata H, Honda S, Maeda S, Chang L, Hirata H, Karin M (2005) Reactive oxygen species promote TNFalpha-induced death and sustained JNK activation by inhibiting MAP kinase phosphatases. Cell 120(5):649–661. doi:10.1016/j.cell.2004.12.041

Kelley EE, Khoo NK, Hundley NJ, Malik UZ, Freeman BA, Tarpey MM (2010) Hydrogen peroxide is the major oxidant product of xanthine oxidase. Free Radic Biol Med 48(4):493–498. doi:10.1016/j.freeradbiomed.2009.11.012

Lebovitz RM, Zhang H, Vogel H, Cartwright J Jr, Dionne L, Lu N, Huang S, Matzuk MM (1996) Neurodegeneration, myocardial injury, and perinatal death in mitochondrial superoxide dismutase-deficient mice. Proc Natl Acad Sci U S A 93(18):9782–9787

Lee YJ, Galoforo SS, Berns CM, Chen JC, Davis BH, Sim JE, Corry PM, Spitz DR (1998) Glucose deprivation-induced cytotoxicity and alterations in mitogen-activated protein kinase activation are mediated by oxidative stress in multidrug-resistant human breast carcinoma cells. J Biol Chem 273(9):5294–5299. doi:10.1074/jbc.273.9.5294

Lee TH, Kim SU, Yu SL, Kim SH, Park DS, Moon HB, Dho SH, Kwon KS, Kwon HJ, Han YH, Jeong S, Kang SW, Shin HS, Lee KK, Rhee SG, Yu DY (2003) Peroxiredoxin II is essential for sustaining life span of erythrocytes in mice. Blood 101(12):5033–5038. doi:10.1182/blood-2002-08-2548

Lenaz G, Bovina C, Castelluccio C, Fato R, Formiggini G, Genova ML, Marchetti M, Pich MM, Pallotti F, Parenti Castelli G, Biagini G (1997) Mitochondrial complex I defects in aging. Mol Cell Biochem 174(1-2):329–333. doi:10.1007/978-1-4615-6111-8_50

Leopold JA, Loscalzo J (2005) Oxidative enzymopathies and vascular disease. Arterioscler Thromb Vasc Biol 25(7):1332–1340. doi:10.1161/01.ATV.0000163846.51473.09

Lesnefsky EJ, Gudz TI, Moghaddas S, Migita CT, Ikeda-Saito M, Turkaly PJ, Hoppel CL (2001) Aging decreases electron transport complex III activity in heart interfibrillar mitochondria by alteration of the cytochrome c binding site. J Mol Cell Cardiol 33(1):37–47. doi:10.1006/jmcc.2000.1273

Levine M, Padayatty SJ, Espey MG (2011) Vitamin C: a concentration-function approach yields pharmacology and therapeutic discoveries. Adv Nutr 2(2):78–88. doi:10.3945/an.110.000109

Li Y, Huang TT, Carlson EJ, Melov S, Ursell PC, Olson JL, Noble LJ, Yoshimura MP, Berger C, Chan PH, Wallace DC, Epstein CJ (1995) Dilated cardiomyopathy and neonatal lethality in mutant mice lacking manganese superoxide dismutase. Nat Genet 11(4):376–381. doi:10.1038/ng1295-376

Liang H, Ran Q, Jang YC, Holstein D, Lechleiter J, McDonald-Marsh T, Musatov A, Song W, Van Remmen H, Richardson A (2009) Glutathione peroxidase 4 differentially regulates the release of apoptogenic proteins from mitochondria. Free Radic Biol Med 47(3):312–320. doi:10.1016/j.freeradbiomed.2009.05.012

Lillig CH, Lonn ME, Enoksson M, Fernandes AP, Holmgren A (2004) Short interfering RNA-mediated silencing of glutaredoxin 2 increases the sensitivity of HeLa cells toward doxorubicin and phenylarsine oxide. Proc Natl Acad Sci U S A 101(36):13227–13232. doi:10.1073/pnas.0401896101

Linares GR, Xing W, Govoni KE, Chen ST, Mohan S (2009) Glutaredoxin 5 regulates osteoblast apoptosis by protecting against oxidative stress. Bone 44(5):795–804. doi:10.1016/j.bone.2009.01.003

Liu P, Qian L, Sung JS, de Souza-Pinto NC, Zheng L, Bogenhagen DF, Bohr VA, Wilson DM 3rd, Shen B, Demple B (2008) Removal of oxidative DNA damage via FEN1-dependent long-patch base excision repair in human cell mitochondria. Mol Cell Biol 28(16):4975–4987. doi:10.1128/MCB.00457-08

Loh K, Deng H, Fukushima A, Cai X, Boivin B, Galic S, Bruce C, Shields BJ, Skiba B, Ooms LM, Stepto N, Wu B, Mitchell CA, Tonks NK, Watt MJ, Febbraio MA, Crack PJ, Andrikopoulos S, Tiganis T (2009) Reactive oxygen species enhance insulin sensitivity. Cell Metab 10(4):260–272. doi:10.1016/j.cmet.2009.08.009

Longo VD, Liou LL, Valentine JS, Gralla EB (1999) Mitochondrial superoxide decreases yeast survival in stationary phase. Arch Biochem Biophys 365(1):131–142. doi:10.1006/abbi.1999.1158

Lundberg M, Johansson C, Chandra J, Enoksson M, Jacobsson G, Ljung J, Johansson M, Holmgren A (2001) Cloning and expression of a novel human glutaredoxin (Grx2) with mitochondrial and nuclear isoforms. J Biol Chem 276(28):26269–26275. doi:10.1074/jbc.M011605200

Maity A, Tuttle SW (2006) 2-Deoxyglucose and radiosensitization: teaching an old DOG new tricks? Cancer Biol Ther 5(7):824–826. doi:10.4161/cbt.5.7.3024

Matsui M, Oshima M, Oshima H, Takaku K, Maruyama T, Yodoi J, Taketo MM (1996) Early embryonic lethality caused by targeted disruption of the mouse thioredoxin gene. Dev Biol 178(1):179–185. doi:10.1006/dbio.1996.0208

McClung JP, Roneker CA, Mu W, Lisk DJ, Langlais P, Liu F, Lei XG (2004) Development of insulin resistance and obesity in mice overexpressing cellular glutathione peroxidase. Proc Natl Acad Sci U S A 101(24):8852–8857. doi:10.1073/pnas.0308096101

Meister A (1995) Mitochondrial changes associated with glutathione deficiency. Biochim Biophys Acta 1271(1):35–42. doi:10.1016/0925-4439(95)00007-Q

Meister A, Tate SS (1976) Glutathione and related gamma-glutamyl compounds: biosynthesis and utilization. Annu Rev Biochem 45:559–604. doi:10.1146/annurev.bi.45.070176.003015

Mitchell P (1979) Keilin's respiratory chain concept and its chemiosmotic consequences. Science 206(4423):1148–1159

Mitchell P (2011) Chemiosmotic coupling in oxidative and photosynthetic phosphorylation. 1966. Biochim Biophys Acta 1807(12):1507–1538. doi:10.1016/j.bbabio.2011.09.018

Moghaddas S, Hoppel CL, Lesnefsky EJ (2003) Aging defect at the QO site of complex III augments oxyradical production in rat heart interfibrillar mitochondria. Arch Biochem Biophys 414(1):59–66. doi:10.1016/S0003-9861(03)00166-8

Muller FL, Liu Y, Van Remmen H (2004) Complex III releases superoxide to both sides of the inner mitochondrial membrane. J Biol Chem 279(47):49064–49073. doi:10.1074/jbc.M407715200

Murphy MP (2009) How mitochondria produce reactive oxygen species. Biochem J 417(1):1–13. doi:10.1042/BJ20081386

Nagy N, Malik G, Fisher AB, Das DK (2006) Targeted disruption of peroxiredoxin 6 gene renders the heart vulnerable to ischemia-reperfusion injury. Am J Physiol Heart Circ Physiol 291(6):H2636–H2640. doi:10.1152/ajpheart.00399.2006

Nelson DL, Cox MM (2000) Lehninger Principles in Biochemistry. Worth Publishers Inc., New York, NY

Nemoto Y, Yamamoto T, Takada S, Matsui Y, Obinata M (1990) Antisense RNA of the latent period gene (MER5) inhibits the differentiation of murine erythroleukemia cells. Gene 91(2):261–265. doi:10.1016/0378-1119(90)90097-B

Neumann CA, Krause DS, Carman CV, Das S, Dubey DP, Abraham JL, Bronson RT, Fujiwara Y, Orkin SH, Van Etten RA (2003) Essential role for the peroxiredoxin Prdx1 in erythrocyte antioxidant defence and tumour suppression. Nature 424(6948):561–565. doi:10.1038/nature01819

Nonn L, Williams RR, Erickson RP, Powis G (2003) The absence of mitochondrial thioredoxin 2 causes massive apoptosis, exencephaly, and early embryonic lethality in homozygous mice. Mol Cell Biol 23(3):916–922. doi:10.1128/MCB.23.3.916-922.2003

Oberley LW, Oberley TD, Buettner GR (1980) Cell differentiation, aging and cancer: the possible roles of superoxide and superoxide dismutases. Med Hypotheses 6(3):249–268. doi:10.1016/0306-9877(80)90123-1

Oberley LW, Oberley TD, Buettner GR (1981) Cell division in normal and transformed cells: the possible role of superoxide and hydrogen peroxide. Med Hypotheses 7(1):21–42. doi:10.1016/0306-9877(81)90018-9

Orrenius S, Gogvadze V, Zhivotovsky B (2007) Mitochondrial oxidative stress: implications for cell death. Annu Rev Pharmacol Toxicol 47:143–183. doi:10.1146/annurev.pharmtox.47.120505.105122

Paradies G, Ruggiero FM, Petrosillo G, Quagliariello E (1997) Age-dependent decline in the cytochrome c oxidase activity in rat heart mitochondria: role of cardiolipin. FEBS Lett 406(1-2):136–138. doi:10.1016/S0014-5793(97)00264-0

Park SH, Ozden O, Jiang H, Cha YI, Pennington JD, Aykin-Burns N, Spitz DR, Gius D, Kim HS (2011) Sirt3, mitochondrial ROS, ageing, and carcinogenesis. Int J Mol Sci 12(9):6226–6239. doi:10.3390/ijms12096226

Pelicano H, Lu W, Zhou Y, Zhang W, Chen Z, Hu Y, Huang P (2009) Mitochondrial dysfunction and reactive oxygen species imbalance promote breast cancer cell motility through a CXCL14-mediated mechanism. Cancer Res 69(6):2375–2383. doi:10.1158/0008-5472.CAN-08-3359

Perez VI, Van Remmen H, Bokov A, Epstein CJ, Vijg J, Richardson A (2009) The overexpression of major antioxidant enzymes does not extend the lifespan of mice. Aging Cell 8(1):73–75. doi:10.1111/j.1474-9726.2008.00449.x

Petros JA, Baumann AK, Ruiz-Pesini E, Amin MB, Sun CQ, Hall J, Lim S, Issa MM, Flanders WD, Hosseini SH, Marshall FF, Wallace DC (2005) mtDNA mutations increase tumorigenicity in prostate cancer. Proc Natl Acad Sci U S A 102(3):719–724. doi:10.1073/pnas.0408894102

Petrosillo G, Matera M, Casanova G, Ruggiero FM, Paradies G (2008) Mitochondrial dysfunction in rat brain with aging Involvement of complex I, reactive oxygen species and cardiolipin. Neurochem Int 53(5):126–131. doi:10.1016/j.neuint.2008.07.001

Phillips JP, Campbell SD, Michaud D, Charbonneau M, Hilliker AJ (1989) Null mutation of copper/zinc superoxide dismutase in Drosophila confers hypersensitivity to paraquat and reduced longevity. Proc Natl Acad Sci U S A 86(8):2761–2765

Polyak K, Li Y, Zhu H, Lengauer C, Willson JK, Markowitz SD, Trush MA, Kinzler KW, Vogelstein B (1998) Somatic mutations of the mitochondrial genome in human colorectal tumours. Nat Genet 20(3):291–293. doi:10.1038/3108

Prosperi MT, Apiou F, Dutrillaux B, Goubin G (1994) Organization and chromosomal assignment of two human PAG gene loci: PAGA encoding a functional gene and PAGB a processed pseudogene. Genomics 19(2):236–241. doi:10.1006/geno.1994.1053

Radi R, Turrens JF, Chang LY, Bush KM, Crapo JD, Freeman BA (1991) Detection of catalase in rat heart mitochondria. J Biol Chem 266(32):22028–22034

Ralph SJ, Rodriguez-Enriquez S, Neuzil J, Saavedra E, Moreno-Sanchez R (2010) The causes of cancer revisited: "mitochondrial malignancy" and ROS-induced oncogenic transformation—why mitochondria are targets for cancer therapy. Mol Aspects Med 31(2):145–170. doi:10.1016/j.mam.2010.02.008

Reeves MA, Hoffmann PR (2009) The human selenoproteome: recent insights into functions and regulation. Cell Mol Life Sci 66(15):2457–2478. doi:10.1007/s00018-009-0032-4

Rhee SG, Chae HZ, Kim K (2005) Peroxiredoxins: a historical overview and speculative preview of novel mechanisms and emerging concepts in cell signaling. Free Radic Biol Med 38(12):1543–1552. doi:10.1016/j.freeradbiomed.2005.02.026

Rigoulet M, Yoboue ED, Devin A (2011) Mitochondrial ROS generation and its regulation: mechanisms involved in H_2O_2 signaling. Antioxid Redox Signal 14(3):459–468. doi:10.1089/ars.2010.3363

Rotruck JT, Pope AL, Ganther HE, Swanson AB, Hafeman DG, Hoekstra WG (1973) Selenium: biochemical role as a component of glutathione peroxidase. Science 179(4073):588–590

Salskov A, Tammisetti VS, Grierson J, Vesselle H (2007) FLT: measuring tumor cell proliferation in vivo with positron emission tomography and 3′-deoxy-3′-[18F]fluorothymidine. Semin Nucl Med 37(6):429–439. doi:10.1053/j.semnuclmed.2007.08.001

Sarsour EH, Kalen AL, Goswami PC (2014) Manganese superoxide dismutase regulates a redox cycle within the cell cycle. Antioxid Redox Signal 20(10):1618–1627. doi:10.1089/ars.2013.5303

Schafer FQ, Buettner GR (2001) Redox environment of the cell as viewed through the redox state of the glutathione disulfide/glutathione couple. Free Radic Biol Med 30(11):1191–1212. doi:10.1016/S0891-5849(01)00480-4

Schriner SE, Linford NJ (2006) Extension of mouse lifespan by overexpression of catalase. Age (Dordr) 28(2):209–218. doi:10.1007/s11357-006-9010-z

Schriner SE, Linford NJ, Martin GM, Treuting P, Ogburn CE, Emond M, Coskun PE, Ladiges W, Wolf N, Van Remmen H, Wallace DC, Rabinovitch PS (2005) Extension of murine life span by overexpression of catalase targeted to mitochondria. Science 308(5730):1909–1911. doi:10.1126/science.1106653

Shau H, Butterfield LH, Chiu R, Kim A (1994) Cloning and sequence analysis of candidate human natural killer-enhancing factor genes. Immunogenetics 40(2):129–134. doi:10.1007/BF00188176

Shields AF, Grierson JR, Dohmen BM, Machulla HJ, Stayanoff JC, Lawhorn-Crews JM, Obradovich JE, Muzik O, Mangner TJ (1998) Imaging proliferation in vivo with [F-18]FLT and positron emission tomography. Nat Med 4(11):1334–1336. doi:10.1038/3337

Sies H (1991) Oxidative stress: from basic research to clinical application. Am J Med 91(3c):31s–38s. doi:10.1016/0002-9343(91)90281-2

Skulachev VP, Anisimov VN, Antonenko YN, Bakeeva LE, Chernyak BV, Erichev VP, Filenko OF, Kalinina NI, Kapelko VI, Kolosova NG, Kopnin BP, Korshunova GA, Lichinitser MR, Obukhova LA, Pasyukova EG, Pisarenko OI, Roginsky VA, Ruuge EK, Senin II, Severina II, Skulachev MV, Spivak IM, Tashlitsky VN, Tkachuk VA, Vyssokikh MY, Yaguzhinsky LS, Zorov DB (2009) An attempt to prevent senescence: a mitochondrial approach. Biochim Biophys Acta 1787(5):437–461. doi:10.1016/j.bbabio.2008.12.008

Smith CV, Jones DP, Guenthner TM, Lash LH, Lauterburg BH (1996) Compartmentation of glutathione: implications for the study of toxicity and disease. Toxicol Appl Pharmacol 140(1):1–12. doi:10.1006/taap.1996.0191

Spitz DR, Sim JE, Ridnour LA, Galoforo SS, Lee YJ (2000) Glucose deprivation-induced oxidative stress in human tumor cells. A fundamental defect in metabolism? Ann N Y Acad Sci 899:349–362. doi:10.1111/j.1749-6632.2000.tb06199

Spitz DR, Azzam EI, Li JJ, Gius D (2004) Metabolic oxidation/reduction reactions and cellular responses to ionizing radiation: a unifying concept in stress response biology. Cancer Metastasis Rev 23(3-4):311–322. doi:10.1023/B:CANC.0000031769.14728.bc

Springer EL (1980) Comparative study of the cytoplasmic organelles of epithelial cell lines derived from human carcinomas and nonmalignant tissues. Cancer Res 40(3):803–817

Stadtman ER, Oliver CN (1991) Metal-catalyzed oxidation of proteins. Physiological consequences. J Biol Chem 266(4):2005–2008

Starke-Reed PE, Oliver CN (1989) Protein oxidation and proteolysis during aging and oxidative stress. Arch Biochem Biophys 275(2):559–567. doi:10.1016/0003-9861(89)90402-5

Sullivan LB, Chandel NS (2014) Mitochondrial reactive oxygen species and cancer. Cancer Metab 2:17. doi:10.1186/2049-3002-2-17

Sundaresan M, Yu ZX, Ferrans VJ, Irani K, Finkel T (1995) Requirement for generation of H_2O_2 for platelet-derived growth factor signal transduction. Science 270(5234):296–299

Szabo C, Ischiropoulos H, Radi R (2007) Peroxynitrite: biochemistry, pathophysiology and development of therapeutics. Nat Rev Drug Discov 6(8):662–680. doi:10.1038/nrd2222

Szczesny B, Tann AW, Longley MJ, Copeland WC, Mitra S (2008) Long patch base excision repair in mammalian mitochondrial genomes. J Biol Chem 283(39):26349–26356. doi:10.1074/jbc.M803491200

Szeto HH, James LP, Atkinson AJ (2014) Mitochondrial pharmacology: its future is now. Clin Pharmacol Ther 96(6):629–633. doi:10.1038/clpt.2014.177

Thomas DD, Heinecke JL, Ridnour LA, Cheng RY, Kesarwala AH, Switzer CH, McVicar DW, Roberts DD, Glynn S, Fukuto JM, Wink DA, Miranda KM (2015) Signaling and stress: The redox landscape in NOS2 biology. Free Radic Biol Med 87:204–225. doi:10.1016/j.freeradbiomed.2015.06.002

Trifunovic A, Wredenberg A, Falkenberg M, Spelbrink JN, Rovio AT, Bruder CE, Bohlooly YM, Gidlof S, Oldfors A, Wibom R, Tornell J, Jacobs HT, Larsson NG (2004) Premature ageing in mice expressing defective mitochondrial DNA polymerase. Nature 429(6990):417–423. doi:10.1038/nature02517

Ussakli CH, Ebaee A, Binkley J, Brentnall TA, Emond MJ, Rabinovitch PS, Risques RA (2013) Mitochondria and tumor progression in ulcerative colitis. J Natl Cancer Inst 105(16):1239–1248. doi:10.1093/jnci/djt167

Van Raamsdonk JM, Hekimi S (2009) Deletion of the mitochondrial superoxide dismutase sod-2 extends lifespan in Caenorhabditis elegans. PLoS Genet 5(2), e1000361. doi:10.1371/journal.pgen.1000361

Vander Heiden MG, Cantley LC, Thompson CB (2009) Understanding the Warburg effect: the metabolic requirements of cell proliferation. Science 324(5930):1029–1033. doi:10.1126/science.1160809

Warburg O (1956) On respiratory impairment in cancer cells. Science 124(3215):269–270. doi:10.1126/science.124.3215.267

Wawryn J, Krzepilko A, Myszka A, Bilinski T (1999) Deficiency in superoxide dismutases shortens life span of yeast cells. Acta Biochim Pol 46(2):249–253

Wells WW, Xu DP, Yang YF, Rocque PA (1990) Mammalian thioltransferase (glutaredoxin) and protein disulfide isomerase have dehydroascorbate reductase activity. J Biol Chem 265(26):15361–15364

Wen ST, Van Etten RA (1997) The PAG gene product, a stress-induced protein with antioxidant properties, is an Abl SH3-binding protein and a physiological inhibitor of c-Abl tyrosine kinase activity. Genes Dev 11(19):2456–2467. doi:10.1101/gad.11.19.2456

Winterbourn CC, Hampton MB (2008) Thiol chemistry and specificity in redox signaling. Free Radic Biol Med 45(5):549–561. doi:10.1016/j.freeradbiomed.2008.05.004

Wood ZA, Poole LB, Karplus PA (2003) Peroxiredoxin evolution and the regulation of hydrogen peroxide signaling. Science 300(5619):650–653. doi:10.1126/science.1080405

Yang W, Li J, Hekimi S (2007) A Measurable increase in oxidative damage due to reduction in superoxide detoxification fails to shorten the life span of long-lived mitochondrial mutants of Caenorhabditis elegans. Genetics 177(4):2063–2074. doi:10.1534/genetics.107.080788

Yant LJ, Ran Q, Rao L, Van Remmen H, Shibatani T, Belter JG, Motta L, Richardson A, Prolla TA (2003) The selenoprotein GPX4 is essential for mouse development and protects from radiation and oxidative damage insults. Free Radic Biol Med 34(4):496–502. doi:10.1016/S0891-5849(02)01360-6

Yu M (2012) Somatic mitochondrial DNA mutations in human cancers. In: Makowski GS (ed) Adv Clin Chem, vol 57. Academic, Elsevier, pp 99–138

Zhou D, Shao L, Spitz DR (2014) Reactive oxygen species in normal and tumor stem cells. Adv Cancer Res 122:1–67. doi:10.1016/b978-0-12-420117-0.00001-3

Chapter 7
Mitochondrial Mutagenesis in Cancer

William J. Valente and Jason H. Bielas

7.1 The Mitochondrial Genome

Mitochondria are unique intracellular organelles that harbor their own genome. The likely product of an early endosymbiotic event, the mitochondrion is indispensable to eukaryotes, as the 16,569 base-pair, circular mitochondrial genome encodes essential elements of respiration. Over evolutionary time, the original proto-mitochondrial genome has gradually migrated to the nucleus, though 37 genes remain in human mitochondrial DNA (mtDNA). These genes underscore the semi-autonomous state of mitochondria, encoding a mitochondrial-specific repertoire of 22 tRNA, 2 rRNA, and 13 of the constituents of the electron transport chain (ETC). The mitochondrial genome resembles its prokaryotic ancestor DNA structurally; in humans, mtDNA is polycistronic, with a transcription start site located on each strand within the noncoding, regulatory displacement loop ("D-loop"). Unequal nucleotide distributions distinguish the guanine-rich, "heavy" strand, which is also more gene dense, encoding 28 genes.

7.1.1 Detection and Definitions of mtDNA Variation

A single human cell can house hundreds of mitochondria, each bearing several copies of the mitochondrial genome. Mitochondrial DNA "homoplasmy" denotes the existence of a single intracellular sequence identity. Conversely, "heteroplasmy" implies multiple unique sets of mtDNA in a given cell or tissue. Homoplasmy, until recently, was considered the normal state of healthy cells. However, with advancements in mtDNA mutation detection technology, mtDNA subpopulations are

W.J. Valente • J.H. Bielas (✉)
Fred Hutchinson Cancer Research Center, Seattle, WA, USA
e-mail: jbielas@fhcrc.org

D.M. Hockenbery (ed.), *Mitochondria and Cell Death*, Cell Death in Biology and Diseases, DOI 10.1007/978-1-4939-3612-0_7

X.-M. Yin and Z. Dong (Series eds.), *Cell Death in Biology and Diseases*

ubiquitously identified in cells and tissues, eroding the previous concept of homoplasmy (Ye et al. 2014; Payne et al. 2013).

The technological limitations found in early studies drastically affected the measurement resolution of mtDNA mutation studies, which erroneously described the mitochondrial genome as homogenous sequences within an individual. In contrast, current-day assays of mtDNA sequence and mutation frequency have enhanced sensitivity that can detect mutant mtDNA at frequencies of 1 mutant in 10^8 base-pairs or less (Taylor et al. 2013; Ericson et al. 2012; Vermulst et al. 2008). With these refined techniques, surveys of mitochondrial genomes have revealed widespread heteroplasmy, indicating "homoplasmy" is dependent on assay measurement resolution (Ye et al. 2014; Payne et al. 2013; He et al. 2010).

This revelation necessitates an adjustment to the definition of homoplasmy. Given the generally accepted Sanger sequencing limit of detection for mtDNA minor variants, typically considered 1 in 20 wild-type base-pairs (Melton et al. 2012), homoplasmy would be better defined as nucleotide concordance of 95 % or greater at a single mtDNA loci. Beneath this threshold, mtDNA would be considered heteroplasmic. This reevaluated definition of homoplasmy not only incorporates the technological limitations of Sanger sequencing (commonly used in mutation research), but also mitochondrial disease genetics. The percent mutant mtDNA required for some pathological manifestations of inherited mitochondrial disorders can exceed 90 % (Yoneda et al. 1995; Santorelli et al. 1993). Thus, the requirement of 95 % sequence identity within our practical definition of homoplasmy unites the detection threshold of a technology frequently used in describing mtDNA mutations and the observed mutant mtDNA fractions in mitochondrial pathologies (Fig. 7.1).

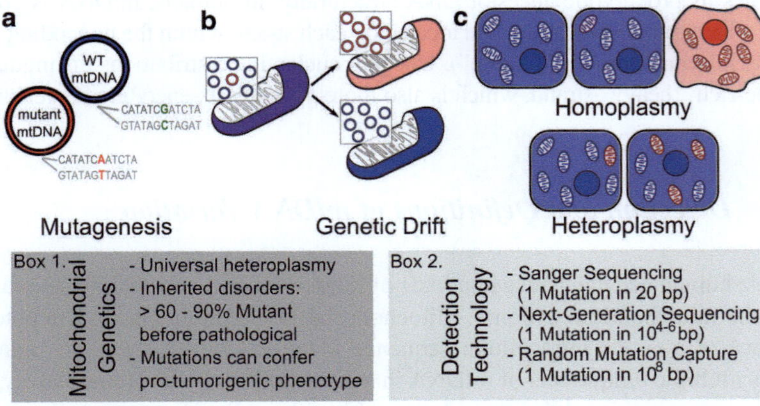

Fig. 7.1 Levels of Mitochondrial DNA Mutagenesis. (**a**) Mutations can occur at the level of individual mtDNA. (**b**) These mutations, within a single mitochondrion, can expand through drift or other factors to encompass the entirety of the sequence diversity within a mitochondrion. (**c**) Through replication, with stochastic processes or selection, cells can become enriched for distinct sequences. This leads to the designations of "homoplasmy" and "heteroplasmy" at the cellular and tissue levels. A practical definition of homoplasmy is a requirement for 95 % concordance at a given loci, which reflects (*Box 1*) technological and (*Box 2*) biological considerations of detection and phenotype penetrance, respectively

7.1.2 *Roles of mtDNA in Metabolism and Human Health*

At 16,569 base-pairs, the human mitochondrial genome may be dwarfed by the comparably expansive nuclear genome, but its impact on cellular processes cannot be overstated. The host of hereditary mtDNA disorders described underscores the health consequences of aberrant mtDNA (Mitomap 2014). Both heteroplasmic and homoplasmic variants in mtDNA have been associated with neurodegenerative disorders, metabolic syndromes, and cancers. However, establishing whether mtDNA mutations are active players or neutral passengers in oncogenesis has proven difficult.

In this chapter, we describe the physiologic consequences of somatic mtDNA mutations in cancer, the spectrum of mtDNA mutations found in malignancy, their pathogenesis, and the utility of mtDNA variation in patient prognosis. Finally, we consider how mitochondrial mutagenesis in cancer should be considered in the context of cellular fitness and its importance to neoplastic change.

7.2 Mitochondrial DNA Mutations in Cancer

Mitochondria have a long-standing association with cancer. In 1930, Otto Warburg first described the aberrant cellular metabolism of tumor cells, noting their reliance on glycolysis even in the presence of oxygen (Warburg et al. 1930). Although the hypothesis that mitochondrial dysfunction causes cancer (Warburg 1956) has not survived the critique of newer lines of evidence (Hanahan et al. 2011), a mitochondrial link to cancer is still intact, and has benefited from contemporary evidence describing mtDNA mutations' possible roles in carcinogenesis. In one of the first surveys of mtDNA in neoplasia, somatically acquired mutations in mtDNA were identified in 7 of 10 human colorectal carcinoma cell lines, the majority being homoplasmic (Polyak et al. 1998). This revelation, amongst others (Horton et al. 1996; Alonso et al. 1997; Habano et al. 1999), reconnected Warburg's decades-old observations of altered cancer cell metabolism with a possible genetic cause, re-instilling interest into the relevance of mitochondrial dysfunction in cancer, and inspiring a wave of mtDNA mutation analysis in tumors.

7.2.1 *Methodological Considerations in mtDNA Analysis*

Many early studies of sequence variation in mtDNA associated with cancer have been questioned, due to fundamental errors in experimental design and execution (Salas et al. 2005). Rampant errors in sample preparation (Fliss et al. 2000; Jerónimo et al. 2001), inappropriate controls (Yeh et al. 2000) (Nishikawa et al. 2001), and interpretation of results (Prior et al. 2006; Tan et al. 2008) cast doubt on the purported role of mtDNA in tumorigenesis (Salas et al. 2005; Bandelt et al. 2009). In this chapter, we have assessed reported studies of mtDNA mutations and cancer for their methodological integrity. We apply quality criteria established by Salas and colleagues (Salas et al. 2005) to ensure cited study designs have included comparisons to normal control tissue for variant analysis, and tests for haplogroup-specific variants to avoid false misclassification of inherited polymorphisms as somatic mutations.

7.2.2 *Widespread mtDNA Alterations in Cancer*

When reporting studies of mtDNA in cancer, we used a set of established criteria (Salas et al. 2005) to screen studies which emphasize proper use of matched normal controls, exclusion of germline polymorphisms, and checks for sample mix-up by matching tumor and normal tissues with shared variants. Including these stringencies, mtDNA mutations have been detected in: breast (Li et al. 2014), colorectal (Ericson et al. 2012; Polyak et al. 1998; Alonso et al. 1997; Habano et al. 1999; Lièvre et al. 2005; Park et al. 2009), prostate (Petros et al. 2005; Kloss-Brandstatter et al. 2010; Gomez-Zaera et al. 2006), pancreatic (Kassauei et al. 2006), gastric (Alonso et al. 1997; Bi et al. 2011; Lee et al. 2014), lung (Yang Ai et al. 2013; Jin et al. 2007), ovarian (Guerra et al. 2012; Liu et al. 2001), brain (Larman et al. 2012), and thyroid (Abu-Amero et al. 2005; Costa-Guda et al. 2007; Gasparre et al. 2007; Maximo et al. 2002) cancers, among others, including non-solid tumors (Carew et al. 2003; He et al. 2003). While these studies all identified the presence of mtDNA mutations in cancer cells, the frequencies of mtDNA mutations found for individual tumors as well as within different study cohorts are highly variable between reports. Even within the same cancer subtype, not only do the reported percentages of tumors harboring mtDNA mutations vary significantly (93 % (Zhu et al. 2005) vs. 30 % (Tseng et al. 2011)), but so do the genomic locations and the types of mutations observed.

7.2.2.1 Mutations by mtDNA Loci

The mitochondrial genome can be subdivided broadly by functionality and transcript identity into regulatory, polypeptide-encoding, and untranslated categories. Within each division, individual mutations may have profoundly different effects (Abu-Amero et al. 2005; Carew et al. 2003; Gasparre et al. 2011; Iommarini et al. 2013), yet this simple organizational strategy enjoys some benefits. The comparable ease in describing mutations in ETC genes, versus the more difficult and perhaps tenuous determination of pathological alterations to regulatory (D-loop) and untranslated (rRNA, tRNA) genes supports our grouping of mutations by their loci and gene product.

D-loop: The mitochondrial D-loop contains the start site for replication, and has been known to forensic science for its two hypervariable regions (HV1 and HV2) that represent common sites of divergence in human mtDNA phylogeny

(Miller et al. 1996; Stoneking et al. 1991). Foundational studies focused on the D-loop as an area of rapid evolution ideal for targeted investigation, leading to observations of the region's high mutation burden in cancers (Alonso et al. 1997; Maximo et al. 2002; Parrella et al. 2001). However, these conclusions were drawn based upon studies with systematic errors in mutation calling (Jerónimo et al. 2001; Yeh et al. 2000; Nishikawa et al. 2001).

Early identification of "cancer-specific mutations" in the D-loop, as elsewhere in the mitochondrial genome, was fraught with faulty assignment of inherited polymorphisms for pathogenic mutants (described by (Salas et al. 2005)). In assigning variant status, inherited polymorphisms should be accounted for by first determining the mtDNA variants present in normal tissue. These inherited variants are summarily discounted from the analysis of somatic mutations.

In contemporary mutation studies with conscious removal of germline polymorphisms, tumor-specific D-loop variants are recurrently identified (Yang Ai et al. 2013; Tseng et al. 2011; Hung et al. 2010; Ding et al. 2010). The functional significance of these noncoding mutations is debated, but is suggested to influence intracellular mtDNA copy number maintenance (Yamada et al. 2006; Yu et al. 2007). However, in a survey of 101 thyroid lesions, similar D-loop point mutations and microsatellite slippage have been observed between malignant thyroid carcinomas and nodular goiters (Ding et al. 2010). It appears D-loop mutations are not unique to neoplasia. However, given that the D-loop contains the start sites of mtDNA replication and transcription, mutations in this region have the potential for drastic effects on the mitochondrion, and may play a role during tumorigenesis.

ETC Subunits: The anticipated consequence of homoplasmic, non-synonymous mutation in a gene encoding any one of the ETC subunits is a potentially catastrophic impairment of cellular metabolism. Focusing on the protein-encoding mtDNA variants in cancer, an early report described a 50 % heteroplasmic deletion in complex I in renal adenocarcinoma (Horton et al. 1996). Interestingly, the authors discovered preferential expression of the truncated transcript via Northern analysis, and did not observe the deletion mutation in any of the five metastases or normal tissue paired to the primary tumor. Mutations in ETC subunits have been found in many cancers, especially in complex I (Gasparre et al. 2007; Maximo et al. 2002; Gasparre et al. 2008; Mayr et al. 2008), and may be characterized by markedly high mutant heteroplasmy, even approaching homoplasmy in some breast cancers (McMahon et al. 2014). Elimination of crucial components of the initial ETC complex would seem antithetical to traditionally associated characteristics of cancer, namely high energy requirements, rapid cellular expansion and dissemination. Interestingly, the degree of dysfunction within respiratory complex I can drastically modify tumor phenotype and progression (Iommarini et al. 2013).

In a study of mtDNA mutations in thyroid carcinoma (Maximo et al. 2002), researchers analyzed 70 % of the mitochondrial genome, including all gene-encoding regions. In the 43 Hürthle and 36 non-Hürthle cell neoplasms assayed, 51.5 % displayed at least one somatic mutation within an ETC polypeptide gene. Of those mutations, more than half resulted in amino acid changes, yet this is less than the percentage expected if there was no selection upon mtDNA mutants within protein-coding regions—evidence of negative selection. Similarly, an investigation of mtDNA mutation frequency in lung cancer identified non-synonymous changes to protein-coding genes in 30.1 % (19/63) of tumor samples (Jin et al. 2007).

tRNA: Variations in tRNA can be associated with non-pathogenic, inherited polymorphisms in addition to mitochondrial disorders such as Myoclonic Epilepsy with Ragged Red Fibers (MERRF) syndrome or Mitochondrial Encephalomyopathy, Lactic Acidosis, and Stroke-like episodes (MELAS) (catalogued in (Mitomap 2014)). Some inherited mutant mitochondrial tRNAs produce debilitating disease, as in MELAS (Corona et al. 2001), whereas others may impart little or no detrimental effect (McFarland et al. 2004). In line with observations in inherited diseases, ascribing pathogenicity, let alone pro-carcinogenic function to tRNA variants is an especially difficult task.

Following on reports of mtDNA susceptibility to nuclear mutagens (Backer et al. 1980; Backer et al. 1982), Taira and colleagues investigated the possibility of tumor-associated mutations in rat cancer cell line mtDNA (Taira et al. 1983). In their study of tumor cell mtDNA, mutations were found in functional regions of three mitochondrial tRNA genes. More recently, in a cell culture model of mutant tRNA, an insertion mutant within the $tRNA^{Arg}$ gene found in murine skin tumors showed enhanced resistance to UV-induced apoptosis, faster proliferation rates, and improved motility (Jandova et al. 2012). This would suggest tRNA mutations influence growth and survival characteristics in cells, and possibly in cancer as well. In humans, sequence alterations in tRNA have been reported in a variety of cancers (Tseng et al. 2011; Dasgupta et al. 2010; Mambo et al. 2003), but their functional significance has been difficult to assess.

7.2.2.2 Mutation Identity

Point Mutations: Single substitutions are the most common variant detected in studies of mtDNA mutations in cancer (Larman et al. 2012; Ju et al. 2014), and are pervasive even in small cohort studies of single tumor types (Kassauei et al. 2006; Tseng et al. 2011; Yin et al. 2010). In an investigation of tumor-specific variants within five tumor types present in The Cancer Genome Atlas (TCGA), paired normal and tumor tissues were assayed for mutations (Larman et al. 2012). In protein coding regions, non-synonymous mutations represented 86 % of somatic mutations, yet only 31 % of inherited variants; and of the non-synonymous variants, only 3 % of inherited versus 50 % of somatic mutations were predicted to have high impact on protein function. This suggests somatic mtDNA point mutations in cancers represent a marked departure from the constraints of germline polymorphisms, existing under separate selection pressures and/or distinct conditions of mutation susceptibility.

Insertions and Deletions: Increased deletions in mtDNA are commonly associated with aging, and clonally expand in both post-mitotic and replicating tissues (Payne et al. 2013; Taylor et al. 2013; Gasparre et al. 2008; Nekhaeva et al. 2002). The mitochondrial "common deletion" represents a region of mtDNA more prone to deletions due to homology-driven lariat formation during replication, and subsequent degradation of the unhybridized ssDNA, excising a 4977 base-pair portion of DNA encoding five tRNA genes and seven ETC constituents. While this large deletion is present in skin squamous cell carcinoma, gastric cancer, colorectal carcinoma and lung carcinoma among others, it must be stressed that mtDNA common deletions were observed at a lower frequency in tumor tissue compared to the surrounding normal tissue (reviewed (Kamalidehghan et al. 2013)).

Smaller deletion events (<1 kb) are apparently not under such strict negative selection in tumor tissue. In hepatocellular carcinoma, two studies have observed tumor mtDNA has a higher prevalence of genomes harboring small deletions (50–70 bp) (Yin et al. 2010; Lee et al. 2004). In a recent cohort study, there were no truncating mutations among the 1262 inherited mtDNA variants observed in a group of 266 cancer patients, whereas deletions represented 39 of the 162 mutations within their tumors (Larman et al. 2012). Interestingly, 65 % of truncating mutations in tumors occurred in *ND5*, often clustered around an 8 base-pair repetitive sequence tract, implicating slippage during mitochondrial replication or sequencing protocol. The expansion of small, rather than large deletions in tumors would suggest some level of selection for mitochondrial function during transformation, perhaps with relaxed pressure in the subsequent neoplastic progression.

7.3 How Mitochondrial DNA Mutations Arise

Many theories have arisen to explain mitochondrial mutagenesis in cancer, and deserve attention for their mechanistic implications. One pivotal consideration that we will discuss is how these theories describe cancer-specific mutants arising from low-level heteroplasmy. As some have suggested, the cancer-specific homoplasmic mutations reported in early studies may represent biased selection and clonal expansion of low-level, mutant mtDNA copies (Salas et al. 2005; Bandelt et al. 2009). We review how, in concert with selection or genetic drift, there is also evidence to suggest mutations arise primarily dependent on spontaneous chemical reactions during mtDNA replication (Ju et al. 2014). Finally, we consider the reputed susceptibility of the mitochondrial genome to genotoxic exposures from endogenous and exogenous sources of mutations.

7.3.1 Selective Pressure

The majority of the somatic mutations observed in cancers are non-synonymous variants when observed in protein coding regions (Larman et al. 2012). Additionally, the proportion of protein-altering mutants is much greater in somatic mutations compared to early developmental or inherited polymorphisms. This finding raises interest in defining how selective pressure could be involved in mtDNA mutagenesis during tumor progression.

Mutation frequencies, and the observed distributions of specific mutation types, for inherited variants and somatic DNA mutations are drastically different. Evidence suggests germline transmission of mtDNA is under strong purifying selection (Stewart et al. 2008). In transmitted mtDNA, mutation frequencies are higher at third codon positions, explained by the variably specific mRNA:anticodon interactions at this position during peptide synthesis (Crick 1966). These results seemingly contrast with the supposed pro-tumorigenic effects of mtDNA mutations observed in many cancers.

In the protein-coding regions of tumor mtDNA, negative selective pressure against mutation is either relaxed or may be positive (Gasparre et al. 2008; Brandon et al. 2006; Shidara et al. 2005). However, negative selection is almost universally present on large deletions. Cell survival depends on mitochondrial function, and mutations with adverse implications for cellular bioenergetics will likely be selected against based on cellular fitness (Mishmar et al. 2003; Ruiz-Pesini et al. 2004; Stafford et al. 2010). For example, a meta-analysis of studies describing the prevalence of the age-associated mitochondrial "common deletion" in breast cancer, which removes 7 peptide-coding genes, revealed it is under significant negative selection pressure compared to normal tissue (Nie et al. 2013). "Common deletion" prevalence is inversely correlated with advancing colorectal cancer stage (Chen et al. 2011), indicating it may pose fitness challenges to neoplasms during tumor progression.

In contrast, many single gene mutations seemingly facilitate aggressive and resilient cancer phenotypes that conceivably increase cancer cell fitness. Complex I mutant tumors, with a heteroplasmic frame-shift in *ND5*, have shown enhanced colony formation and in vivo tumor growth when injected into nude mice (Park et al. 2009). A homoplasmic point mutation in the ATP synthase subunit 6 (ATP6), typically associated with inherited mitochondrial encephalomyopathy, conferred apoptotic resistance to a human ovarian cancer cell line (Shidara et al. 2005). Notably, expression of wild-type *ATP6* abrogated growth advantages in the mutant cells, and expression of mutant *ATP6* in a wild-type background accelerated tumor growth and diminished respiration (Shidara et al. 2005). Much of the enrichment for mutant mtDNA in cancers appears predicated on functional consequences of these mutations, implying selection primarily occurs at the cellular level at a point when mutations have aggregated in sufficient concentration to alter tumor phenotype. Greater discussion of the physiological consequences of mtDNA mutations in cancer will be presented later in this chapter.

7.3.2 *Mitochondrial Genomic Drift*

Mutations in mtDNA, even pathogenic mutations, are relatively common in the human population at low levels (Nekhaeva et al. 2002; Elliott et al. 2008). Tumor mtDNA mutants, in models lacking selection bias, can originate from low-level heteroplasmic mutants (Nekhaeva et al. 2002; Coller et al. 2001; Taylor et al. 2003).

Given the stem cell-like nature of many tumors, the many cellular divisions preceding transformation may allow for stochastic isolation and enrichment of a unique mtDNA genome without any need for growth advantages or tumorigenic phenotypes derived from that mtDNA mutant (Coller et al. 2001). In mathematical models of this phenomenon, random processes predict that the majority of tumors (58 %) will harbor homoplasmic mtDNA mutations given the growth parameters of many tumors. Patient-derived samples support this model with instances of germ-line mutant mtDNA expanding to homoplasmic abundance in head and neck squamous cancer (Gekeler et al. 2009). However, mtDNA mutations' random drift in tumorigenesis has been called into question recently (Stafford et al. 2010).

Sequencing datasets offering comparisons across tumors show predominantly relaxed negative selection for somatic mtDNA mutations relative to the germline, but possible positive selection at residues of high functionality (McMahon et al. 2014). A large cohort analysis observed strong negative selection pressure on truncating mtDNA mutations and on tRNA, especially within anticodon sequences (Ju et al. 2014). Despite the many descriptions of individual mtDNA mutations facilitating cancers with ostensible fitness benefits, selection pressures within tumors are predominantly negative or neutral for many mtDNA mutations (Ju et al. 2014; Stewart et al. 2008; Stafford et al. 2010), and positive selection has not been demonstrated in a large study (>1000 patients) (McMahon et al. 2014).

7.3.3 *The Role of Mutagen Exposure*

The primary mtDNA mutations observed in cancer and normal tissues are C→T and T→C transitions, with higher proportion of the former on the heavy strand and the latter on the light strand (Ju et al. 2014). The mtDNA leading strand during replication is the heavy strand. The strand-biased distribution of mutations could be coupled to events during replication. Lending credence to this assertion, C→T transitions are consistent with base mis-pairing after deamination of cytosine, a reaction 140-fold faster in single-stranded DNA (Frederico et al. 1990). C→T/G→A transitions are common in noncancerous tissue, as seen in human colon (Taylor et al. 2003; Greaves et al. 2014) and in colorectal adenocarcinoma (Ericson et al. 2012). However, in colorectal cancer, the frequency of mtDNA mutations is reduced when compared to normal or adenomatous tissue, and most notably C→T mutations. This mutation type is associated with ROS DNA damage, and as mitochondria are a primary source of ROS from OXPHOS, this leads to the hypothesis that metabolic changes during cancer progression (e.g., the reduction in OXPHOS represented by the Warburg effect) contribute to mtDNA mutation. In support of this hypothesis is the concomitant decrease of OXPHOS and mtDNA mutagenesis (Ericson et al. 2012), whereby decreased OXPHOS-derived ROS results in increased mitochondrial genome fidelity.

In addition to endogenous sources of mutagens, namely ROS, the mitochondrial genome is noted for its sensitivity to environmental toxins (Backer et al. 1980; Backer et al. 1982) and genotoxic agents employed in chemotherapy (LeDoux et al. 1992; Murata et al. 1990; Wisnovsky et al. 2013; Yang et al. 2006; Gogvadze et al. 2009). By extension, nuclear carcinogens might play a role in the genesis of mtDNA mutations. In early reports derived from observations in buccal epithelium swabs of patients with smoking-associated lung carcinomas, the frequency of mtDNA mutations was significantly elevated over normal tissue (Prior et al. 2006; Tan et al. 2008). These results in have attracted scrutiny (Bandelt et al. 2009), and the inability to observe mtDNA mutational susceptibility in smokers (Lai et al. 2013) has since obscured the possible role of tobacco-derived carcinogens in mtDNA mutagenesis. In a recent study of oral squamous cell carcinoma in Taiwanese patients, neither smoking status nor areca quid chewing had a significant effect on the presence of mutations across the mitochondrial genome (Lai et al. 2013). Considering other classic nuclear mutagens, there has been some suggestion that UV exposure may precipitate neoplastic change and mtDNA mutation in squamous carcinoma of the skin (Berneburg et al. 1999). A mutation in mtDNA derived from murine skin cancer produced enhanced cellular growth kinetics, but whether UV irradiation or the transformation process caused the mutation is unclear (Jandova et al. 2012). Thus far few studies link known nuclear mutagens to increased mitochondrial mutagenesis (Mambo et al. 2003; Stumpf et al. 2014). Thus it appears, despite the elevated spontaneous mitochondrial genome mutation frequency relative to nDNA, mtDNA is remarkably resistant to induced mutation when exposed to established nDNA mutagenic compounds (Ju et al. 2014; Mita et al. 1988).

7.4 Tumorigenic Effects of mtDNA Mutations

The prevalence of mtDNA mutations in cancers suggests that these mutations might play a causative role in tumorigenesis or contribute to cancer cell phenotypes through mitochondrial dysfunction (Iommarini et al. 2013; Sharma et al. 2011), destabilization of the respiration/glycolysis balance (Lee et al. 2014; Jandova et al. 2012; Hu et al. 2013), and dysregulation of apoptosis (Park et al. 2009; Gasparre et al. 2011; Jandova et al. 2012; Lemarie et al. 2011). Mutations in cancer mtDNA are common (Larman et al. 2012; McMahon et al. 2014; Ju et al. 2014), and advances in detection technologies have solidified the association of cancers with mtDNA mutations. However, doubt remains regarding mtDNA mutations' specific relationship to the physiologic alterations observed in neoplastic change.

Compounding the difficulties with assigning function to mtDNA variants is the uniquely mitochondrial consideration of heteroplasmic proportion. Gasparre and colleagues have shown that the presence of an insertion mutant into *ND1* carries a concentration-dependent role in cancer cells. For tumors that displayed above 83 % mutant heteroplasmy the pro-tumorigenic effects of slightly less heteroplasmic complex I mutation were lost, resulting in a less aggressive, oncocytic phenotype (Gasparre et al. 2011). These observations underscore the importance of heteroplasmic proportion to mtDNA mutations' role in tumorigenesis.

7.4.1 Cancer Establishment and Adaptation

Cybrids represent a widely used tool for exploring the phenotypic consequences of pathogenic mtDNA mutations. Briefly, the process involves depleting the native mtDNA populations of a cell using mtDNA-specific toxins such as ethidium bromide. These ρ_0 cells are fused with enucleated cells, cytoplasts, derived from a cellular source known to harbor an mtDNA mutation. In this way, the nuclear backgrounds of test cells are standardized, and the mtDNA effects can be discerned in the absence of possible confounding factors. By creating cybrids harboring pathogenic variants in mtDNA associated with Leigh Syndrome (Dahl 1998) in HeLa cells, Shidara and colleagues were able to demonstrate the mutant mtDNA conferred both early growth advantages, as well as a lowered frequency of apoptosis, both in vitro and in vivo (Shidara et al. 2005). Similar growth advantages have been discovered in other cybrids (Jandova et al. 2012), indicating mtDNA mutations may play roles in both early and adaptive stages of tumorigenesis.

In prostate cancer, mtDNA mutations may encourage treatment-resistant adaptations. Initially, prostate tumors are androgen-dependent for growth and progress to androgen-independence after treatment with androgen-ablation therapy. Large deletions have been identified at higher frequency in the mtDNA of androgen-independent prostate cancer cell lines (Higuchi et al. 2005). Mutations in mtDNA have also been associated with chemoresistance and anchorage-independent growth, both clinically relevant cancer adaptations correlated with poorer patient outcomes.

7.4.2 Metastatic Disease and mtDNA Mutation

Experiments that utilize cybrids have demonstrated that mtDNA mutations can be pro-metastatic (Ishikawa et al. 2008; Ohta 2006), and suggests their involvement in carcinogenesis. MtDNA mutations that conferred high metastatic potential in these studies displayed concomitant resistance to chemotherapy-induced apoptosis, highlighting the clinical relevance of specific mtDNA mutations. Ishikawa and colleagues used this technique to transfer the mtDNA background of a highly metastatic cancer cell line into a cell line with low metastatic potential when injected into nude mice. The resultant increase in metastatic potential in the previously low metastatic potential cell line highlighted the role pathogenic mutations in mtDNA can play in tumor progression. This same group also found that replacing the mtDNA of a highly metastatic breast cancer cell line, MDA-MB-231, with normal mtDNA restored normal respiratory capacity and suppressed metastatic potential; though, critically, without affecting the ROS production observed in the cell line (Imanishi et al. 2011). This observation suggests that mtDNA-mediated metastatic pathways can exist in a ROS-independent function. Thus, there may be multiple mechanisms by which mtDNA mutations' augment metastatic potential.

7.4.3 Resistance to Apoptosis

Mitochondria are intrinsically linked to apoptosis, and so it is not surprising that mutations in mtDNA can modulate cellular death. Mutations in *ND4* have notably produced apoptotic resistance to chemotherapy in ovarian carcinoma (Guerra et al. 2012). The proposed mechanism of action for this cancer phenotype is explored in Chap. 7 in more detail, though it bears mention here that mtDNA mutations can produce profound effects on cancer metabolism, some of which may be protective. Many chemotherapeutics have been shown to preferentially interact with mtDNA over nDNA (Gogvadze et al. 2009; Montero et al. 2011; Palorini et al. 2013; Qian et al. 2005), as well as increase ROS, which could explain why changes to intracellular ROS secondary to mtDNA mutations are associated with chemoresistance (Yu et al. 2009).

7.5 Mutations as Cancer Biomarkers

Relating findings to disease status and patient outcomes motivates the analysis of mtDNA mutations in cancer (Kulawiec et al. 2010). The ability to use mtDNA mutations as biomarkers during the diagnosis, prognosis, and treatment of cancer could bring significant benefits to clinical practice and patient management.

As mentioned previously, mtDNA mutations can impact chemosensitivity (reviewed in (Yu 2011)), as well as promote aggressive behavior (Maybury 2013). Changes in mtDNA sequence associated with specific clinical manifestations and patient prognosis are numerous. Patients with type III or IV gastric cancers bearing D-loop mutations were found to have poor prognoses and lower 5-year survival rates after gastric resection (Wu et al. 2005). Increased mutations in mtDNA are associated with higher histological grade of breast cancer (Yu et al. 2007) and advanced staging in lung cancer (Matsuyama et al. 2003). The presence of pathogenic mtDNA mutations, defined as gross changes to sequence (≥50 bp insertions and deletions) or alterations to evolutionarily conserved nucleotides and encoded amino acids with computationally predicted deleterious effects, in oral squamous cell carcinoma (OSCC) was associated with poorer disease-free survival in those patients that received adjuvant chemotherapy after surgery (Lai et al. 2013). Similar to the findings in OSCC, patients with colorectal tumors bearing D-loop mutations had poorer 3-year survival and did not benefit from adjuvant chemotherapy (Lièvre et al. 2005).

However, an analysis of head and neck cancer found D-loop mutations to be infrequent, present in only 18 % (6 of 34) of tissue samples, and not correlated with patient outcomes (Challen et al. 2011). Yet a similar study in head and neck cancer found mtDNA mutation enrichment in patient samples, with almost half displaying some mtDNA mutation (Dasgupta et al. 2010). Currently, clinical implications of mtDNA mutation presence in esophageal or colon cancer are debatable (Larman et al. 2012; Hung et al. 2010). Moreover, despite the numerous observations of mtDNA mutations fostering pro-tumorigenic phenotypes, data describing positive selection of mutations across cancers is distinctly lacking (McMahon et al. 2014; Ju et al. 2014; Stafford et al. 2010). Thus the true utility of mtDNA mutations as a cancer biomarker is a contentious assertion, though it clearly warrants further investigation.

7.6 Concluding Remarks and Future Directions

From its early associations with cancer, through a period of relative stagnation before a recent explosion in interest, the study of mitochondria and mtDNA in cancer has benefitted from marked improvements in detection technologies, methods for phenotypic characterization, and data processing. However, a critical separation exists between the study of mtDNA and nDNA: thus far, there is no reliable methodology for site-directed mutagenesis in the mammalian mitochondrial genome. This singular deficiency in the field significantly hampers the capacity to validate the functional consequences of the diverse set of mtDNA mutations found in cancer. With an established technique for directed mutagenesis, the cellular effects of individual mutations could be evaluated without reliance on patient-derived mitochondria, as is requisite for cybrid technology. A mutation observed in a tumor could be thus evaluated in a cell line with defined physiology, revealing the nuclear genome-independent phenotype conferred by an mtDNA mutation. Additionally, the dynamics of an mtDNA mutation in a population of cells could be assayed in a controlled environment, which could reveal the role of selection in tumor progression or adaptation.

Studies have shown the ratio of somatic missense to silent substitutions in cancer is between 1.4 and 2.1 times higher in mtDNA than in nDNA (Greenman et al. 2006; Nik-Zainal et al. 2012). This is even more impressive considering the number of mutations relative to the disparate genomic size and gene density between mtDNA and nDNA. Averaging even just one somatic mutation within 16,569 base-pairs, extending the mutation frequency of mtDNA in many cancers to just the nuclear exome (~30 million base-pairs) would yield more than 1800 observed mutations. This mutant burden is nearly 40 times greater than the frequency found in whole-exome deep sequencing of tumors (Li et al. 2014; Zang et al. 2012). The biological processes that link the rise in mtDNA mutations during oncogenesis are currently unresolved. To answer these questions, future investigative efforts must progress from correlative studies to identification of causal mechanisms in order to rectify the conflicted understanding of mtDNA's role in cancer.

References

Abu-Amero K et al (2005) High frequency of somatic mitochondrial DNA mutations in human thyroid carcinomas and complex I respiratory defect in thyroid cancer cell lines. Oncogene 24(8):1455–1460

Alonso A et al (1997) Detection of somatic mutations in the mitochondrial DNA control region of colorectal and gastric tumors by heteroduplex and single-strand conformation analysis. Electrophoresis 18(5):682–685

Backer JM et al (1980) Mitochondrial DNA is a major cellular target for a dihydrodiol-epoxide derivative of benzo[a]pyrene. Science 209:297–299
Backer JM et al (1982) Interaction of benzo(a)pyrene and its dihydrodiol-epoxide derivative with nuclear and mitochondrial DNA in C3h10t½ cell cultures. Cancer Res 42(7):2764–2769
Bandelt H et al (2009) Contamination and sample mix-up can best explain some patterns of mtDNA instabilities in buccal cells and oral squamous cell carcinoma. BMC Cancer 9:113
Berneburg M et al (1999) Singlet oxygen mediates the UVA-induced generation of the photoaging-associated mitochondrial common deletion. J Biol Chem 274(22):15345–15349
Bi R et al (2011) Rapid identification of mtDNA somatic mutations in gastric cancer tissues based on the mtDNA phylogeny. Mutat Res 709–710:15–20
Brandon M et al (2006) Mitochondrial mutations in cancer. Oncogene 25(34):4647–4662
Carew JS et al (2003) Mitochondrial DNA mutations in primary leukemia cells after chemotherapy: clinical significance and therapeutic implications. Leukemia 17(8):1437–1447
Challen C et al (2011) Mitochondrial DNA mutations in head and neck cancer are infrequent and lack prognostic utility. Br J Cancer 104(8):1319–1324
Chen T et al (2011) The mitochondrial DNA 4,977-Bp deletion and its implication in copy number alteration in colorectal cancer. BMC Med Genet 12:8
Coller H et al (2001) High frequency of homoplasmic mitochondrial DNA mutations in human tumors can be explained without selection. Nat Genet 28(2):147–150
Corona P et al (2001) A novel mtDNA mutation in the Nd5 subunit of complex I in two MELAS patients. Ann Neurol 49(1):106–110
Costa-Guda J et al (2007) Mitochondrial DNA mutations in oxyphilic and chief cell parathyroid adenomas. BMC Endocr Disord 7:8
Crick FHC (1966) Codon—anticodon pairing: the wobble hypothesis. J Mol Biol 19(2):548–555
Dahl H-HM (1998) Getting to the nucleus of mitochondrial disorders: identification of respiratory chain–enzyme genes causing Leigh syndrome. Am J Hum Genet 63(6):1594–1597
Dasgupta S et al (2010) Mitochondrial DNA mutation in normal margins and tumors of recurrent head and neck squamous cell carcinoma patients. Cancer Prev Res 3(9):1205–1211
Ding Z et al (2010) Analysis of mitochondrial DNA mutations in D-loop region in thyroid lesions. Biochim Biophys Acta Gen Subj 1800(3):271–274
Elliott HR et al (2008) Pathogenic mitochondrial DNA mutations Are common in the general population. Am J Hum Genet 83(2):254–260
Ericson NG et al (2012) Decreased mitochondrial DNA mutagenesis in human colorectal cancer. PLoS Genet 8(6), e1002689
Fliss M et al (2000) Facile detection of mitochondrial DNA mutations in tumors and bodily fluids. Science 287(5460):2017–2019
Frederico LA et al (1990) A sensitive genetic assay for the detection of cytosine deamination: determination of rate constants and the activation energy. Biochemistry 29(10):2532–2537
Gasparre G et al (2007) Disruptive mitochondrial DNA mutations in complex I subunits Are markers of oncocytic phenotype in thyroid tumors. Proc Natl Acad Sci U S A 104(21): 9001–9006
Gasparre G et al (2008) Clonal expansion of mutated mitochondrial DNA is associated with tumor formation and complex I deficiency in the benign renal oncocytoma. Hum Mol Genet 17(7):986–995
Gasparre G et al (2011) A mutation threshold distinguishes the antitumorigenic effects of the mitochondrial gene Mtnd1, an oncojanus function. Cancer Res 71(19):6220–6229
Gekeler J et al (2009) Clonal expansion of different mtDNA variants without selective advantage in solid tumors. Mutat Res 662(1-2):28–32
Gogvadze V et al (2009) Mitochondria as targets for cancer chemotherapy. Semin Cancer Biol 19(1):57–66
Gomez-Zaera M et al (2006) Identification of somatic and germline mitochondrial DNA sequence variants in prostate cancer patients. Mutat Res 595(1-2):42–51
Greaves LC et al (2014) Clonal expansion of early to mid-life mitochondrial DNA point mutations drives mitochondrial dysfunction during human ageing. PLoS Genet 10(9), e1004620

Greenman C et al (2006) Statistical analysis of pathogenicity of somatic mutations in cancer. Genetics 173(4):2187–2198

Guerra F et al (2012) Mitochondrial DNA mutation in serous ovarian cancer: implications for mitochondria-coded genes in chemoresistance. J Clin Oncol 30(36):e373–e378

Habano W et al (1999) Mitochondrial gene mutation, but not large-scale deletion, is a feature of colorectal carcinomas with mitochondrial microsatellite instability. Int J Cancer 83(5):625–629

Hanahan D et al (2011) Hallmarks of cancer: the next generation. Cell 144(5):646–674

He L et al (2003) Somatic mitochondrial DNA mutations in adult-onset leukaemia. Leukemia 17(12):2487–2491

He Y et al (2010) Heteroplasmic mitochondrial DNA mutations in normal and tumour cells. Nature 464(7288):610–614

Higuchi M et al (2005) Mitochondrial DNA determines androgen dependence in prostate cancer cell lines. Oncogene 25(10):1437–1445

Horton TM et al (1996) Novel mitochondrial DNA deletion found in a renal cell carcinoma. Genes Chromosomes Cancer 15(2):95–101

Hu J et al (2013) Heterogeneity of tumor-induced gene expression changes in the human metabolic network. Nat Biotechnol 31(6):522–529

Hung WY et al (2010) Somatic mutations in mitochondrial genome and their potential roles in the progression of human gastric cancer. Biochim Biophys Acta 1800(3):264–270

Imanishi H et al (2011) Mitochondrial DNA mutations regulate metastasis of human breast cancer cells. PLoS One 6(8), e23401

Iommarini L et al (2013) Different mtDNA mutations modify tumor progression in dependence of the degree of respiratory complex I impairment. Hum Mol Genet 23(6):1453–1466

Ishikawa K et al (2008) Ros-generating mitochondrial DNA mutations can regulate tumor cell metastasis. Science 320(5876):661–664

Jandova J et al (2012) Somatic alterations in mitochondrial DNA produce changes in cell growth and metabolism supporting a tumorigenic phenotype. Biochim Biophys Acta (BBA)—Mol Basis Dis 1822(2):293–300

Jerónimo C et al (2001) Mitochondrial mutations in early stage prostate cancer and bodily fluids. Oncogene 20(37):5195–5198

Jin X et al (2007) Relationship between mitochondrial DNA mutations and clinical characteristics in human lung cancer. Mitochondrion 7(5):347–353

Ju YS et al. (2014) Origins and Functional Consequences of Somatic Mitochondrial DNA Mutations in Human Cancer. Elife. doi: 10.7554/eLife.02935.

Kamalidehghan B et al (2013) Pitfalls for common mitochondrial DNA deletion (DeltamtDNA4977) as a biomarker of cancer. Arch Med Res 44(1):79–80

Kassauei K et al (2006) Mitochondrial DNA mutations in pancreatic cancer. Int J Gastrointest Cancer 37(2-3):57–64

Kloss-Brandstatter A et al (2010) Somatic mutations throughout the entire mitochondrial genome Are associated with elevated Psa levels in prostate cancer patients. Am J Hum Genet 87(6):802–812

Kulawiec M et al (2010) Generation, function, and prognostic utility of somatic mitochondrial DNA mutations in cancer. Environ Mol Mutagen 51(5):427–439

Lai CH et al (2013) Clinical significance in oral cavity squamous cell carcinoma of pathogenic somatic mitochondrial mutations. PLoS One 8(6), e65578

Larman TC et al (2012) Spectrum of somatic mitochondrial mutations in five cancers. Proc Natl Acad Sci 109(35):14087–14091

LeDoux SP et al (1992) Repair of mitochondrial DNA after various types of DNA damage in Chinese hamster ovary cells. Carcinogenesis 13(11):1967–1973

Lee H-C et al (2004) Somatic mutations in the D-loop and decrease in the copy number of mitochondrial DNA in human hepatocellular carcinoma. Mutat Res 547(1–2):71–78

Lee HC et al (2014) Somatic alterations in mitochondrial DNA and mitochondrial dysfunction in gastric cancer progression. World J Gastroenterol 20(14):3950–3959

Lemarie A et al (2011) Mitochondrial respiratory chain complexes: apoptosis sensors mutated in cancer[quest]. Oncogene 30(38):3985–4003
Li LH et al (2014) Detection of mitochondrial DNA mutations by high-throughput sequencing in the blood of breast cancer patients. Int J Mol Med 33(1):77–82
Lièvre A et al (2005) Clinical value of mitochondrial mutations in colorectal cancer. J Clin Oncol 23(15):3517–3525
Liu V et al (2001) High incidence of somatic mitochondrial DNA mutations in human ovarian carcinomas. Cancer Res 61(16):5998–6001
Mambo E et al (2003) Electrophile and oxidant damage of mitochondrial DNA leading to rapid evolution of homoplasmic mutations. Proc Natl Acad Sci 100(4):1838–1843
Matsuyama W et al (2003) Mitochondrial DNA mutation correlates with stage progression and prognosis in non-small cell lung cancer. Hum Mutat 21(4):441–443
Maximo V et al (2002) Mitochondrial DNA somatic mutations (point mutations and large deletions) and mitochondrial DNA variants in human thyroid pathology: a study with emphasis on Hürthle cell tumors. Am J Pathol 160(5):1857–1865
Maybury BD (2013) Mitochondrial DNA damage is uncommon in cancer but can promote aggressive behaviour. Anticancer Res 33(9):3543–3552
Mayr JA et al (2008) Loss of complex I Due to mitochondrial DNA mutations in renal oncocytoma. Clin Cancer Res 14(8):2270–2275
McFarland R et al (2004) Assigning pathogenicity to mitochondrial trna mutations: when 'definitely Maybe' is Not good enough. Trends Genet 20(12):591–596
McMahon S et al (2014) Mutational patterns in the breast cancer mitochondrial genome, with clinical correlates. Carcinogenesis 35(5):1046–1054
Melton T, Holland C, Holland M (2012) Forensic mitochondrial DNA analysis: current practice and future potential. Forensic Sci Rev 24(101):102–122
Miller KWP et al (1996) A concordance of nucleotide substitutions in the first and second hypervariable segments of the human mtDNA control region. Int J Legal Med 109(3):107–113
Mishmar D et al (2003) Natural selection shaped regional mtDNA variation in humans. Proc Natl Acad Sci U S A 100(1):171–176
Mita S et al (1988) Resistance of Hela cell mitochondrial DNA to mutagenesis by chemical carcinogens. Cancer Res 48(16):4578–4583
Mitomap (2014) A human mitochondrial genome database. http://www.mitomap.org. Accessed 31 Oct 2014
Montero AJ et al (2011) Cellular redox pathways as a therapeutic target in the treatment of cancer. Drugs 71(11):1385–1396
Murata T et al (1990) Preferential binding of cisplatin to mitochondrial DNA and suppression of Atp generation in human malignant melanoma cells. Biochem Int 20(5):949–955
Nekhaeva E et al (2002) Clonally expanded mtDNA point mutations are abundant in individual cells of human tissues. Proc Natl Acad Sci 99(8):5521–5526
Nie H et al (2013) Mitochondrial common deletion, a potential biomarker for cancer occurrence, is selected against in cancer background: a meta-analysis of 38 studies. PLoS One 8(7), e67953
Nik-Zainal S et al (2012) Mutational processes molding the genomes of 21 breast cancers. Cell 149(5):979–993
Nishikawa M et al (2001) Somatic mutation of mitochondrial DNA in cancerous and noncancerous liver tissue in individuals with hepatocellular carcinoma. Cancer Res 61(5):1843–1845
Ohta S (2006) Contribution of somatic mutations in the mitochondrial genome to the development of cancer and tolerance against anticancer drugs. Oncogene 25(34):4768–4776
Palorini R et al (2013) Mitochondrial complex I inhibitors and forced oxidative phosphorylation synergize in inducing cancer cell death. Int J Cell Biol 2013:243876
Park JS et al (2009) A heteroplasmic, Not homoplasmic, mitochondrial DNA mutation promotes tumorigenesis Via alteration in reactive oxygen species generation and apoptosis. Hum Mol Genet 18(9):1578–1589
Parrella P et al (2001) Detection of mitochondrial DNA mutations in primary breast cancer and fine-needle aspirates. Cancer Res 61(20):7623–7626

Payne BA et al (2013) Universal heteroplasmy of human mitochondrial DNA. Hum Mol Genet 22(2):384–390

Petros J et al (2005) MtDNA mutations increase tumorigenicity in prostate cancer. Proc Natl Acad Sci U S A 102(3):719–724

Polyak K et al (1998) Somatic mutations of the mitochondrial genome in human colorectal tumours. Nat Genet 20(3):291–293

Prior SL et al (2006) Mitochondrial DNA mutations in oral squamous cell carcinoma. Carcinogenesis 27(5):945–950

Qian W et al (2005) Mitochondrial density determines the cellular sensitivity to cisplatin-induced cell death. Am J Physiol Cell Physiol 289(6):C1466–C1475

Ruiz-Pesini E et al (2004) Effects of purifying and adaptive selection on regional variation in human mtDNA. Science 303(5655):223–226

Salas A et al (2005) A critical reassessment of the role of mitochondria in tumorigenesis. PLoS Med 2(11), e296

Santorelli FM et al (1993) The mutation at Nt 8993 of mitochondrial DNA is a common cause of Leigh's syndrome. Ann Neurol 34(6):827–834

Sharma LK et al (2011) Mitochondrial respiratory complex I dysfunction promotes tumorigenesis through Ros alteration and Akt activation. Hum Mol Genet 20(23):4605–4616

Shidara Y et al (2005) Positive contribution of pathogenic mutations in the mitochondrial genome to the promotion of cancer by prevention from apoptosis. Cancer Res 65(5):1655–1663

Stafford P et al (2010) The pattern of natural selection in somatic cancer mutations of human mtDNA. J Hum Genet 55(9):605–612

Stewart JB et al (2008) Strong purifying selection in transmission of mammalian mitochondrial DNA. PLoS Biol 6(1), e10

Stoneking M et al (1991) Population variation of human mtDNA control region sequences detected by enzymatic amplification and sequence-specific oligonucleotide probes. Am J Hum Genet 48(2):370–382

Stumpf JD et al (2014) Mms exposure promotes increased mtDNA mutagenesis in the presence of replication-defective disease-associated DNA polymerase Γ variants. PLoS Genet 10(10), e1004748

Taira M et al (1983) Tumor-associated mutations of Rat mitochondrial transfer Rna genes. Nucleic Acids Res 11(6):1635–1644

Tan D et al (2008) Associations between cigarette smoking and mitochondrial DNA abnormalities in buccal cells. Carcinogenesis 29(6):1170–1177

Taylor RW et al (2003) Mitochondrial DNA Mutations in Human Colonic Crypt Stem Cells. J Clin Invest 112(9):1351–1360

Taylor SD et al (2013) Targeted enrichment and high-resolution digital profiling of mitochondrial DNA deletions in human brain. Aging Cell 13(1):29–38

Tseng LM et al (2011) Somatic mutations of the mitochondrial genome in human breast cancers. Genes Chromosomes Cancer 50(10):800–811

Vermulst M et al (2008) Quantification of random mutations in the mitochondrial genome. Methods 46(4):263–268

Warburg O (1956) On the origin of cancer cells. Science 123(3191):309–314

Warburg O, Posener K, Negelein E (1930) Ueber Den stoffwechsel Der tumoren. Biochem Z 152:319–344

Wisnovsky SP et al (2013) Targeting mitochondrial DNA with a platinum-based anticancer agent. Chem Biol 20(11):1323–1328

Wu C-W et al (2005) Mitochondrial DNA mutations and mitochondrial DNA depletion in gastric cancer. Genes Chromosom Cancer 44(1):19–28

Yamada S et al (2006) Correlation between copy number of mitochondrial DNA and clinico-pathologic parameters of hepatocellular carcinoma. Eur J Surg Oncol 32(3):303–307

Yang Ai SS et al (2013) Mitochondrial DNA mutations in exhaled breath condensate of patients with lung cancer. Respir Med 107(6):911–918

Yang Z et al (2006) Cisplatin preferentially binds mitochondrial DNA and voltage-dependent anion channel protein in the mitochondrial membrane of head and neck squamous cell carcinoma: possible role in apoptosis. Clin Cancer Res 12(19):5817–5825

Ye K et al (2014) Extensive pathogenicity of mitochondrial heteroplasmy in healthy human individuals. Proc Natl Acad Sci 111(29):10654–10659

Yeh J et al (2000) Somatic mitochondrial DNA (mtDNA) mutations in papillary thyroid carcinomas and differential mtDNA sequence variants in cases with thyroid tumours. Oncogene 19(16):2060–2066

Yin P et al (2010) Somatic mutations of mitochondrial genome in hepatocellular carcinoma. Mitochondrion 10(2):174–182

Yoneda M et al (1995) Heteroplasmic mitochondrial trna(Lys) mutation and its complementation in merrf patient-derived mitochondrial transformants. Muscle Nerve Suppl 3:S95–S101

Yu M (2011) Generation, function and diagnostic value of mitochondrial DNA copy number alterations in human cancers. Life Sci 89(3–4):65–71

Yu M et al (2007) Reduced mitochondrial DNA copy number is correlated with tumor progression and prognosis in Chinese breast cancer patients. IUBMB Life 59(7):450–457

Yu M et al (2009) Mitochondrial DNA depletion promotes impaired oxidative status and adaptive resistance to apoptosis in T47d breast cancer cells. Eur J Cancer Prev 18(6):445–457

Zang ZJ et al (2012) Exome sequencing of gastric adenocarcinoma identifies recurrent somatic mutations in cell adhesion and chromatin remodeling genes. Nat Genet 44(5):570–574

Zhu W et al (2005) Mitochondrial DNA mutations in breast cancer tissue and in matched nipple aspirate fluid. Carcinogenesis 26(1):145–152

Chapter 8
Mitochondrial UPR in Cancer

Doris Germain

8.1 Introduction

The identification of the UPR^{ER} arose from the observation that accumulation of misfolded proteins in the lumen of the endoplasmic reticulum in yeast is immediately followed by the transcriptional upregulation of the endoplasmic reticulum chaperone BiP (Mori et al. 1992). This observation raised the question as to how the detection of proteotoxic stress in the lumen of the organelle is communicated to the nucleus. Using a genetic screen, the first sensor of the UPR^{ER} was identified as the inositol-requiring transmembrane kinase/endonuclease, (IRE1) (Mori et al. 1993; Cox et al. 1993). In yeast, IRE-1 is the only signaling cascade that is activated. In higher eukaryotes however, the complexity of the UPR^{ER} expanded to three distinct branches; PERK, ATF6, and IRE, all of which are bound to the chaperone BiP and kept inactive in the absence of misfolded proteins (Schroder 2008). Upon accumulation of misfolded proteins, BiP is sequestered away from PERK, ATF6, and IRE by the misfolded proteins. As a result, PERK, ATF6, and IRE are activated. PERK is a transmembrane protein with a kinase domain facing the cytoplasm. Upon its activation, PERK phosphorylates eIF2α, leading to the repression of translation. ATF6 is a transcription factor, which is released from the endoplasmic reticulum when unbound to BiP and is transported to the Golgi where proteases cleave its transmembrane domain. The resulting cytoplasmic form of ATF6 then translocates to the nucleus to induce transcription of genes required to decrease stress in the endoplasmic reticulum such as chaperones. IRE is an endoribonuclease that cleaves the mRNA of transcription factor XBP1 allowing its activation. IRE therefore converts XBP-1U (unspliced) into XBP-1S. XBP-1S much like ATF6 activates the

D. Germain (✉)
Division of Hematology/Oncology Tisch Cancer Institute,
Icahn School of Medicine at Mount Sinai, New York, NY, USA
e-mail: doris.germain@mssm.edu

D.M. Hockenbery (ed.), *Mitochondria and Cell Death*, Cell Death in Biology and Diseases, DOI 10.1007/978-1-4939-3612-0_8

X.-M. Yin and Z. Dong (Series eds.), *Cell Death in Biology and Diseases*

transcription of genes to reduce stress in the endoplasmic reticulum. IRE-1 also has a kinase activity, which can lead to its autophosphorylation and promote the formation of protein complexes with TRAF2 and ASK1 leading to apoptosis or IKK leading to survival.

Therefore, the UPR^{ER} consists of three branches responsible for the activation of distinct pathways, which collectively aim at reducing proteotoxic stress in the organelle.

In addition to the UPR^{ER}, accumulation of misfolded proteins in this organelle is linked to endoplasmic reticulum associated degradation (ERAD). ERAD acts by promoting the retrotranslocation of misfolded proteins from the lumen of the endoplasmic reticulum to the cytoplasm for their degradation by the proteasome (Romisch 2005). Inhibition of the proteasome therefore results in sustained proteotoxic stress in the endoplasmic reticulum.

Since their discovery, the UPR^{ER} and ERAD have been implicated in a wide array of diseases. Our understanding of the mitochondrial UPR (UPR^{mt}) is in its infancy in comparison to the UPR^{ER}. However, the picture that emerges from what we have learned so far suggests that, similarly to the UPR^{ER}, the UPR^{mt} may also consists of multiple branches leading to an array of cytoprotective outcomes. In term of cancer biology, we propose that the distinct branches provide flexibility to the activation of the UPR^{mt} so that cancer cells can adapt to varying stress conditions.

8.2 The Mitochondria of Cancer Cells

The discovery 72 years ago by Otto Warburg that cancer cells produce excessive lactate in the presence of oxygen, as a result of their shift to aerobic glycolysis for the production of ATP, has led to the misconception that mitochondria are defective in cancer cells. In contrast, however, it is now clear that cancer cells are heavily dependent on their mitochondrial network for cell viability. For instance, deletion of mitochondrial DNA severely reduces their growth. Therefore, rather than being defective, mitochondria of cancer cells are reprogrammed. The metabolic reprogramming is achieved through several mechanisms: mutation in nuclear DNA-encoded mitochondrial enzymes such as succinate dehydrogenase, isocitrate dehydrogenase 1 and 2, or fumarate hydratase results in elevation in metabolites that contribute to cellular proliferation (Wallace 2012), and mutation in mitochondrial DNA affects the activity of complex I, which increases ROS that then act as signaling molecules to also promote proliferation (Wallace 2012; McMahon and LaFramboise 2014). Further, mutations in the mitochondrial DNA affecting specifically NADH dehydrogenase 6 subunit have been shown to promote metastasis (He et al. 2013; Ishikawa et al. 2008; Santidrian et al. 2013), although the mechanism remains unclear. Mutations in mitochondrial DNA also affect several tRNAs required for mitochondrial translation (Wallace 2012; McMahon and LaFramboise 2014). Defects in the function of the super-complexes leads to formation of ROS, which in turn promotes more mitochondrial DNA mutations and accumulation of misfolded proteins. Adding to this complexity is the fact that the vast majority of mitochondrial proteins is encoded by nuclear genes, with only 13 mitochondrial proteins are encoded by the mitochondrial genome. Since the electron transport chain is composed of proteins encoded by both genomes, in order to achieve stoichiometry of the complexes, the correct balance between the two genomes must be created. Failure to do so leads to the presence of unassembled proteins also contributing to proteotoxic stress. Therefore, while mutations in both nuclear and mitochondrial DNA-encoded mitochondrial proteins and elevated ROS cooperate to promote proliferation, they also lead to proteotoxic stress. As such, mechanisms must exist to limit ROS and the accumulation of misfolded proteins. In order to do so the level of chaperones must match the level of unfolded proteins so as to reach protein homeostasis and maintain the integrity of the organelle. Importantly, as the mitochondria are divided into the matrix and the intermembrane space (IMS), but the chaperone capacity of each subcompartment is drastically different, the effect of ROS on the accumulation of misfolded proteins is likely very different between the two subcompartments. The following sections focus on highlighting these differences.

8.2.1 The Protein Quality Control of the Matrix

The matrix contains several proteases and chaperones responsible for the degradation of unassembled or misfolded proteins (Arnold and Langer 2002; Bota and Davies 2002; Altieri 2013; Tatsuta and Langer 2008). The LONP, the CLPP, and the matrix AAA-proteases promote the degradation of these proteins in the matrix (Arnold and Langer 2002). LONP is overexpressed in cancer cells (Zhu et al. 2002; Bernstein et al. 2012) and CLPP has recently been identified in an shRNA screen for genes required for the viability of leukemia cells (Cole et al. 2015). While the substrates of the human CLPP protease remain poorly characterized, CLPP appears to play an important role in the UPR^{mt} as it has been identified in a genetic screen in *C. elegans* for genes that are activated by mitochondrial stress. This aspect is discussed in more detail later in this chapter.

The heat shock protein HSP60 is of particular interest to this chapter since it has been used extensively in genetic screens to identify genes implicated in the UPR^{mt}. HSP90 is another chaperone that has been linked to the UPR^{mt} and shown to localize to the matrix of cancer cells specifically (Altieri 2013; Siegelin et al. 2011). The selective accumulation of HSP90 in the mitochondria of cancer cells raises the possibility of exploiting HSP90 as a novel therapeutic option with low toxicity to normal tissues (Siegelin et al. 2011). In agreement with this possibility, disruption of mitochondrial HSP^{90} was shown to promote cell death (Siegelin et al. 2011).

Taken together, the upregulation of the proteases and chaperones of the mitochondria in cancer cells suggest that they play an important function in the maintenance of the integrity of the mitochondria in cancer cells. This aspect has been described in detail in a recent review (Goard and Schimmer 2014).

The elevation of proteases and chaperones raises the possibility that their expression may be coordinated by a transcriptional program activated in mitochondrial proteotoxic stress conditions. The UPR^{mt} is a prime candidate toward achieving this goal. The importance of these proteases and chaperones is further supported by the fact that, as described in the following section, the antioxidant machinery of the matrix is downregulated in cancer.

8.2.2 The Antioxidant Machinery of the Matrix

In the matrix, the manganese-superoxide dismutase SOD2 promotes the conversion of superoxide to hydrogen peroxide, which can then be converted to water by catalase. The activity of SOD2 is regulated by acetylation (Tao et al. 2010). In the acetylated form, SOD2 is inactive. The matrix deacetylase SIRT3, deacetylates SOD2 and promotes its activation (Tao et al. 2010). Therefore, SIRT3 and SOD2 represent an important mechanism, which limits the accumulation of superoxide in the matrix and, as a result, limits the oxidation of proteins (Lombard et al. 2011). While the antioxidant machinery of the matrix comprises several additional components, only SIRT3 and SOD2 are discussed here as they are the only element of the antioxidant machinery that, so far at least, has been linked to the UPR^{mt}.

Recently, the expression of SIRT3 has been reported to be reduced in 87 % of breast cancers and the authors suggested that the resulting elevation in ROS is critical in the stabilization of HIF1α and the switch to glycolysis (Finley et al. 2011). Subsequently, we reported that SOD2 is also downregulated in breast cancers (Papa et al. 2014a, b). Therefore, the expression of both SIRT3 and SOD2 is inhibited during transformation to contribute to the elevation in ROS levels observed in cancer cells. As ROS act as signaling molecules, their elevation may be a mean by which mitochondrial proteotoxic stress is communicated to increase the transcription of proteases and chaperones by the UPR^{mt}. This possibility is discussed later in this chapter.

8.3 The Intermembrane Space of Cancer Cells

Compared to the intensive studies of the inner membrane (electron transport chain), outer membrane (permeabilization during apoptosis) and matrix (Kreb cycle, etc.), the IMS has largely been overlooked. The general view is that the IMS is a passive subcompartment, which acts mainly as a storage space for a few pro-apoptotic proteins. This view cannot be further from reality.

The IMS contains over 100 proteins (Herrmann and Riemer 2010a). Collectively, the various functions of these proteins indicate that the IMS acts as a logistic hub that orchestrates metabolic processes, import of proteins, oxidative protein folding, protein degradation, transport of metabolites, lipids, and metal ions, export of ferrous precursors, assembly of the respiratory chain, detoxification of reactive oxygen species (ROS), and ROS-mediated signaling (Herrmann and Riemer 2010a).

Oxidative protein folding in the IMS is of particular interest, since the IMS is one of only two cellular compartments in which this process takes place; the other is the endoplasmic reticulum. Each cycle of folding generates one molecule of ROS. Oxidation of cysteines leads to the formation of disulfide bonds, which if inappropriate, will lead to misfolding and protein aggregation. While ROS generated by oxidative protein folding is the only source of ROS in the endoplasmic reticulum, the activity of the electron transport chain is another major source of ROS in the IMS. Therefore, the IMS can be considered as a highly oxidative cellular compartment, implying that IMS proteins are at high risk of misfolding (Herrmann and Riemer 2010a). This fact highlights the crucial importance of detecting and signaling the presence of misfolded proteins in the IMS to the nucleus in order to reduce IMS stress (IMSS) and to preserve the integrity of the mitochondria.

The IMS has also an important role in the assembly of the electron transport chain complexes. The inner membrane is considered the most protein-enriched membrane of the cell, with up to 46 subunits per complex, including some soluble subunits facing the IMS. Oxidation and misfolding of these subunits represent another mechanism leading to defects in the super-complexes and ROS generation.

Therefore, contrary to the general perception, the IMS represents an important logistic hub that orchestrates essential mitochondrial functions but is likely to suffer from chronic proteotoxic stress as a result of excessive ROS production in the IMS of cancer cells. In support of this possibility, both the protein quality control (PQC) and the antioxidant machinery of the IMS appear deficient in comparison to the matrix.

8.3.1 The Protein Quality Control of the IMS

The IMS contains several proteases, however, most are involved either in the processing of mitochondrial targeting signals in an ATP-dependent manner or cleavage of peptides generated by the processing proteases (Herrmann and Riemer 2010a). The i-AAA-protease acts mainly on protein assembly of the respiratory chain complex in the inner membrane (Herrmann and Riemer 2010a). Therefore, only OMI/HTRA2 appears to have the ability to cleave misfolded proteins, which suggests that in comparison to the matrix, the IMS has limited capacity to degrade misfolded proteins (Radke et al. 2008). Since the outer membrane of the mitochondria contains several ubiquitin ligases (Germain 2008), it is tempting to postulate that, as for the endoplasmic reticulum, a ERAD-like retrotranslocation of misfolded proteins from the lumen of the IMS into the cytoplasm for their degradation by the proteasome may take place. While evidence supporting the possibility of MAD (mitochondria-associated degradation) has been reported (Bragoszewski et al. 2013), data obtained in our group favor the view that the effect of the proteasome on the protein quality control of the IMS proteins take place prior to their import (Radke et al. 2008). Since, we also found that OMI/HTRA2 promotes the degradation of misfolded proteins in the IMS, we proposed a two step checkpoint, where the proteasome plays a major primary role in limiting the import of misfolded proteins in the IMS but if proteins escape this first step and enter the IMS, then as a secondary step OMI promotes their degradation (Radke et al. 2008). It remains entirely possible, however, that the proteasome plays a role in both pre- and post-import mechanisms such as MAD. In support of this possibility, a recent study reported the retro-translocation of proteins from the IMS into the cytosol (Bragoszewski et al. 2015).

Further, in contrast to the matrix, the IMS does not contain any heat shock proteins. While folding activity toward selected proteins has been associated with Mia40 (Banci et al. 2009), collectively the lack of chaperones combined with the limited proteolytic activity against misfolded proteins in the IMS support the notion that the IMS is prone to their accumulation.

8.3.2 The Antioxidant Machinery of the IMS

While superoxide produced by complex I localizes to the matrix, complex III produces superoxide in both the matrix and the IMS. The observation that the IMS is more oxidized than either the cytoplasm or the matrix suggests that, as for the protein quality control, the antioxidant capacity of the IMS is limited. Several factors culminate to make the environment of the IMS more oxidized. First, only a small fraction of total copper/zinc-dismutase SOD1 localizes to the IMS, as the vast majority resides in the cytoplasm. Second, the diffusion of glutathione across the outer membrane is limited. Third, the oxidative folding machinery of the ERV1-MIA40 disulfide relay leads to the generation of additional ROS (Herrmann and Riemer 2010a). Fourth, the peroxidase activity of cytochrome c allows the conversion of hydrogen peroxide produced by SOD1 to the highly toxic peroxidase compound 1 (Goldsteins et al. 2008). Fifth, oxidative folding of proteins in the IMS also promotes the formation of hydrogen peroxide (Sideris and Tokatlidis 2010; Deponte and Hell 2009) and stimulates the peroxidase activity of cytochrome c (Goldsteins et al. 2008; Diederix et al. 2002). These observations offer a mechanistic basis to explain the more oxidized environment of the IMS. They also highlight the possibility that proteins residing in the IMS may be especially prone to oxidation and misfolding. In agreement with this possibility, numerous proteins that localize to the IMS contain multiple cysteine residues and therefore are prone to the formation of abnormal disulfide bonds (Herrmann and Riemer 2010a). This characteristic of the IMS is of particular importance in the context of the UPR^{mt} as it suggests that the IMS may be a potent activator of the UPR^{mt} by having a lower threshold of tolerance for the accumulation of misfolded proteins.

8.4 Mitophagy

While a detailed review of mitophagy is beyond the focus of this chapter, it is nevertheless essential to emphasis the importance of mitophagy in the context of its role in the quality control of the mitochondrial network. At the molecular level, the selective elimination of irreversibly damaged mitochondria from the mitochondrial network can be view as a complementary mechanism to the protein quality control and the antioxidant machinery. As discussed in the following sections, the UPR^{mt} acts to activate simultaneously all of these mechanisms.

8.5 Summary of the Mechanisms Limiting Accumulation of Misfolded Proteins in the Matrix and IMS Respectively

Figure 8.1a illustrates the multiple mechanisms by which accumulation of misfolded proteins can be limited in the mitochondrial matrix. The antioxidant machinery can do so by limiting oxidation of proteins and mutations in mitochondrial DNA encoding mitochondrial proteins. However, this machinery is severely reduced in cancer cells. The protein quality control directly promotes the degradation of damaged or unassembled proteins in the matrix. Several of the proteases and chaperones are indeed upregulated in cancer cells, which has led to the hypothesis that targeting the chaperones and proteases of the matrix may represent a potent anticancer strategy (Siegelin et al. 2011; Goard and Schimmer 2014). Mitophagy indirectly

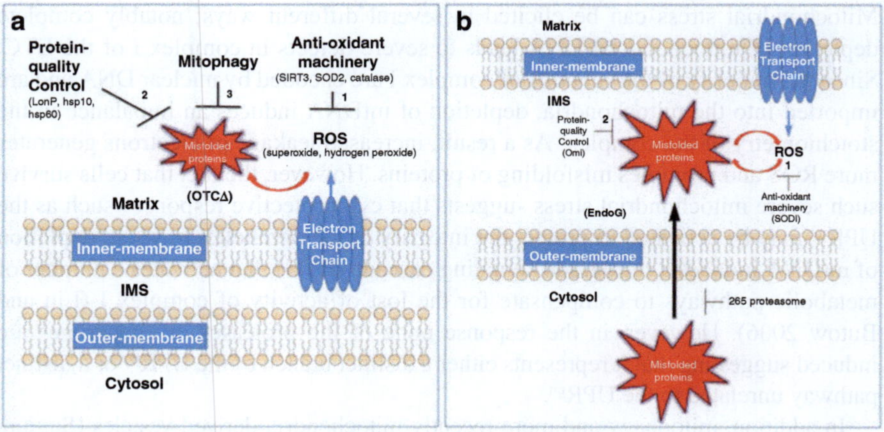

Fig. 8.1 Matrix and IMS differ in their ability to cope with proteotoxic stress. (**a**) Summary of the mechanisms contributing to the elimination of misfolded proteins in the matrix. (**b**) Summary of the mechanisms contributing to the elimination of misfolded proteins in the IMS. See text for details

contributes to limiting the accumulation of misfolded proteins by eliminating individual mitochondria with excessive proteotoxic stress from the mitochondrial network. Further, as described in the following section, the UPR^{mt} allows the transcriptional activation of all of these mechanisms. In mammalian cells, the matrix protein OTCΔ has been used as a tool to monitor the effect of accumulation of misfolded proteins in the matrix.

Figure 8.1b illustrates the more limited ability of the IMS to cope with the accumulation of misfolded proteins. The protease OMI/HTRA2 and the superoxide dismutase SOD1 contribute to lower proteotoxic stress in the lumen of the IMS. However, as attempts to fold proteins through oxidative folding can contribute to the generation of ROS, the combined effect of oxidative folding and the electron transport chain leads to a more oxidized environment in the IMS. Upregulation of OMI/HTRA2 and the activity of the proteasome to inhibit the import of misfolded proteins in the IMS, by the UPR^{mt} are discussed in the following sections. Lastly, in mammalian cells, the IMS protein ENDOG has been used as a tool to monitor the effect of accumulation of misfolded proteins in the IMS. While the use of OTCΔ and ENDOG has allowed the dissection of the relative contribution of each subcompartment of the mitochondria to the activation of the UPR^{mt}, in reality the accumulation of misfolded proteins is simultaneously observed in both the matrix and the IMS under oxidative stress conditions.

8.6 The UPR^{mt} Coordinates Protein and Organelle Quality Controls

8.6.1 Mitochondrial Stress Responses and the UPR^{mt}

Mitochondrial stress can be elicited in several different ways; notably complete depletion of mitochondrial DNA leads to severe defects in complex I of the ETC. Since the majority of the subunits of complex I are encoded by nuclear DNA and are imported into the mitochondria, depletion of mtDNA induces an imbalance in the stoichiometry of the complex. As a result, increased leakage of electrons generates more ROS and promotes misfolding of proteins. However, the fact that cells survive such severe mitochondrial stress suggests that cytoprotective responses such as the UPR^{mt} may be activated to protect the integrity of the organelle. In yeast, depletion of mtDNA induces a retrograde signaling cascade (RTG) that activates a number of metabolic pathways to compensate for the lost of activity of complex I (Liu and Butow 2006). However, in the response none of the proteases or chaperones are induced suggesting that it represents either a distinct branch of the UPR^{mt} or a distinct pathway unrelated to the UPR^{mt}.

In addition, mitophagy and more recently mitochondria-derived vesicles (Sugiura et al. 2014) represent mitochondrial stress responses that are up to now considered independent of the UPR^{mt}. Work from our group however raises the possibility that accumulation of misfolded proteins in the mitochondria may induce some or perhaps all of the responses simultaneously but at various degrees to orient the fate of individual mitochondria within the mitochondrial network depending on the levels of stress experienced by the mitochondrion. If so, it would challenge the current view of parallel mitochondrial stress pathways and favor a view where upregulation of the PQC, the antioxidant machinery and mitophagy represent different branches of the UPR^{mt}.

If correct, the parallel and gradual induction of distinct branches of the UPR^{mt} relative to mitochondrial stress levels would mimic the UPR^{ER}, where a gradation of responses is observed relative to ER stress (Wang and Kaufman 2014).

The following section summarizes some of the work supporting such a model.

8.6.2 *The UPR^{mt}, A Complex Transcriptional Program?*

The effect of the accumulation of misfolded proteins in the matrix of the mitochondria was first described by the Hoogenraad group (Martinus et al. 1996). Their pioneer study of the UPR^{mt} used the expression of OTCΔ, which misfolded and formed aggregates in the mitochondrial matrix. They reported that the resulting proteotoxic stress leads to the activation of the transcription factor CHOP and the subsequent expression of matrix LONP protease and the hsp10 and 60 chaperones (Zhao et al. 2002). They further described that CHOP binding to the promoter of these genes required the presence of AP-1 (Aldridge et al. 2007; Horibe and Hoogenraad 2007; Ryan and Hoogenraad 2007). Since CHOP is implicated in ER-stress, but none of the UPR^{ER} genes are activated by matrix proteotoxic stress, the binding to AP-1 offers a potential explanation for the dual function of CHOP in both mitochondrial and ER-stress.

The Hoogenraad group is also the first to have identified endonucleaseG (ENDOG), which resides in the intermembrane space of the mitochondria as a gene upregulated by the UPR^{mt}. While the identification of ENDOG remained enigmatic at that time, subsequently ENDOG has been identified as a transcriptional target of ERRα (Ranhotra 2014). As ERRα/PGC1α is a major transcriptional complex involved in the regulation of a large number of mitochondrial genes involved in biogenesis and oxidative phosphorylation, their findings suggested that ERRα/PGC1α is also activated by mitochondrial proteotoxic stress.

This possibility is supported by our more recent finding that the transcription factor FOXO3A is activated by the UPR^{mt}, since PGC1α is a transcriptional target of FOXO3A. We identified FOXO3A using the same OTCΔ model as the Hoogenraad group and also confirmed that ENDOG is upregulated by mitochondrial proteotoxic stress. While the role of ENDOG in the UPR^{mt} remains enigmatic, other targets of FOXO3A and PGC1α were also found to be activated, namely the antioxidants SOD2 and catalase (Papa and Germain 2014). We further found that mitochondrial stress led to the nuclear accumulation of FOXO3A and its deacetylation. This finding led us to investigate whether the deacetylase SIRT1, which is well documented to activate both FOXO3A and PGC1α was affected by the UPR^{mt}. However, we found no evidence for upregulation of SIRT1. In contrast, we found that the matrix SIRT3 is upregulated. In fact our results indicate that nuclear translocation of FOXO3A and the induction of SOD2 and catalase are dependent on SIRT3. Since SIRT3 is in the mitochondria and FOXO3A in the cytoplasm, how SIRT3 can affect FOXO3A is unclear. The mitochondrial localization of FOXO3A and the cytoplasmic localization of SIRT3 have both been reported, but are very controversial (Jacobs et al. 2008; Sundaresan et al. 2009). Therefore, whether SIRT3 affects FOXO3A directly or indirectly remains to be determined.

The dependency of the induction of SOD2 and catalase on SIRT3 was specific, since neither were affected by the inhibition of CHOP. Conversely, inhibition of SIRT3 had no effect on the level of LONP, HSP10 or HSP60. We therefore concluded that SIRT3 and CHOP represent distinct branches of the UPR^{mt} upon accumulation of mutant OTC in the matrix. Furthermore, the CHOP branch is responsible for promoting PQC of the matrix, while the SIRT3 branch promotes the antioxidant machinery of the matrix, therefore offering complementary cytoprotective effects.

This work also led to the finding, using electron microscopy of cells undergoing matrix proteotoxic stress, that a fraction of the mitochondria were targeted for elimination by mitophagy (Papa and Germain 2014). In agreement with this finding, we found that several markers of mitophagy were induced under these conditions. Since SIRT3, but not CHOP, was found to be essential for the induction of these markers, we conclude that the SIRT3 branch of the UPR^{mt} has the dual role of activating the antioxidant machinery and mitophagy. This finding raises the possibility that SIRT3 may play an important role in a mitochondrial checkpoint where moderately stressed mitochondria may be repaired through an upregulation of the antioxidant machinery, while irreversibly damaged organelles are selectively eliminated from the network by mitophagy. If so, the question that emerges is what distinguish moderately from irreversibly damaged mitochondria. Work from Youle's group indicates that accumulation of misfolded OTC in the matrix of the mitochondria leads to the accumulation of PINK and the translocation of the ubiquitin ligase PARK2 to the mitochondria, offering therefore a potential mechanism of selection (Jin and Youle 2013).

Considering that the expression of SIRT3 is downregulated in cancer, and is essential for the shift to glycolysis (Finley et al. 2011; Haigis et al. 2012) the finding that SIRT3 regulates at least one branch of the UPR^{mt} raises a paradox. In order to resolve this paradox, we propose that the expression of SIRT3 acts as rheostat of ROS in cancer cells (Fig. 8.2). In this model, a decrease in SIRT3 is necessary during transformation to elevate ROS levels and promote the shift to glycolysis. However, in cells where ROS levels reach excessive levels, the expression of SIRT3 may remain inducible by the UPR^{mt} and allow decreased ROS in these cells to levels that are compatible with the viability of the organelle. In addition, the inducible expression of SIRT3 in these cells would allow the elimination of irreversibly damaged mitochondria from the mitochondrial network and restoration of a healthier and uniform network (Fig. 8.2).

The notion that ROS levels must be maintained within the elevated range but below a certain threshold to maintain the integrity of the mitochondria in cancer cells is reminiscent of mitochondrial hormesis, or dual effect of ROS, during aging. In this setting, it is now recognized that moderate levels of ROS are protective against aging by activating cytoprotective responses but failure of these responses leads to aging (for an extensive review (Ristow 2014; Ristow and Schmeisser 2014)). Considering that expression of SIRT3 decreases in both aging cells and cancer cells, it is tempting to speculate that cancer cells may extend their ability to activate the UPR^{mt} and maintain viability, while the UPR^{mt} fails in aging cells.

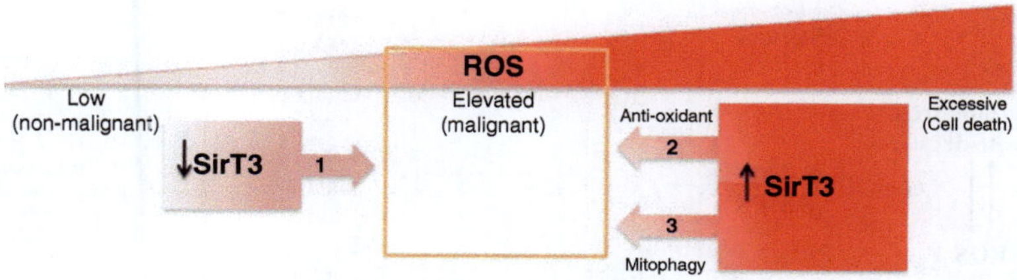

Fig. 8.2 Proposed dual roles of Sirt3 in cancer. A reduction in SIRT3 contributes to the elevation in ROS levels in cancer cells and contributes to the switch to glycolysis. However, when ROS reach excessive levels, the induction of Sirt3 by the UPR^{mt} will promote the activation of FOXO3A and the transcription of the antioxidant machinery as well as removal of irreversibly damaged organelles by mitophagy. As a result, ROS levels are reduced within a window that is compatible with cellular viability

8.6.3 The IMS, An Active Player in the UPR^{mt}

The IMS appears to be deficient in terms of antioxidant machinery as well as PQC relative to the matrix. In addition, being a more oxidized environment (Herrmann and Riemer 2010a, b), the folding machinery of the IMS is also reported to be affected by hypoxia (Fukuda et al. 2007). Collectively, these weaknesses in the capacity of the IMS culminate to argue that it may represent a subcompartment particularly prone to the accumulation of misfolded proteins and therefore to the activation of the UPR^{mt}. In order to test this possibility, we aimed at inducing proteotoxic stress in the IMS. This line of investigation has led to the discovery of a third branch of the UPR^{mt}, which is dependent on the transcriptional activity of the estrogen receptor alpha (Papa and Germain 2011). Estrogen receptors (ER) are implicated in the regulation of mitochondrial functions (Pedram et al. 2006), although the mechanism remains unclear. Our data suggest that ROS generated by the IMS proteotoxic stress promote the activation of AKT, which than phosphorylates the estrogen receptor alpha (ERα), leading to its activation in a ligand-independent fashion (Papa and Germain 2011). The transcriptional activity of ERα stimulates the expression of OMI/HTRA2 and leads to stimulation of the activity of the proteasome, therefore allowing a reduction in the import of misfolded proteins in the IMS and their degradation in the IMS. In addition, since the transcription factor NRF1, a key regulation of several genes of the ETC, was reported to be dependent on the estrogen receptor (Klinge 2008; Mattingly et al. 2008), we also monitored the level of NRF1 and found that it is upregulated (Papa and Germain 2011). Importantly, in breast cancer cells that do not express the estrogen receptor, proteotoxic stress in the IMS was found to be a potent activator of CHOP and SIRT3.

8.6.4 Summary of the UPR^{mt}

Figure 8.3 shows a diagram to summarize our current model of the UPR^{mt} in mammalian cells. We propose that mitochondrial proteotoxic stress and ROS are intimately linked. On one hand, oxidation of proteins can promote their misfolding

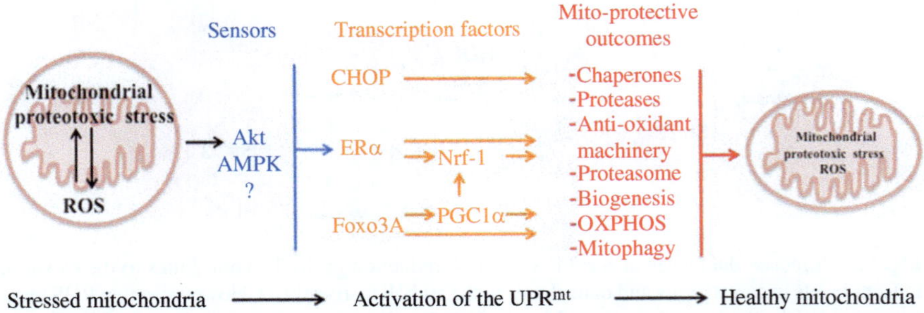

Fig. 8.3 Summary of the UPRmt. Stressed mitochondria due to accumulation of misfolded proteins activate sensors such as AKT and AMPK. However, whether additional sensors able to act as bridge to communicate directly mitochondrial proteotoxic stress to the cytoplasm remain to be determined. The sensors then act to activate a number of transcription factors, CHOP, ERα, FOXO3A, PCG1α and NRF-1, which lead to the up-regulation of protective outcomes to reduce mitochondrial stress and restore mitochondrial health

and, at least in the IMS, oxidative folding also generates ROS. On the other hand, ROS can promote mutations in mtDNA and defects in the electron transport chain complexes that may not only lead to misfolding of subunits encoded by the mitochondrial genome but also lead to more ROS. Therefore, in our view, inducing mitochondrial stress either by the deletion of mtDNA, pharmacological inhibition of the ETC complex or causing proteotoxic stress directly by the expression of misfolded proteins such as OTCΔ are likely to overlap in their outcome. If so, the distinction between the UPR^{mt} and the retrograde signaling cascade may be subtle under physiological conditions rather than experimental conditions (overexpression of misfolded protein and depletion of mtDNA respectively) that were used for their discoveries.

So far AKT and AMPK have been shown to be activated by ROS, and therefore, we propose that they act as sensors of mitochondrial stress (Fig. 8.3). Akt can phosphorylate both ERα and FOXO3A. We show that the activation of the ERα leads to an elevation in the proteasome and the protease OMI/HTRA2 as well as NRF1 to protect the mitochondria against proteotoxic stress. AKT also phosphorylates FOXO3A and prevents it accumulation in the nucleus. Therefore, AKT opposes the effect of AMPK on FOXO3A (for recent reviews (Taylor et al. 2015; Wu et al. 2014)). Since FOXO3A leads to the transcription of the antioxidant machinery and PGC1α but also of pro-apoptotic genes, FOXO3A holds the potential to have a dual effect on the mitochondria. This observation raises the possibility that limited accumulation of FOXO3A in the nucleus promotes the activation of protective genes against mitochondrial stress while avoiding the activation of pro-apoptotic genes. However, if the stress becomes too severe, the level of FOXO3A could act as a defining threshold to push cells toward apoptosis. This possibility remains to be tested.

We propose that AKT and AMPK act as sensors of ROS and therefore contribute to communicate mitochondrial stress to the transcriptional factors implicated in the UPR^{mt}. Whether other sensors are present at the mitochondrial membrane itself remains a question. By similarity to ATF6, IRE, and PERK in the UPR^{ER}, such sensors would also be transmembrane proteins with a domain facing the IMS to detect the presence of misfolded proteins and another domain facing the cytosol in order to activate a signal cascade. The discovery of such sensors would represent a major advance in the elucidation of the UPR^{mt}.

So far, CHOP, ERα, NRF1, FOXO3A, and PGC1α have been linked to the UPR^{mt} in mammalian cells (Fig. 8.3). Since NRF1 is activated by both the ERα and PGC1α, cross talk between the targets of the various transcription factors involved may provide a potent mechanism to allow the sustained activation of the UPR^{mt} despite loss of expression of one of the transcription factors. Collectively, the "mito-protective" outcomes of this transcriptional program cover all aspects of proteostasis and organellar repair to allow the maintenance of a healthy mitochondrial network in conditions of oxidative stress (Fig. 8.3). In addition, the dissection of the respective contribution of the matrix and the IMS has highlighted the possibility that the apparent weaknesses of the IMS in term of protein quality control and antioxidant machinery may result in the activation of the UPR^{mt} at a lower threshold of proteotoxic stress than in the matrix. As such, the IMS may be seen as a potent activator of the UPR^{mt}.

8.7 The UPR^{mt} in Aging and Development

While the UPR^{mt} has initially been described in mammals (Martinus et al. 1996), at the molecular level, it has been most extensively dissected using model organisms such as *C. elegans* in the context of aging and development. Notably, genetic and pharmacological alteration of the sirtuin-2.1-SOD2 axis of the UPR^{mt} in *C. elegans* affects life-span, where reduction in sirtuin-2.1 accelerates aging, while its activation expanded life-span (Mouchiroud et al. 2013). Considering that aging is one of the highest risk factors of developing cancer, it raises the possibility that the reduction in SIRT3 during aging contributes to the transformation process by increasing ROS and stabilizing HIF1α.

The elegant work on the UPR^{mt} from the Haynes group in *C. elegans* has led to the identification of two important transcriptional regulators of the UPR^{mt}, ATFS-1 and the DVE-1/UBL5 complex (Baker et al. 2012; Nargund et al. 2012; Pellegrino et al. 2013; Haynes and Ron 2010; Haynes et al. 2010; Haynes et al. 2007; Benedetti et al. 2006). In addition, mitochondrial proteotoxic stress was found to regulate protein translation through the inactivation of eIF2α (Baker et al. 2012). A decrease in protein translation diminishes the import of mitochondrial proteins and therefore results in reduction in protein load on the mitochondrial chaperones. Interestingly, since induction of eIF2α was found to be independent of ATFS-1, eIF2α may represent a different branch of the UPR^{mt}.

More recently, one of the transcription factors they originally identified in worms has been validated in mammalian cells. Using ChIP-sequencing, they reported that the transcription factor ATFS-1 binds directly to the promoters of oxidative phosphorylation genes in both the nuclear and mitochondrial genomes and therefore promotes oxidative phosphorylation (Nargund et al. 2015). They found that ATFS-1 also upregulates chaperones, antioxidant machinery, and glycolysis (Nargund et al. 2015).

However, since these pathways have been recently and extensively reviewed elsewhere,(Pellegrino et al. 2013; Haynes and Ron 2010) they are not further discussed in this chapter. Whether these cascades are conserved in humans and play a role in cancer represent an important area of research. In agreement with the conservation of the pathway is the observation that activation of the UPR^{mt} in both *C. elegans* and mice correlates with longevity (Houtkooper et al. 2013). Strikingly, mitochondrial ribosomal protein Mrps5 was found to map to the longevity locus in mice (Houtkooper et al. 2013). Additionally, induction of the UPR^{mt} by ROS in *C. elegans* has recently led to the identification of 54 novel regulators (Runkel et al. 2013). Therefore, there is no doubt that the understanding of the role of the UPR^{mt} in aging will lead to major advances in our understanding of its role in cancer.

8.8 Toward the Validation of the UPR^{mt} in Cancer

Our current understanding of the UPR^{mt} suggests that oxidative phosphorylation, mitochondrial biogenesis and other outcomes of the activation may be reversible depending on the level of local oxidative stress. Consistent with this possibility, several studies have now reported the upregulation of PGC1α and SOD2 in recurrent tumors and metastasis.

A study by the Draetta group in pancreatic cancer indicated that oxidative phosphorylation, autophagy, and mitochondrial biogenesis are upregulated in recurrent pancreatic tumors (Viale et al. 2014). In this study inducible mutant Ras was used to promote the formation of pancreatic tumors in mice. Upon inhibition of Ras expression, tumors regress rapidly; however, 4–5 months later, Ras-negative tumors emerged. Their characterization revealed that these tumors selectively increased mitochondrial biogenesis, were not dependent on glycolysis but rather required oxidative phosphorylation and the antioxidant machinery (Viale et al. 2014).

Likewise, the Depinho group found that tumors that recurred after surviving telomere-shortening crisis by activating the ALT pathway are characterized by an upregulation of mitochondrial biogenesis, oxidative phosphorylation, and antioxidant machinery (Hu et al. 2012). These tumors were found to be drastically more sensitive to SOD2 inhibition (Hu et al. 2012).

In addition, analysis of circulating cancer cells showed that they specifically upregulate mitochondrial biogenesis and oxidative phosphorylation through PGC1α (LeBleu et al. 2014). Notably, this study also found a strong positive correlation between PGC1α expression and distant metastasis (LeBleu et al. 2014).

Therefore, while switch from oxidative phosphorylation to glycolysis characterizes primary tumors, more recent data suggest that disseminated cells, circulating cancer cells and metastatic cells reverse to oxidative phosphorylation and may be highly dependent on increased antioxidant machinery. Therefore, the activation of the UPR^{mt} is likely to play a major role in the changes in mitochondrial activity in these lesions.

8.9 Therapeutic Impact and Future Directions

Extensive analysis of the mitochondrial genome in breast cancer cells has revealed that the vast majority contains mtDNA mutations and that the frequency of mutation correlates with clinical outcome (McMahon and LaFramboise 2014). This recent study adds to the now undeniable link between mitochondrial reprogramming and cancer. These findings argue that activation of the UPR^{mt} is likely essential to prevent cancer cell death due to failure of mitochondria to cope with proteotoxic stress.

Work from the Haigis and Gius labs (Tao et al. 2010; Finley et al. 2011; Kim et al. 2009) has established that SIRT3 acts as a mitochondrial tumor suppressor (Finley et al. 2011). Therefore, lost of SIRT3 increases ROS and contributes to the transformation. Based on our findings, the resulting increase in ROS is predicted to lead to the activation of the UPR^{mt}, but in absence of SIRT3, only the CHOP and ER branches of the pathway can protect the mitochondria from proteotoxic stress. Theoretically, activating the UPR^{mt} through separate branches allows flexibility and facilitates the adaptation to the loss of one branch. In support of this possibility, in the situation where SIRT3 is entirely lost, the proliferation of ER-positive breast cancer cells is predicted to be favored over the proliferation of ER-negative cells. Consistent with this prediction, genetic deletion of SIRT3 in mice leads to the formation of exclusively ER positive mammary tumors (Kim et al. 2009). As the vast majority of breast cancer models in mice develop ER-negative mammary tumors, the absence of SIRT3 imposes a potent selective pressure for the formation of ER-positive tumors.

Further relevance of ER-positive mammary tumors in SIRT3 knockout mice to the etiology of breast cancer arose from the observation that the SIRT3 knockout mice develop mammary tumors only in older mice. In humans, the vast majority of breast cancers are ER-positive and develop in older, post-menopausal women, therefore, the SIRT3 knockout mouse can be viewed as one of the best models of breast cancer, since it reflects more accurately the disease in humans.

In addition, since the activation of ERα by mitochondrial proteotoxic stress is independent of ligand, the UPR^{mt} may contribute to resistance to endocrine therapy. Therefore, the finding of the UPR^{mt} strongly supports that targeting chaperones and proteases of the mitochondria may be an important therapeutic avenue for the future (Siegelin et al. 2011; Goard and Schimmer 2014).

Compared to the UPR^{ER}, our knowledge of the UPR^{mt} is in its infancy and much remains to be discovered about not only the identification of all of the key players it engages, but also the extend of its role in several mitochondria-related diseases. In terms of cancer therapeutics, the development of matrix protease inhibitors as well as chaperone inhibitors offers an exciting area of research to prevent their cytoprotective effect on the mitochondria. However, as the differential bioenergetic characteristics of stem cells become more evident, how stem cells regulate the UPR^{mt} remains entirely unknown. Further, the UPR^{mt} may be non-cell autonomous (Durieux et al. 2011) and be activated in one tissue by mitochondrial stress in a distant tissue. Lastly, evidence is emerging of a cross-talk between the UPR^{mt} and UPR^{ER} (Runkel et al. 2013). Therefore, the complexity of the UPR^{mt} is likely to offer many surprises and levels of complexity that are yet to be discovered. Nevertheless given the anticipated complexity of this pathway, the study of the UPR^{mt} represents an exciting new field of research likely to have profound impact on several diseases including cancer.

References

Aldridge JE, Horibe T, Hoogenraad NJ (2007) Discovery of genes activated by the mitochondrial unfolded protein response (mtUPR) and cognate promoter elements. PLoS One 2:e874. doi:10.1371/journal.pone.0000874

Altieri DC (2013) Mitochondrial, HSP90s and tumor cell metabolism. Autophagy 9:244–245. doi:10.4161/auto.22527

Arnold I, Langer T (2002) Membrane protein degradation by AAA proteases in mitochondria. Biochim Biophys Acta 1592:89–96

Baker BM, Nargund AM, Sun T, Haynes CM (2012) Protective coupling of mitochondrial function and protein synthesis via the eIF2alpha kinase GCN-2. PLoS Genet 8, e1002760. doi:10.1371/journal.pgen.1002760

Banci L et al (2009) MIA40 is an oxidoreductase that catalyzes oxidative protein folding in mitochondria. Nat Struct Mol Biol 16:198–206. doi:10.1038/nsmb.1553

Benedetti C, Haynes CM, Yang Y, Harding HP, Ron D (2006) Ubiquitin-like protein 5 positively regulates chaperone gene expression in the mitochondrial unfolded protein response. Genetics 174:229–239

Bernstein SH et al (2012) The mitochondrial ATP-dependent Lon protease: a novel target in lymphoma death mediated by the synthetic triterpenoid CDDO and its derivatives. Blood 119:3321–3329. doi:10.1182/blood-2011-02-340075

Bota DA, Davies KJ (2002) Lon protease preferentially degrades oxidized mitochondrial aconitase by an ATP-stimulated mechanism. Nat Cell Biol 4:674–680

Bragoszewski P, Gornicka A, Sztolsztener ME, Chacinska A (2013) The ubiquitin-proteasome system regulates mitochondrial intermembrane space proteins. Mol Cell Biol 33:2136–2148. doi:10.1128/MCB.01579-12

Bragoszewski P et al (2015) Retro-translocation of mitochondrial intermembrane space proteins. Proc Natl Acad Sci U S A 112:7713–7718. doi:10.1073/pnas.1504615112

Cole A et al (2015) Inhibition of the mitochondrial protease ClpP as a therapeutic strategy for human acute myeloid leukemia. Cancer Cell 27:864–876. doi:10.1016/j.ccell.2015.05.004

Cox JS, Shamu CE, Walter P (1993) Transcriptional induction of genes encoding endoplasmic reticulum resident proteins requires a transmembrane protein kinase. Cell 73:1197–1206

Deponte M, Hell K (2009) Disulphide bond formation in the intermembrane space of mitochondria. J Biochem 146:599–608. doi:10.1093/jb/mvp133, mvp133 [pii]

Diederix RE, Ubbink M, Canters GW (2002) Peroxidase activity as a tool for studying the folding of c-type cytochromes. Biochemistry 41:13067–13077, doi:bi0260841 [pii]

Durieux J, Wolff S, Dillin A (2011) The cell-non-autonomous nature of electron transport chain-mediated longevity. Cell 144:79–91. doi:10.1016/j.cell.2010.12.016

Finley LW et al (2011) SIRT3 opposes reprogramming of cancer cell metabolism through HIF1alpha destabilization. Cancer Cell 19:416–428. doi:10.1016/j.ccr.2011.02.014

Fukuda R et al (2007) HIF-1 regulates cytochrome oxidase subunits to optimize efficiency of respiration in hypoxic cells. Cell 129:111–122. doi:10.1016/j.cell.2007.01.047

Germain D (2008) Ubiquitin-dependent and -independent mitochondrial protein quality controls: implications in ageing and neurodegenerative diseases. Mol Microbiol 70:1334–1341. doi:10.1111/j.1365-2958.2008.06502.x

Goard CA, Schimmer AD (2014) Mitochondrial matrix proteases as novel therapeutic targets in malignancy. Oncogene 33:2690–2699. doi:10.1038/onc.2013.228

Goldsteins G et al (2008) Deleterious role of superoxide dismutase in the mitochondrial intermembrane space. J Biol Chem 283:8446–8452. doi:10.1074/jbc.M706111200, M706111200 [pii]
Haigis MC, Deng CX, Finley LW, Kim HS, Gius D (2012) SIRT3 is a mitochondrial tumor suppressor: a scientific tale that connects aberrant cellular ROS, the Warburg effect, and carcinogenesis. Cancer Res 72:2468–2472. doi:10.1158/0008-5472.CAN-11-3633
Haynes CM, Ron D (2010) The mitochondrial UPR - protecting organelle protein homeostasis. J Cell Sci 123:3849–3855. doi:10.1242/jcs.075119
Haynes CM, Petrova K, Benedetti C, Yang Y, Ron D (2007) ClpP mediates activation of a mitochondrial unfolded protein response in C. elegans. Dev Cell 13:467–480
Haynes CM, Yang Y, Blais SP, Neubert TA, Ron D (2010) The matrix peptide exporter HAF-1 signals a mitochondrial UPR by activating the transcription factor ZC376.7 in C. elegans. Mol Cell 37:529–540. doi:10.1016/j.molcel.2010.01.015
He X et al (2013) Suppression of mitochondrial complex I influences cell metastatic properties. PLoS One 8:e61677. doi:10.1371/journal.pone.0061677
Herrmann JM, Riemer J (2010a) The intermembrane space of mitochondria. Antioxid Redox Signal 13:1341–1358. doi:10.1089/ars.2009.3063
Herrmann JM, Riemer J (2010b) Oxidation and reduction of cysteines in the intermembrane space of mitochondria: multiple facets of redox control. Antioxid Redox Signal 13:1323–1326. doi:10.1089/ars.2010.3270
Horibe T, Hoogenraad NJ (2007) The chop gene contains an element for the positive regulation of the mitochondrial unfolded protein response. PLoS One 2:e835. doi:10.1371/journal.pone.0000835
Houtkooper RH et al (2013) Mitonuclear protein imbalance as a conserved longevity mechanism. Nature 497:451–457. doi:10.1038/nature12188
Hu J et al (2012) Antitelomerase therapy provokes ALT and mitochondrial adaptive mechanisms in cancer. Cell 148:651–663. doi:10.1016/j.cell.2011.12.028
Ishikawa K et al (2008) ROS-generating mitochondrial DNA mutations can regulate tumor cell metastasis. Science 320:661–664. doi:10.1126/science.1156906
Jacobs KM et al (2008) SIRT3 interacts with the daf-16 homolog FOXO3A in the mitochondria, as well as increases FOXO3A dependent gene expression. Int J Biol Sci 4:291–299
Jin SM, Youle RJ (2013) The accumulation of misfolded proteins in the mitochondrial matrix is sensed by PINK1 to induce PARK2/Parkin-mediated mitophagy of polarized mitochondria. Autophagy 9:1750–1757. doi:10.4161/auto.26122
Kim HS et al (2009) SIRT3 is a mitochondria-localized tumor suppressor required for maintenance of mitochondrial integrity and metabolism during stress. Cancer Cell 17:41–52. doi:10.1016/j.ccr.2009.11.023, S1535-6108(09)00428-0 [pii]
Klinge CM (2008) Estrogenic control of mitochondrial function and biogenesis. J Cell Biochem 105:1342–1351. doi:10.1002/jcb.21936
LeBleu VS et al (2014) PGC1alpha mediates mitochondrial biogenesis and oxidative phosphorylation in cancer cells to promote metastasis. Nat Cell Biol 16(992–1003):1001–1015. doi:10.1038/ncb3039
Liu Z, Butow RA (2006) Mitochondrial retrograde signaling. Annu Rev Genet 40:159–185. doi:10.1146/annurev.genet.40.110405.090613
Lombard DB, Tishkoff DX, Bao J (2011) Mitochondrial sirtuins in the regulation of mitochondrial activity and metabolic adaptation. Handb Exp Pharmacol 206:163–188. doi:10.1007/978-3-642-21631-2_8
Martinus RD et al (1996) Selective induction of mitochondrial chaperones in response to loss of the mitochondrial genome. Eur J Biochem 240:98–103
Mattingly KA et al (2008) Estradiol stimulates transcription of nuclear respiratory factor-1 and increases mitochondrial biogenesis. Mol Endocrinol 22:609–622. doi:10.1210/me.2007-0029
McMahon S, LaFramboise T (2014) Mutational patterns in the breast cancer mitochondrial genome, with clinical correlates. Carcinogenesis 35:1046–1054. doi:10.1093/carcin/bgu012
Mori K et al (1992) A 22 bp cis-acting element is necessary and sufficient for the induction of the yeast KAR2 (BiP) gene by unfolded proteins. EMBO J 11:2583–2593

Mori K, Ma W, Gething MJ, Sambrook J (1993) A transmembrane protein with a cdc2+/CDC28-related kinase activity is required for signaling from the ER to the nucleus. Cell 74:743–756
Mouchiroud L et al (2013) The NAD(+)/Sirtuin pathway modulates longevity through activation of mitochondrial UPR and FOXO signaling. Cell 154:430–441. doi:10.1016/j.cell.2013.06.016
Nargund AM, Pellegrino MW, Fiorese CJ, Baker BM, Haynes CM (2012) Mitochondrial import efficiency of ATFS-1 regulates mitochondrial UPR activation. Science 337:587–590. doi:10.1126/science.1223560
Nargund AM, Fiorese CJ, Pellegrino MW, Deng P, Haynes CM (2015) Mitochondrial and nuclear accumulation of the transcription factor ATFS-1 promotes OXPHOS recovery during the UPR(mt). Mol Cell 58:123–133. doi:10.1016/j.molcel.2015.02.008
Papa L, Germain D (2011) Estrogen receptor mediates a distinct mitochondrial unfolded protein response. J Cell Sci 124:1396–1402. doi:10.1242/jcs.078220
Papa L, Germain D (2014) SirT3 regulates the mitochondrial unfolded protein response. Mol Cell Biol 34:699–710. doi:10.1128/MCB.01337-13
Papa L, Manfredi G, Germain D (2014a) SOD1, an unexpected novel target for cancer therapy. Genes Cancer 5:15–21
Papa L, Hahn M, Marsh EL, Evans BS, Germain D (2014b) SOD2 to SOD1 switch in breast cancer. J Biol Chem 289:5412–5416. doi:10.1074/jbc.C113.526475
Pedram A, Razandi M, Wallace DC, Levin ER (2006) Functional estrogen receptors in the mitochondria of breast cancer cells. Mol Biol Cell 17:2125–2137
Pellegrino MW, Nargund AM, Haynes CM (2013) Signaling the mitochondrial unfolded protein response. Biochim Biophys Acta 1833:410–416. doi:10.1016/j.bbamcr.2012.02.019
Radke S et al (2008) Mitochondrial protein quality control by the proteasome involves ubiquitination and the protease Omi. J Biol Chem 283:12681–12685
Ranhotra HS (2014) Estrogen-related receptor alpha and mitochondria: tale of the titans. J Recept Signal Transduct Res 1–5, doi: 10.3109/10799893.2014.959592
Ristow M (2014) Unraveling the truth about antioxidants: mitohormesis explains ROS-induced health benefits. Nat Med 20:709–711. doi:10.1038/nm.3624
Ristow M, Schmeisser K (2014) Mitohormesis: promoting health and lifespan by increased levels of Reactive Oxygen Species (ROS). Dose Response 12:288–341. doi:10.2203/dose-response.13-035. Ristow
Romisch K (2005) Endoplasmic reticulum-associated degradation. Annu Rev Cell Dev Biol 21:435–456
Runkel ED, Liu S, Baumeister R, Schulze E (2013) Surveillance-activated defenses block the ROS-induced mitochondrial unfolded protein response. PLoS Genet 9:e1003346. doi:10.1371/journal.pgen.1003346
Ryan MT, Hoogenraad NJ (2007) Mitochondrial-nuclear communications. Annu Rev Biochem 76:701–722. doi:10.1146/annurev.biochem.76.052305.091720
Santidrian AF et al (2013) Mitochondrial complex I activity and NAD+/NADH balance regulate breast cancer progression. J Clin Invest 123:1068–1081. doi:10.1172/JCI64264
Schroder M (2008) Endoplasmic reticulum stress responses. Cell Mol Life Sci 65:862–894. doi:10.1007/s00018-007-7383-5
Sideris DP, Tokatlidis K (2010) Trapping oxidative folding intermediates during translocation to the intermembrane space of mitochondria: in vivo and in vitro studies. Methods Mol Biol 619:411–423. doi:10.1007/978-1-60327-412-8_25
Siegelin MD et al (2011) Exploiting the mitochondrial unfolded protein response for cancer therapy in mice and human cells. J Clin Invest 121:1349–1360. doi:10.1172/JCI44855
Sugiura A, McLelland GL, Fon EA, McBride HM (2014) A new pathway for mitochondrial quality control: mitochondrial-derived vesicles. EMBO J doi: 10.15252/embj.201488104
Sundaresan NR et al (2009) Sirt3 blocks the cardiac hypertrophic response by augmenting FOXO3A-dependent antioxidant defense mechanisms in mice. J Clin Invest 119:2758–2771. doi:10.1172/JCI39162
Tao R et al (2010) Sirt3-mediated deacetylation of evolutionarily conserved lysine 122 regulates MnSOD activity in response to stress. Mol Cell 40:893–904. doi:10.1016/j.molcel.2010.12.013, S1097-2765(10)00969-X [pii]

Tatsuta T, Langer T (2008) Quality control of mitochondria: protection against neurodegeneration and ageing. EMBO J 27:306–314. doi:10.1038/sj.emboj.7601972

Taylor S et al (2015) Evaluating the evidence for targeting FOXO3A in breast cancer: a systematic review. Cancer Cell Int 15:1. doi:10.1186/s12935-015-0156-6

Viale A et al (2014) Oncogene ablation-resistant pancreatic cancer cells depend on mitochondrial function. Nature 514:628–632. doi:10.1038/nature13611

Wallace DC (2012) Mitochondria and cancer. Nat Rev Cancer 12:685–698. doi:10.1038/nrc3365

Wang M, Kaufman RJ (2014) The impact of the endoplasmic reticulum protein-folding environment on cancer development. Nat Rev Cancer 14:581–597. doi:10.1038/nrc3800

Wu SB, Wu YT, Wu TP, Wei YH (2014) Role of AMPK-mediated adaptive responses in human cells with mitochondrial dysfunction to oxidative stress. Biochim Biophys Acta 1840:1331–1344. doi:10.1016/j.bbagen.2013.10.034

Zhao Q et al (2002) A mitochondrial specific stress response in mammalian cells. EMBO J 21:4411–4419

Zhu Y et al (2002) Epidermal growth factor up-regulates the transcription of mouse lon homology ATP-dependent protease through extracellular signal-regulated protein kinase- and phosphatidylinositol-3-kinase-dependent pathways. Exp Cell Res 280:97–106

Chapter 9
Mitochondrial Signaling

Nuno Raimundo, Lorena Fernandez-Mosquera, and King Faisal Yambire

9.1 Introduction

The roles of mitochondria in the cell go well beyond the classic image of "energy factory" portrayed in introductory biology textbooks. The discovery of mutations in mitochondrial DNA (mtDNA) as a cause of disease (Holt et al. 1988; Wallace et al. 1988) opened the door to a new paradigm in which mitochondria were no longer seen as passive ATP-generating entities but rather as active participants in the biology of cells and tissues (Nunnari and Suomalainen 2012; Pagliarini and Rutter 2013; Raimundo 2014). Just 2 years later, the existence of a mechanism of communication from mitochondria to the nucleus was demonstrated in the Butow lab, using yeast as a model (Liao and Butow 1993). Shortly after, the identification of nuclear respiratory factor 1 (NRF1) as a major mediator of mitochondrial biogenesis by the Scarpulla lab established a mechanism of communication between the nucleus and mitochondria (Virbasius et al. 1993). These findings establish the proof-of-principle of bidirectional communication between mitochondria and the nucleus, underlying the need for a strict coordination between the two cellular genomes.

The mitochondria-to-nucleus pathway identified in yeast was referred as mitochondrial retrograde response (as opposed to the anterograde nucleus-to-mitochondria signaling). While the molecular mechanism involved in this response in yeast is not conserved in higher eukaryotes, the principle is valid, since other pathways evolved to mediate the "retrograde" communication between mitochondria and nucleus (Butow and Avadhani 2004). There is some confusion regarding to the terms used to describe the communication originating in mitochondria and leading to changes in nuclear gene expression in higher eukaryotes, with some

N. Raimundo (✉) • L. Fernandez-Mosquera • K.F. Yambire
Institute for Cellular Biochemistry, Universitätsmedizin Göttingen,
Humboldtallee 23, Room 01.423, 37073 Göttingen, Germany
e-mail: nuno.raimundo@med.uni-goettingen.de

D.M. Hockenbery (ed.), *Mitochondria and Cell Death*, Cell Death
in Biology and Diseases, DOI 10.1007/978-1-4939-3612-0_9

X.-M. Yin and Z. Dong (Series eds.), *Cell Death in Biology and Diseases*

researchers referring to it as mitochondrial stress signaling, and others reserving the term "retrograde response" for experimental systems based only on cell lines with mtDNA depletion by chemical treatments. Since the communication of mitochondria with the rest of the cell happens both under basal (unstressed) and stressed conditions, the terminology used in this chapter is rather "mitochondrial signaling." In the remainder of this chapter, this term does not refer specifically to the communication between mitochondria and the nucleus, but to the signaling originating in mitochondria and sensed outside mitochondria, either by cytoplasmic signaling pathways, eventually affecting nuclear gene expression, or by other cellular organelles.

The role that mitochondrial signaling plays in cells and tissues goes beyond academic interest. The widespread assumption that mitochondrial malfunction triggers pathology due to a decrease in ATP availability has been defied—blockage of the signaling pathways triggered by mitochondrial malfunction ablates pathology both at cellular and tissue level, despite the presence of the malfunctioning mitochondria (Raimundo 2014). The clear understanding of which signaling pathways are activated by different mitochondrial stresses presents therefore a clear path for the definition of therapeutic targets. Eventually, all signaling pathways triggered by mitochondrial malfunction start at the same point: a signal that is released from mitochondria and conveys stress.

9.2 The Conceptual Framework of Mitochondrial Signaling

In order to understand mitochondrial signaling, several aspects need to be addressed: the origin of the signals, their identity both inside and outside mitochondria, and their effect outside mitochondria. While these are essential to characterize the signals, the context must also be taken into account. The "signaling context" refers to the biological event that causes the signal (for example, is the nature of the mitochondrial perturbation a genetic defect directly affecting a mitochondrial protein or an event that was triggered outside mitochondria but impacts the function of the organelle?), but also to the intensity, duration, and frequency of the signal and, finally, to how that signal is interpreted by the cellular signaling environment. The interpretation of the signals is particularly important when the tissue-specificity of the phenotypes of mitochondrial diseases is taken into account, and it is complex to define. For example, if mitochondria in different tissues generate superoxide at the same rate, then tissues that have higher expression of SOD2 (superoxide dismutase 2, which converts superoxide into O_2 and hydrogen peroxide) will be able to maintain the steady-state levels of superoxide stable, and below a threshold to trigger signaling, while tissues where SOD2 expression is lower will be more likely to trigger superoxide-dependent signaling. This example illustrates how the same signal may be interpreted in a tissue-specific form. Furthermore, the tissue-specific signaling environment affects how mitochondrial signals are interpreted. For example, a mitochondrial signal that activates AMPK will have different consequences depending on the expression of AMPK subunits and basal AMPK activity in each tissue, as well as signaling pathways downstream of AMPK (e.g., mTORC1).

9.2.1 Definition

Despite several signaling pathways having been identified as responsive to mitochondrial signaling, the signals that trigger them remain often elusive. Part of the problem is the difficulty in establishing what exactly constitutes a signal, which is compounded by the inability of many molecules to cross the mitochondrial inner membrane and by the fact that the nature of the signal can change from inside to outside mitochondria (e.g., superoxide being converted to H_2O_2, which can diffuse across the mitochondrial membranes).

An effective mitochondrial signal has to present several characteristics. First, it needs to be generated by mitochondria. Second, its generation and elimination should occur rapidly to increase the levels of the signal upon certain conditions and reciprocally decrease them once the situation returns to normal. This aspect is particularly important for physiological signals responding to sudden changes in the environment (e.g., decrease in O_2 levels, change in catabolic substrates) compared to pathological signals which are chronic. Third, it needs to reach the cytoplasm and be readable by cellular signaling cascades. There are multiple entities matching these properties, of which ADP, AMP, and Ca^{2+} are pivotal examples: these molecules can be released by mitochondria and trigger AMPK and calcineurin signaling in the cytoplasm (Raimundo 2014).

Another group of molecules that fits the profile are the mitochondrial metabolites. They are generated inside mitochondria, mostly in the mitochondrial matrix, and have dedicated carriers to transport them across the inner membrane to the intermembrane space, from where they can easily diffuse across the outer membrane. While many of these metabolites can also be consumed or produced in the cytoplasm (Raimundo et al. 2011), their accumulation in mitochondria will result in a corresponding increase in their cytoplasmic levels (Isaacs et al. 2005; Raimundo et al. 2008; Selak et al. 2005). Metabolites have been seen for decades as mere intermediaries in a complex network of chemical reactions, but in the recent years their roles as signaling mediators have started to be unveiled.

Other entities, such as peptides and mitochondrial DNA (mtDNA), have been proposed as mitochondrial signals. Unfolded proteins in mitochondria can result in the efflux of peptides from mitochondria, in which the peptides constitute a signal that triggers a response akin to the endoplasmic reticulum unfolded protein response (Nargund et al. 2012). Instability in mtDNA has recently been shown to result in the release of mtDNA to the cytoplasm and activation of the antiviral innate immune response (West et al. 2015).

9.2.2 Origin

The origin of the signal conveying a particular type of mitochondrial stress depends on the nature of that stress, which of course depends on the primary cause being intrinsic or extrinsic to mitochondria. Intrinsic mitochondrial causes of stress would be mutations that affect nuclear genes encoding for proteins functioning in mitochondria, or mutations in mtDNA. Extrinsic causes could be cytoplasmic signaling pathways that perturb mitochondrial function, for example as it has been recently shown for the ERK pathway (Pyakurel et al. 2015).

The topology of signal generation is fundamentally important in mitochondria due to the impermeability of the inner mitochondrial membrane. The relevance of this barrier is best evidenced by the effect of superoxide production. Superoxide is a by-product of the respiratory chain generated in complexes I and III (Cadenas et al. 1977; Drose and Brandt 2012). It is currently assumed that it cannot diffuse through the membranes (although this is a contested matter), and the circumstances of its possible transport through the inner mitochondrial membrane remain unclear (Budzinska et al. 2009; Hawkins et al. 2007; Lustgarten et al. 2012; Mumbengegwi et al. 2008). While complex I releases superoxide only to the mitochondrial matrix, complex III releases it both to the mitochondrial matrix and the intermembrane space (Muller et al. 2004), most likely to the intra-cristae space where the respiratory chain complexes are known to preferentially locate (Scorrano 2013). The detoxification of superoxide is carried out by enzymes called superoxide dismutases, which convert superoxide into H_2O_2 and O_2 (Murphy 2009). Unlike superoxide, H_2O_2 can diffuse through membranes (Cadenas and Davies 2000). The two cellular superoxide dismutases, SOD1 and SOD2, have different intracellular locations. SOD1 is cytoplasmic and also found in the intermembrane space, and SOD2 is found in the mitochondrial matrix, but their relative amounts in the local compartments remain unclear (Missirlis et al. 2003; Zelko et al. 2002). It is also not clear in which area of the intermembrane space is SOD1 located, in particular if it i in the intracristae space or in the more diffuse areas between the inner and the outer membrane. Therefore, the place where superoxide is released is extremely important because it will influence the speed of its detoxification and the species that it can encounter. Particularly, superoxide released into the mitochondrial matrix can eventually react with numerous metabolic enzymes, eventually directly affecting metabolic fluxes, or with mtDNA, affecting the quality and/or quantity of mitochondrial gene expression (Murphy 2009).

Furthermore, even if superoxide would be eliminated with the same efficiency from the matrix and the intracristae space, the increase in the resulting product, H_2O_2, would have different consequences in the two compartments. Despite the ability of H_2O_2 to cross membranes, the availability of enzymes able to process H_2O_2 (e.g., glutathione peroxidases, peroxiredoxins) and their corresponding co-substrates (e.g., GSH) is different in the matrix and in the intermembrane space (Burdon et al. 1996; Franco et al. 1999).

9.2.3 Identity

The original signal triggered by mitochondrial malfunction is not necessarily the signal that is released by mitochondria. As discussed in the section above, an increase in superoxide generation in mitochondria can result in mitochondrial release of H_2O_2. But a burst of superoxide in the matrix can also result in the increase of cytoplasmic citrate: superoxide can inactivate the mitochondrial aconitase, thereby increasing the levels of citrate in the mitochondrial matrix, which will then result in release of citrate to the cytoplasm (Castro et al. 1994; Hausladen and Fridovich 1994; Murphy 2009).

9.2.4 Interpretation Outside Mitochondria

The fundamental objective of any signal is to be read and trigger the necessary response. The question of how mitochondrial signals are read in the cytoplasm naturally trails the identification of the actual signals. Some signals, such as H_2O_2, ADP, or AMP are known to activate AMPK signaling (Hardie 2011; Sena and Chandel 2012), and Ca^{2+} triggers calcineurin-dependent signaling (Szabadkai and Rizzuto 2013). Some metabolites with signaling properties, such as fumarate and succinate, trigger the hypoxia response through stabilization of hypoxia-inducible factor 1α (HIF1α) (Raimundo et al. 2011). Other metabolites affect the signaling environment in a way more related to their metabolic roles: acetyl-CoA is the donor of acetyl moieties for acetylation reactions, so an accumulation of Ac-CoA in the cytoplasm modifies the acetylation state of several proteins (Wellen et al. 2009; Zhao et al. 2010).

9.2.5 Termination

Any physiological signal must be eventually terminated. This can happen once the originating stress ends (e.g., hypoxia returned to normoxia). The termination of the signal can also be due to the transformation of the signal into another species that does not have the same signaling properties (e.g., the reduction of H_2O_2 to water by glutathione peroxidases or peroxiredoxins in mitochondria and cytoplasm or by catalase in the peroxisomes), by direct removal (e.g., uptake by another organelle of Ca^{2+} released by mitochondria) or by selective removal of the individual mitochondria which are producing the pathological signaling. The issue of selective autophagy of damaged mitochondria is worthy of a long discussion on its own, and therefore will not be addressed in detail here (Youle and Narendra 2011).

Chronic mitochondrial stress is less likely to include signal termination, since the stress persists for a long period of time. Eventually, the inability to terminate the signal contributes to transform a stress signal into a pathological signal, of which the sustained hyperactivation of AMPK in chronic models of mitochondrial malfunction is a pivotal example (Bokko et al. 2007; Raimundo et al. 2012).

9.3 Mitochondrial Signals

This section aims to discuss many of the signals that are known to convey information about the mitochondrial compartment to the cytoplasm. It is naturally impossible to address them all in detail, but the goal is to provide an overview of the different strategies that mitochondria can use to communicate their status to the rest of the cell. Below, four examples of mitochondrial signaling are presented in more detail in order to illustrate the concept in more practical terms.

Example 1. Mitochondrial depolarization by triggering the permeability transition pore or by treating cells with a mitochondrial uncoupler results in an increase in cytoplasmic Ca^{2+} concentration. This triggers the Ca^{2+}-dependent phosphatase calcineurin, which dephosphorylates the protein dynamin-related protein 1 (DRP1), allowing it to relocate from the cytoplasm to the mitochondrial outer membrane, where it triggers mitochondrial fission, causing a marked fragmentation of the mitochondrial network (Cereghetti et al. 2008).

Example 2. Treatment of cultured cells with an uncoupler leads to a rapid accumulation of the kinase PINK1 in mitochondria. Under basal conditions, PINK1 is imported by the mitochondrial protein import machinery all the way to the inner membrane, where it is constitutively degraded by the protease PARL. However, the import of PINK1 to the inner membrane is dependent on the mitochondrial membrane potential (Jin et al. 2010). Upon treatment with the uncoupler, PINK1 can no longer be imported to the inner membrane and is mistargeted to the outer mitochondrial membrane, where it accumulates. This constitutes a signal that the E3 ubiquitin ligase Parkin recognizes, leading to the ubiquitination of several proteins in the outer mitochondrial membrane, marking those uncoupled mitochondria for autophagic degradation.

Example 3. Perturbations of the respiratory chain function can result in increased production of superoxide, resulting in the release of a superoxide-dependent signal that activates AMPK (Emerling et al. 2009; Raimundo et al. 2012). The sustained activation of AMPK upon mitochondrial stress leads to increased cell death propensity, under some conditions (Raimundo et al. 2012). The inhibition of AMPK or the ablation of its pro-apoptotic downstream targets results in the removal of the pathological signaling both in cultured cells and in vivo (Bokko et al. 2007; Raimundo et al. 2012).

Example 4. Mutations in genes encoding subunits of the respiratory chain may cause increased superoxide production in the matrix (Lenaz et al. 2006). Superoxide has affinity for Fe-S centers, and it can react with mitochondrial aconitase, inactivating it (Castro et al. 1994; Hausladen and Fridovich 1994; Missirlis et al. 2003; Muller et al. 2004). This would lead to the inability of converting citrate to isocitrate in the matrix, with the consequent accumulation of citrate, which would be exported to the cytoplasm. Citrate is a known allosteric regulator of a number of enzymes (Srere 1992), thus transmitting a message of stress from mitochondria and coordinating the corresponding adaptation in cytoplasm (e.g., glycolysis inhibition).

9.3.1 ATP/ADP

The role of the balance between ATP and ADP is known for several decades as a major allosteric regulator of many enzymes involved in energy metabolism and biosynthesis. Rather than the amount of ATP or ADP specifically, those enzymes respond to what is recognized as the "energy charge," which is represented by the ratio ([ATP] + 0.5[ADP])/([ATP] + [ADP] + [AMP]). Besides their allosteric effects, these three molecules also directly regulate a major signaling pathway in cells, the AMPK pathway (Hardie 2011). Increases in ADP or AMP, usually reflecting corresponding decreases in ATP, result in the activation of the AMPK pathway (Hardie 2011).

9.3.2 NAD^+/NADH

Changes in the ratio between NAD^+ and NADH reflect an evident signal of mitochondrial malfunction: the respiratory chain is not transferring electrons from NADH at the same speed as the molecule is being reduced, resulting in a decrease of the NAD^+/NADH ratio. Since glycolysis requires NAD^+ to proceed, and the cell cannot afford to completely stop ATP production, when the levels of NAD^+ decrease below a critical threshold, pyruvate oxidizes NADH, generating lactate and NAD^+, which can then be used by glycolytic enzymes. For this reason, increases in lactate, or in the ratio lactate/pyruvate, are widely used as markers for mitochondrial malfunction, including for diagnosis of many mitochondrial diseases (DiMauro and Schon 2003; Robinson 2006).

The NAD^+ molecule is a long-known component of many key catabolic reactions, but it also has fundamental roles besides redox chemistry: NAD^+ is a co-substrate of sirtuin deacetylases, which have widespread roles in the regulation of key cellular processes (Hall et al. 2013).

9.3.3 Ca^{2+}

Mitochondria is one of the organelles with major roles in cellular Ca^{2+} buffering, together with lysosomes and the endoplasmic reticulum. Upon mitochondrial malfunction, mitochondria will be less able to uptake Ca^{2+} and often release Ca^{2+} to the cytoplasm. This will result in increased cytoplasmic Ca^{2+} concentration, which triggers the activation of the Ca^{2+}-dependent phosphatase calcineurin. This protein affects a wide array of targets, regulating processes ranging from mitochondrial fragmentation (Cereghetti et al. 2008) to lysosomal biogenesis and autophagy (Medina et al. 2015). It is not yet completely clear what are the specificities of Ca^{2+} signaling that are elicited by each of the organelles. Importantly, changes of Ca^{2+} concentration inside mitochondria will trigger further changes at a metabolic level, since Ca^{2+} is an important regulator of key metabolic pathways such as the citrate cycle (Srere 1992). This situation illustrates how often there is not "one" mitochondrial signal but rather a set of changes that together signal mitochondrial stress to other organelles and to cytoplasmic signaling pathways.

9.3.4 Reactive Oxygen Species

While reactive oxygen species (ROS) were seen for a long time as toxic byproducts of the respiratory chain, their roles as signaling mediators is now established (Raimundo 2014; Rhee et al. 2005; Sena and Chandel 2012). ROS comprise the molecules in which one oxygen atom has one unpaired electron (e.g., superoxide, hydroxyl), but the definition is usually loosened to include H_2O_2. H_2O_2 has no unpaired electrons but can generate other ROS through Fenton chemistry (Cadenas and Davies 2000). The major source of ROS in the mitochondrial respiratory chain are complexes I and III (Murphy 2009). One interesting aspect of ROS signaling is that superoxide generated by respiratory chain complex I is released exclusively to the mitochondrial matrix, while the complex III-derived superoxide is released both to the matrix and the intra-cristae space (Missirlis et al. 2003; Muller et al. 2004). This is the specialized area of the intermembrane space that is localized "inside" the cristae, since it is now evident that the cristae are a partly impermeable compartment (Frezza et al. 2006; Scorrano 2013).

The role of ROS as signaling mediators has been studied for about 20 years now, and the volume of knowledge generated is too vast to address in detail here.

9.3.5 Reactive Nitrogen Species

Nitric oxide (NO) is the flagship of the reactive nitrogen species (RNS). RNS have one unpaired electron in one nitrogen atom. Besides NO, the other RNS with major relevance is peroxynitrite, which results from the reaction of NO with superoxide. Both NO and peroxynitrite can be generated in mitochondria and diffuse across membranes (Marla et al. 1997). Furthermore, they both have the capacity to inhibit the mitochondrial respiratory chain (Antunes et al. 2004) and to trigger signaling leading to mitochondrial biogenesis (Nisoli et al. 2003)—another example of how one signal triggers other signals. Additionally, NO can react with cysteine residues of proteins and modify them by a process of S-nitrosylation, which affects their signaling properties (Nakamura et al. 2013).

9.3.6 *Metabolites*

Metabolites present many of the key characteristics expected from a mitochondrial signal—particularly, those that are generated within mitochondria and can easily shuttle to the cytoplasm through dedicated carriers. Acetyl-CoA (Ac-CoA) presents a good example of mitochondrial metabolite with signaling potential. It is energetically expensive to make and it can be readily convertible to citrate. Ac-CoA cannot cross the mitochondrial membranes, but it can signal outside mitochondria through citrate: mitochondrial Ac-CoA is converted to citrate, which can shuttle out of mitochondria though its carrier, and be reconverted to Ac-CoA by ATP-citrate lyase in the cytoplasm (Wellen et al. 2009). Cytoplasmic Ac-CoA levels are major mediators of protein acetylation, which has widespread impacts on cellular signaling pathways as well as biological processes such as autophagy (Eisenberg et al. 2014; Marino et al. 2014).

The role of metabolites as signaling mediators does not end with Ac-CoA. In fact, the whole concept of metabolites affecting cellular signaling was boosted by the genetic mutations in fumarate hydratase and succinate dehydrogenase which lead to tumor formation (Baysal et al. 2000; Baysal et al. 2002; Tomlinson et al. 2002). These mutations lead to the accumulation of fumarate and succinate, which have the ability to inhibit the alpha-ketoglutarate-dependent dioxygenase superfamily (Isaacs et al. 2005; Selak et al. 2005). This impacts several signaling pathways, of which hypoxia signaling has particular importance for the tumorigenesis process in the patients with those mutations (Raimundo et al. 2011). However, it should not be understood that the metabolites can only regulate signaling through allosteric regulation. Fumarate accumulation can also repress the differentiation of smooth muscle progenitor cells, through repression of serum response factor (SRF)-dependent signaling (Raimundo et al. 2009). Fumarate can react with lysine residues of proteins resulting in a process called succinylation, in which one adduct between fumarate and the lysine residue is formed (Zhang et al. 2011). This process is likely to have a wide impact in many signaling pathways, and is demonstrated to affect NRF2 (nuclear factor erythroid 2-like) signaling, the major regulator of antioxidant responses (Adam et al. 2011).

As demonstrated for Ac-CoA and fumarate, the reaction between metabolites and proteins, provides another layer of interaction between metabolism and the regulation of the signaling environment. Another line of evidence of this interaction comes from arginine. Arginine can react with lysine residues and modify proteins by "arginylation," with enzyme catalysis by arginyl transferase 1 (ATE1) (Hu et al. 2006; Saha and Kashina 2011). Arginylation can modify the half-life of proteins, thus providing yet another way for metabolites to regulate signaling pathways (Tasaki et al. 2012).

9.3.7 Mitochondria-Derived Peptides

Mitochondria import over 1000 proteins translated in the cytoplasmic ribosomes. The last step of the complex import pathway is the folding of the proteins to their final conformation. Upon stress, the proteins in charge of protein folding can be overloaded and thus lead to the accumulation of unfolded proteins in the mitochondrial matrix, which would trigger a stress pathway designated mitochondrial unfolded protein response (mitoUPR) (Haynes et al. 2013; Pellegrino et al. 2013). MitoUPR has many similarities with the endoplasmic reticulum UPR. Of particular significance, mitoUPR seems to be signaled to the cytoplasm by peptides, at least in *C. elegans*. The unfolded proteins get degraded by quality control proteases, yielding peptides of 8–20 amino acids in length, which are then exported from the mitochondrial matrix to the intermembrane space by an ATP-binding cassette transporter, and then diffuse across the outer membrane to the cytoplasm (Nargund et al. 2012). Once in the cytoplasm, these peptides activate a transcription factor which drives the mitoUPR.

9.3.8 Mitochondria-Derived Vesicles

Mitochondria emit vesicular trafficking directed either to the peroxisomes or to the lysosomes under regulated circumstances (Braschi et al. 2010; Neuspiel et al. 2008; Soubannier et al. 2012). The contents of the vesicles directed to each of these organelles is different. While the vesicles directed to the lysosomes seem to be associated with a mechanism of mitochondrial quality control, it remains to be determined how these mitochondria-derived vesicles contribute to the communication between mitochondria and other organelles. The effects of these vesicles on the target organelles, still unknown, are particularly tantalizing.

9.3.9 Mitochondria-Derived mtDNA

The maintenance of mtDNA is performed by nuclear-encoded proteins (Bonawitz et al. 2006). No histones are found in mitochondria, but the protein TFAM functions as a *bona fide* histone, conferring protection to mtDNA (TFAM is also necessary for mtDNA transcription) by packaging it (Ekstrand et al. 2004). The levels of TFAM are directly proportional to the levels of mtDNA, confirmed both by overexpression and knockdown and knockout of TFAM (Larsson et al. 1998; Ylikallio et al. 2010). Upon partial loss of TFAM (TFAM+/– mouse embryonic fibroblasts and macrophages), the packaging of mtDNA becomes aberrant, resulting in mtDNA degradation but also in the release of some mtDNA fragments to the cytoplasm, where it activates an antiviral response (West et al. 2015). A similar response was observed upon infection of the cells with herpesvirus, implying a role for mtDNA as a mitochondrial signal in innate immunity (West et al. 2015).

9.4 Signaling Threshold and Tissue-Specificity

Mitochondrial diseases have been the subject of solid research efforts, which led to the characterization of many syndromes and the identification of many disease genes (Koopman et al. 2012). One particular aspect of mitochondrial diseases is simultaneously clear and obscure: the tissue-specificity (DiMauro 2004; DiMauro and Schon 2003; Koopman et al. 2012). One particular mutation can result in different syndromes, particularly more significant when considering mtDNA mutations, since then a number of nuclear modifiers, many of them still unknown, also contributes to the penetration of the mutation. The mtDNA mutation A3243G, in the gene encoding for one of two $tRNA^{Leu}$ is the most prevalent mtDNA mutation, and can result in two different syndromes: mitochondrial encephalopathy, lactic acidosis, and stroke-like episodes (MELAS) or maternally inherited diabetes and deafness (DiMauro 2004; DiMauro and Schon 2003; Koopman et al. 2012). It remains unclear how the same mutation can result in two different diseases, affecting different tissues. This is further complicated by the fact that mutations in other genes (most of them mtDNA-encoded tRNAs, but also three subunits of complex I, ND1, ND5, and ND6) also cause MELAS, while some other mutations in nuclear-encoded complex I subunits cause a different clinical presentation (Leigh syndrome) (Crimi et al. 2003; Lamperti et al. 2012; Malfatti et al. 2007; Mancuso et al. 2014). The objective of this section, and chapter, is not to dwell into the complexity of mitochondrial diseases, but to highlight that it remains unclear why certain mitochondrial diseases manifest in some tissues, but not in others. This can be illustrated by three mtDNA depletion syndromes, all characterized by a decrease in the amount of mtDNA. MNGIE (mitochondrial neurogastrointestinal encephalopathy syndrome) caused by mutations in thymidine phosphorylase leads to encephalopathy without affecting the liver, while mutations in DGUOK cause hepatic dysfunction and encephalopathy (Mandel et al. 2001), and mutations in TK2 affect predominantly the muscle (Saada et al. 2001).

The present postulate is that when mitochondrial malfunction reaches a certain threshold in a certain tissue it will trigger pathology. For mtDNA mutations this is certainly true, since the amount of mutant mtDNA determines the occurrence of pathology and its severity. For example, early-onset MELAS is associated with higher amounts of mutant mtDNA while later-onset has slightly lower mutant mtDNA. However, it is not linear to transfer this notion to nuclear-encoded genes, since 0 (control individuals), 50 or 100 % of the protein product can present the mutation. This gets further complicated if the mutation affects the stability of the RNA or protein, in which case there may be no protein product at all.

In conceptual terms, the activation of a signaling pathway in response to mitochondrial stress represents an attempt of the cells and tissues to adapt to the defect. In the case of mutations causing disease, it is possible that some tissues are more able than others to terminate the mitochondrial signals that convey certain stresses. Naturally, this depends on the stress, the signal and the tissue. If the signal is superoxide, tissues that have higher expression of superoxide dismutases will be more likely to quench it and the downstream signaling than tissues with lower expression of the enzymes. In another level of regulation, since damaged mitochondria are removed by selective autophagy (Novak and Dikic 2011; Youle and Narendra 2011), tissues in which the autophagic capacity is higher can probably deal with the problem, while others in which there is not so much availability of lysosomes would undergo saturation of the autophagic pathway, resulting in the accumulation of damaged mitochondria and continued pathological signaling (Raimundo 2014), as observed in the absence of autophagy (Tal et al. 2009).

9.5 Inter-organelle Signaling

The mitochondrial stress responses have traditionally been studied as if the cells contained only mitochondria, the nuclear gene expression machinery and the signaling pathways that regulate them. However, in the last years a wealth of information has accumulated demonstrating that mitochondria have interactions, some involving physical contacts, with many other cellular organelles. For example, mitochondria have direct contacts with endoplasmic reticulum (ER), which regulate Ca^{2+} homeostasis and ER stress response (de Brito and Scorrano 2008; Murley et al. 2013; Rowland and Voeltz 2012). ER also delivers several lipid and protein components to peroxisomes, which share a key fission protein (Drp1) with mitochondria (Thoms et al. 2009). The signaling via PGC-1α and PPARγ drives the biogenesis of both mitochondria and peroxisomes (Thoms et al. 2009). Lysosomes are essential for the removal of damaged mitochondria and peroxisomes (Nixon 2013). Moreover, upon destabilization of the lysosomal membrane, the cross-talk between lysosomes and mitochondria generates a feed-forward loop that promotes apoptosis (Aits and Jaattela 2013). Furthermore, mitochondria release vesicles directed to peroxisomes and lysosomes, under regulated conditions and with defined composition, further highlighting the relationship of these three organelles (Braschi et al. 2010; Neuspiel et al. 2008; Soubannier et al. 2012). It has also been recently shown that in yeast mitochondria and lysosomes have physical contacts (Elbaz-Alon et al. 2014), and while the molecular regulators are not conserved it is possible that the existence of these contacts exists also in higher eukaryotes.

The importance of interactions between mitochondria and other organelles is further evidenced by disease phenotypes, as primary defects in peroxisomes or lysosomes often result in secondary mitochondrial perturbations. Lysosomal storage diseases (e.g., Pompe's disease) show abnormalities in peroxisomal metabolism, as well as in mitochondrial function (Raben et al. 2012; Selak et al. 2000; Takikita et al. 2010). Peroxisomal diseases (e.g., Zellweger syndrome) affect mitochondrial structure, redox balance, and metabolism (Dirkx et al. 2005; Wanders 2013). Interestingly, this seems to be a two-way street, since mitochondrial myopathies (e.g., progressive external ophthalmoplegia) seem to perturb lysosomal autophagic capacity leading to the accumulation of autophagosomes (Tyynismaa et al. 2005). Additionally, mitochondrial β-oxidation disorders trigger peroxisomal biogenesis (Janssen and Stoffel 2002), showing that rather than resulting in simple metabolic blockage, defects in one organelle can induce the biogenesis of another.

The reductionist view that these organelles are isolated entities in the cell is therefore no longer acceptable (McBride 2015; Raimundo 2014). While the contact sites between mitochondria and endoplasmic reticulum have received a lot of attention from the scientific community, it remains poorly understood how mitochondria interact with peroxisomes and lysosomes, but also how mitochondrial perturbations affect the formation and function of mitochondria-ER contact sites. Most importantly, it remains to be determined how the effect of mitochondrial stress on other organelles contributes to the global cellular response to mitochondrial malfunction.

9.6 Concluding Remarks

Mitochondrial signaling is an exciting place to be, full of mysteries that drive the imagination of scientists, full of surprises that overturn concepts and long-held paradigms. Several major questions are being intensively addressed, and poised to make leaps in the current knowledge.

First, the mitochondrial signals, and especially the roles of metabolites as such, including lipids, which have received far less attention. The emerging methodologies of metabolomics have reached unprecedented levels of detail, particularly NMR-based metabolomics that have temporal resolution. It is likely that the list of metabolites with signaling roles will increase, and the role of metabolite-mediated posttranslational modifications is likely to take center stage, given its over-reaching implications for cellular biology and molecular medicine.

Next, the signaling pathways. Given the recent advances in technologies for gene editing, it is expected that new animal models of mitochondrial malfunction will be generated in the coming years, with particular incidence in mouse models of disease. The study of signaling pathways in organisms, possibly taking advantage of new methodologies of RNA sequencing and quantitative proteomics, is likely to provide key developments towards the understanding of tissue-specific signaling and pathology. Furthermore, the use of animal models will allow the understanding of mitochondrial signaling at a systems level—is one given tissue developing pathology because of intrinsic or extrinsic reasons (e.g., due to endocrine mediators released by another tissue, in a similar pattern that explains the loss of muscle mass in cancer patients)? A better understanding of the signaling mechanisms will also lead to more targeted approaches for therapeutic strategies.

Finally, the biology of mitochondrial signaling. In recent years it was shown that peptides, mtDNA fragments and mitochondria-derived vesicles can function as mitochondrial signals. It would be hard to imagine that there are no more surprises in the treasure chest. The scientific community will certainly learn a great deal, in the near future, in how these "new" signals function, how they are regulated, and what is their impact in physiology and pathology. And, of course, the interaction between organelles. How does malfunction of one organelle affect others, how do the other organelles signal their mitochondria-induced stress stage, what is the contribution of the network between mitochondria and ER and lysosomes and peroxisomes (and others?) for cellular life and death.

References

Adam J, Hatipoglu E, O'Flaherty L, Ternette N, Sahgal N, Lockstone H, Baban D, Nye E, Stamp GW, Wolhuter K et al (2011) Renal cyst formation in Fh1-deficient mice is independent of the Hif/Phd pathway: roles for fumarate in KEAP1 succination and Nrf2 signaling. Cancer Cell 20:524–537

Aits S, Jaattela M (2013) Lysosomal cell death at a glance. J Cell Sci 126:1905–1912

Antunes F, Boveris A, Cadenas E (2004) On the mechanism and biology of cytochrome oxidase inhibition by nitric oxide. Proc Natl Acad Sci U S A 101:16774–16779

Baysal BE, Ferrell RE, Willett-Brozick JE, Lawrence EC, Myssiorek D, Bosch A, van der Mey A, Taschner PE, Rubinstein WS, Myers EN et al (2000) Mutations in SDHD, a mitochondrial complex II gene, in hereditary paraganglioma. Science 287:848–851

Baysal BE, Willett-Brozick JE, Lawrence EC, Drovdlic CM, Savul SA, McLeod DR, Yee HA, Brackmann DE, Slattery WH 3rd, Myers EN et al (2002) Prevalence of SDHB, SDHC, and SDHD germline mutations in clinic patients with head and neck paragangliomas. J Med Genet 39:178–183

Bokko PB, Francione L, Bandala-Sanchez E, Ahmed AU, Annesley SJ, Huang X, Khurana T, Kimmel AR, Fisher PR (2007) Diverse cytopathologies in mitochondrial disease are caused by AMP-activated protein kinase signaling. Mol Biol Cell 18:1874–1886

Bonawitz ND, Clayton DA, Shadel GS (2006) Initiation and beyond: multiple functions of the human mitochondrial transcription machinery. Mol Cell 24:813–825

Braschi E, Goyon V, Zunino R, Mohanty A, Xu L, McBride HM (2010) Vps35 mediates vesicle transport between the mitochondria and peroxisomes. Curr Biol 20:1310–1315

Budzinska M, Galganska H, Karachitos A, Wojtkowska M, Kmita H (2009) The TOM complex is involved in the release of superoxide anion from mitochondria. J Bioenerg Biomembr 41:361–367

Burdon RH, Gill V, Alliangana D (1996) Hydrogen peroxide in relation to proliferation and apoptosis in BHK-21 hamster fibroblasts. Free Radic Res 24:81–93

Butow RA, Avadhani NG (2004) Mitochondrial signaling: the retrograde response. Mol Cell 14:1–15

Cadenas E, Davies KJ (2000) Mitochondrial free radical generation, oxidative stress, and aging. Free Radic Biol Med 29:222–230

Cadenas E, Boveris A, Ragan CI, Stoppani AO (1977) Production of superoxide radicals and hydrogen peroxide by NADH-ubiquinone reductase and ubiquinol-cytochrome c reductase from beef-heart mitochondria. Arch Biochem Biophys 180:248–257

Castro L, Rodriguez M, Radi R (1994) Aconitase is readily inactivated by peroxynitrite, but not by its precursor, nitric oxide. J Biol Chem 269:29409–29415

Cereghetti GM, Stangherlin A, Martins de Brito O, Chang CR, Blackstone C, Bernardi P, Scorrano L (2008) Dephosphorylation by calcineurin regulates translocation of Drp1 to mitochondria. Proc Natl Acad Sci U S A 105:15803–15808

Crimi M, Galbiati S, Moroni I, Bordoni A, Perini MP, Lamantea E, Sciacco M, Zeviani M, Biunno I, Moggio M et al (2003) A missense mutation in the mitochondrial ND5 gene associated with a Leigh-MELAS overlap syndrome. Neurology 60:1857–1861

de Brito OM, Scorrano L (2008) Mitofusin 2 tethers endoplasmic reticulum to mitochondria. Nature 456:605–610

DiMauro S (2004) Mitochondrial diseases. Biochim Biophys Acta 1658:80–88

DiMauro S, Schon EA (2003) Mitochondrial respiratory-chain diseases. N Engl J Med 348:2656–2668

Dirkx R, Vanhorebeek I, Martens K, Schad A, Grabenbauer M, Fahimi D, Declercq P, Van Veldhoven PP, Baes M (2005) Absence of peroxisomes in mouse hepatocytes causes mitochondrial and ER abnormalities. Hepatology 41:868–878

Drose S, Brandt U (2012) Molecular mechanisms of superoxide production by the mitochondrial respiratory chain. Adv Exp Med Biol 748:145–169

Eisenberg T, Schroeder S, Andryushkova A, Pendl T, Kuttner V, Bhukel A, Marino G, Pietrocola F, Harger A, Zimmermann A et al (2014) Nucleocytosolic depletion of the energy metabolite acetyl-coenzyme a stimulates autophagy and prolongs lifespan. Cell Metab 19:431–444

Ekstrand MI, Falkenberg M, Rantanen A, Park CB, Gaspari M, Hultenby K, Rustin P, Gustafsson CM, Larsson NG (2004) Mitochondrial transcription factor A regulates mtDNA copy number in mammals. Hum Mol Genet 13:935–944

Elbaz-Alon Y, Rosenfeld-Gur E, Shinder V, Futerman AH, Geiger T, Schuldiner M (2014) A dynamic interface between vacuoles and mitochondria in yeast. Dev Cell 30:95–102

Emerling BM, Weinberg F, Snyder C, Burgess Z, Mutlu GM, Viollet B, Budinger GR, Chandel NS (2009) Hypoxic activation of AMPK is dependent on mitochondrial ROS but independent of an increase in AMP/ATP ratio. Free Radic Biol Med 46:1386–1391

Franco AA, Odom RS, Rando TA (1999) Regulation of antioxidant enzyme gene expression in response to oxidative stress and during differentiation of mouse skeletal muscle. Free Radic Biol Med 27:1122–1132

Frezza C, Cipolat S, Martins de Brito O, Micaroni M, Beznoussenko GV, Rudka T, Bartoli D, Polishuck RS, Danial NN, De Strooper B et al (2006) OPA1 controls apoptotic cristae remodeling independently from mitochondrial fusion. Cell 126:177–189

Hall JA, Dominy JE, Lee Y, Puigserver P (2013) The sirtuin family's role in aging and age-associated pathologies. J Clin Invest 123:973–979

Hardie DG (2011) AMP-activated protein kinase: an energy sensor that regulates all aspects of cell function. Genes Dev 25:1895–1908

Hausladen A, Fridovich I (1994) Superoxide and peroxynitrite inactivate aconitases, but nitric oxide does not. J Biol Chem 269:29405–29408

Hawkins BJ, Madesh M, Kirkpatrick CJ, Fisher AB (2007) Superoxide flux in endothelial cells via the chloride channel-3 mediates intracellular signaling. Mol Biol Cell 18:2002–2012

Haynes CM, Fiorese CJ, Lin YF (2013) Evaluating and responding to mitochondrial dysfunction: the mitochondrial unfolded-protein response and beyond. Trends Cell Biol 23:311–318

Holt IJ, Harding AE, Morgan-Hughes JA (1988) Deletions of muscle mitochondrial DNA in patients with mitochondrial myopathies. Nature 331:717–719

Hu RG, Brower CS, Wang H, Davydov IV, Sheng J, Zhou J, Kwon YT, Varshavsky A (2006) Arginyltransferase, its specificity, putative substrates, bidirectional promoter, and splicing-derived isoforms. J Biol Chem 281:32559–32573

Isaacs JS, Jung YJ, Mole DR, Lee S, Torres-Cabala C, Chung YL, Merino M, Trepel J, Zbar B, Toro J et al (2005) HIF overexpression correlates with biallelic loss of fumarate hydratase in renal cancer: novel role of fumarate in regulation of HIF stability. Cancer Cell 8:143–153

Janssen U, Stoffel W (2002) Disruption of mitochondrial beta -oxidation of unsaturated fatty acids in the 3,2-trans-enoyl-CoA isomerase-deficient mouse. J Biol Chem 277:19579–19584

Jin SM, Lazarou M, Wang C, Kane LA, Narendra DP, Youle RJ (2010) Mitochondrial membrane potential regulates PINK1 import and proteolytic destabilization by PARL. J Cell Biol 191:933–942

Koopman WJ, Willems PH, Smeitink JA (2012) Monogenic mitochondrial disorders. N Engl J Med 366:1132–1141

Lamperti C, Diodato D, Lamantea E, Carrara F, Ghezzi D, Mereghetti P, Rizzi R, Zeviani M (2012) MELAS-like encephalomyopathy caused by a new pathogenic mutation in the mitochondrial DNA encoded cytochrome c oxidase subunit I. Neuromuscul Disord 22:990–994

Larsson NG, Wang J, Wilhelmsson H, Oldfors A, Rustin P, Lewandoski M, Barsh GS, Clayton DA (1998) Mitochondrial transcription factor A is necessary for mtDNA maintenance and embryogenesis in mice. Nat Genet 18:231–236

Lenaz G, Baracca A, Fato R, Genova ML, Solaini G (2006) New insights into structure and function of mitochondria and their role in aging and disease. Antioxid Redox Signal 8:417–437

Liao X, Butow RA (1993) RTG1 and RTG2: two yeast genes required for a novel path of communication from mitochondria to the nucleus. Cell 72:61–71

Lustgarten MS, Bhattacharya A, Muller FL, Jang YC, Shimizu T, Shirasawa T, Richardson A, Van Remmen H (2012) Complex I generated, mitochondrial matrix-directed superoxide is released from the mitochondria through voltage dependent anion channels. Biochem Biophys Res Commun 422:515–521

Malfatti E, Bugiani M, Invernizzi F, de Souza CF, Farina L, Carrara F, Lamantea E, Antozzi C, Confalonieri P, Sanseverino MT et al (2007) Novel mutations of ND genes in complex I deficiency associated with mitochondrial encephalopathy. Brain 130:1894–1904

Mancuso M, Orsucci D, Angelini C, Bertini E, Carelli V, Comi GP, Donati A, Minetti C, Moggio M, Mongini T et al (2014) The m.3243A>G mitochondrial DNA mutation and related phenotypes. A matter of gender? J Neurol 261:504–510

Mandel H, Szargel R, Labay V, Elpeleg O, Saada A, Shalata A, Anbinder Y, Berkowitz D, Hartman C, Barak M et al (2001) The deoxyguanosine kinase gene is mutated in individuals with depleted hepatocerebral mitochondrial DNA. Nat Genet 29:337–341

Marino G, Pietrocola F, Eisenberg T, Kong Y, Malik SA, Andryushkova A, Schroeder S, Pendl T, Harger A, Niso-Santano M et al (2014) Regulation of autophagy by cytosolic acetyl-coenzyme A. Mol Cell 53:710–725

Marla SS, Lee J, Groves JT (1997) Peroxynitrite rapidly permeates phospholipid membranes. Proc Natl Acad Sci U S A 94:14243–14248

McBride HM (2015) Open questions: seeking a holistic approach for mitochondrial research. BMC Biol 13:8

Medina DL, Di Paola S, Peluso I, Armani A, De Stefani D, Venditti R, Montefusco S, Scotto-Rosato A, Prezioso C, Forrester A et al (2015) Lysosomal calcium signalling regulates autophagy through calcineurin and TFEB. Nat Cell Biol 17:288–299

Missirlis F, Hu J, Kirby K, Hilliker AJ, Rouault TA, Phillips JP (2003) Compartment-specific protection of iron-sulfur proteins by superoxide dismutase. J Biol Chem 278:47365–47369

Muller FL, Liu Y, Van Remmen H (2004) Complex III releases superoxide to both sides of the inner mitochondrial membrane. J Biol Chem 279:49064–49073

Mumbengegwi DR, Li Q, Li C, Bear CE, Engelhardt JF (2008) Evidence for a superoxide permeability pathway in endosomal membranes. Mol Cell Biol 28:3700–3712

Murley A, Lackner LL, Osman C, West M, Voeltz GK, Walter P, Nunnari J (2013) ER-associated mitochondrial division links the distribution of mitochondria and mitochondrial DNA in yeast. eLife 2:e00422

Murphy MP (2009) How mitochondria produce reactive oxygen species. Biochem J 417:1–13

Nakamura T, Tu S, Akhtar MW, Sunico CR, Okamoto S, Lipton SA (2013) Aberrant protein s-nitrosylation in neurodegenerative diseases. Neuron 78:596–614

Nargund AM, Pellegrino MW, Fiorese CJ, Baker BM, Haynes CM (2012) Mitochondrial import efficiency of ATFS-1 regulates mitochondrial UPR activation. Science 337:587–590

Neuspiel M, Schauss AC, Braschi E, Zunino R, Rippstein P, Rachubinski RA, Andrade-Navarro MA, McBride HM (2008) Cargo-selected transport from the mitochondria to peroxisomes is mediated by vesicular carriers. Curr Biol 18:102–108

Nisoli E, Clementi E, Paolucci C, Cozzi V, Tonello C, Sciorati C, Bracale R, Valerio A, Francolini M, Moncada S et al (2003) Mitochondrial biogenesis in mammals: the role of endogenous nitric oxide. Science 299:896–899

Nixon RA (2013) The role of autophagy in neurodegenerative disease. Nat Med 19:983–997

Novak I, Dikic I (2011) Autophagy receptors in developmental clearance of mitochondria. Autophagy 7:301–303

Nunnari J, Suomalainen A (2012) Mitochondria: in sickness and in health. Cell 148:1145–1159

Pagliarini DJ, Rutter J (2013) Hallmarks of a new era in mitochondrial biochemistry. Genes Dev 27:2615–2627

Pellegrino MW, Nargund AM, Haynes CM (2013) Signaling the mitochondrial unfolded protein response. Biochim Biophys Acta 1833:410–416

Pyakurel A, Savoia C, Hess D, Scorrano L (2015) Extracellular regulated kinase phosphorylates mitofusin 1 to control mitochondrial morphology and apoptosis. Mol Cell 58(2):244–254

Raben N, Wong A, Ralston E, Myerowitz R (2012) Autophagy and mitochondria in Pompe disease: nothing is so new as what has long been forgotten. Am J Med Genet C: Semin Med Genet 160:13–21

Raimundo N (2014) Mitochondrial pathology: stress signals from the energy factory. Trends Mol Med 20:282–292

Raimundo N, Ahtinen J, Fumic K, Baric I, Remes AM, Renkonen R, Lapatto R, Suomalainen A (2008) Differential metabolic consequences of fumarate hydratase and respiratory chain defects. Biochim Biophys Acta 1782:287–294

Raimundo N, Vanharanta S, Aaltonen LA, Hovatta I, Suomalainen A (2009) Downregulation of SRF-FOS-JUNB pathway in fumarate hydratase deficiency and in uterine leiomyomas. Oncogene 28:1261–1273

Raimundo N, Baysal BE, Shadel GS (2011) Revisiting the TCA cycle: signaling to tumor formation. Trends Mol Med 17:641–649

Raimundo N, Song L, Shutt TE, McKay SE, Cotney J, Guan MX, Gilliland TC, Hohuan D, Santos-Sacchi J, Shadel GS (2012) Mitochondrial stress engages E2F1 apoptotic signaling to cause deafness. Cell 148:716–726

Rhee SG, Kang SW, Jeong W, Chang TS, Yang KS, Woo HA (2005) Intracellular messenger function of hydrogen peroxide and its regulation by peroxiredoxins. Curr Opin Cell Biol 17:183–189

Robinson BH (2006) Lactic acidemia and mitochondrial disease. Mol Genet Metab 89:3–13

Rowland AA, Voeltz GK (2012) Endoplasmic reticulum-mitochondria contacts: function of the junction. Nat Rev Mol Cell Biol 13:607–625

Saada A, Shaag A, Mandel H, Nevo Y, Eriksson S, Elpeleg O (2001) Mutant mitochondrial thymidine kinase in mitochondrial DNA depletion myopathy. Nat Genet 29:342–344

Saha S, Kashina A (2011) Posttranslational arginylation as a global biological regulator. Dev Biol 358:1–8

Scorrano L (2013) Keeping mitochondria in shape: a matter of life and death. Eur J Clin Investig 43:886–893

Selak MA, de Chadarevian JP, Melvin JJ, Grover WD, Salganicoff L, Kaye EM (2000) Mitochondrial activity in Pompe's disease. Pediatr Neurol 23:54–57

Selak MA, Armour SM, MacKenzie ED, Boulahbel H, Watson DG, Mansfield KD, Pan Y, Simon MC, Thompson CB, Gottlieb E (2005) Succinate links TCA cycle dysfunction to oncogenesis by inhibiting HIF-alpha prolyl hydroxylase. Cancer Cell 7:77–85

Sena LA, Chandel NS (2012) Physiological roles of mitochondrial reactive oxygen species. Mol Cell 48:158–167

Soubannier V, McLelland GL, Zunino R, Braschi E, Rippstein P, Fon EA, McBride HM (2012) A vesicular transport pathway shuttles cargo from mitochondria to lysosomes. Curr Biol 22:135–141

Srere PA (1992) The molecular physiology of citrate. Curr Top Cell Regul 33:261–275

Szabadkai G, Rizzuto R (2013) Kalphalambdaovarsigma kappaalphaiota Agammaalphathe taovarsigma: how mitochondrial beauty translates into biological virtue. Curr Opin Cell Biol 25:477–482

Takikita S, Schreiner C, Baum R, Xie T, Ralston E, Plotz PH, Raben N (2010) Fiber type conversion by PGC-1alpha activates lysosomal and autophagosomal biogenesis in both unaffected and Pompe skeletal muscle. PLoS ONE 5, e15239

Tal MC, Sasai M, Lee HK, Yordy B, Shadel GS, Iwasaki A (2009) Absence of autophagy results in reactive oxygen species-dependent amplification of RLR signaling. Proc Natl Acad Sci U S A 106:2770–2775

Tasaki T, Sriram SM, Park KS, Kwon YT (2012) The N-end rule pathway. Annu Rev Biochem 81:261–289

Thoms S, Gronborg S, Gartner J (2009) Organelle interplay in peroxisomal disorders. Trends Mol Med 15:293–302

Tomlinson IP, Alam NA, Rowan AJ, Barclay E, Jaeger EE, Kelsell D, Leigh I, Gorman P, Lamlum H, Rahman S et al (2002) Germline mutations in FH predispose to dominantly inherited uterine fibroids, skin leiomyomata and papillary renal cell cancer. Nat Genet 30:406–410

Tyynismaa H, Mjosund KP, Wanrooij S, Lappalainen I, Ylikallio E, Jalanko A, Spelbrink JN, Paetau A, Suomalainen A (2005) Mutant mitochondrial helicase Twinkle causes multiple mtDNA deletions and a late-onset mitochondrial disease in mice. Proc Natl Acad Sci U S A 102:17687–17692

Virbasius CA, Virbasius JV, Scarpulla RC (1993) NRF-1, an activator involved in nuclear-mitochondrial interactions, utilizes a new DNA-binding domain conserved in a family of developmental regulators. Genes Dev 7:2431–2445

Wallace DC, Singh G, Lott MT, Hodge JA, Schurr TG, Lezza AM, Elsas LJ 2nd, Nikoskelainen EK (1988) Mitochondrial DNA mutation associated with Leber's hereditary optic neuropathy. Science 242:1427–1430

Wanders RJ (2013) Peroxisomes in human health and disease: metabolic pathways, metabolite transport, interplay with other organelles and signal transduction. Sub Cell Biochem 69:23–44

Wellen KE, Hatzivassiliou G, Sachdeva UM, Bui TV, Cross JR, Thompson CB (2009) ATP-citrate lyase links cellular metabolism to histone acetylation. Science 324:1076–1080

West AP, Khoury-Hanold W, Staron M, Tal MC, Pineda CM, Lang SM, Bestwick M, Duguay BA, Raimundo N, MacDuff DA et al (2015) Mitochondrial DNA stress primes the antiviral innate immune response. Nature 520(7548):553–557

Ylikallio E, Tyynismaa H, Tsutsui H, Ide T, Suomalainen A (2010) High mitochondrial DNA copy number has detrimental effects in mice. Hum Mol Genet 19:2695–2705

Youle RJ, Narendra DP (2011) Mechanisms of mitophagy. Nat Rev Mol Cell Biol 12:9–14

Zelko IN, Mariani TJ, Folz RJ (2002) Superoxide dismutase multigene family: a comparison of the CuZn-SOD (SOD1), Mn-SOD (SOD2), and EC-SOD (SOD3) gene structures, evolution, and expression. Free Radic Biol Med 33:337–349

Zhang Z, Tan M, Xie Z, Dai L, Chen Y, Zhao Y (2011) Identification of lysine succinylation as a new post-translational modification. Nat Chem Biol 7:58–63

Zhao S, Xu W, Jiang W, Yu W, Lin Y, Zhang T, Yao J, Zhou L, Zeng Y, Li H et al (2010) Regulation of cellular metabolism by protein lysine acetylation. Science 327:1000–1004

Chapter 10
Mitochondria and Antiviral Immunity

Sunil Thomas and Michael Gale Jr.

10.1 Introduction

Mitochondria are unique dynamic organelles that evolved from free-living bacteria into endosymbionts of mammalian hosts (Sagan 1967; Hatefi 1985). They have a distinct ~16.6 kb closed circular DNA genome coding for 13 polypeptides (Taanman 1999). In addition, a majority of the ~1500 mitochondrial proteins are encoded in the nucleus and transported to the mitochondria (Bonawitz et al. 2006). Mitochondria have two membranes: an outer smooth membrane and a highly folded inner membrane called cristae, which encompasses the matrix that houses the enzymes of the tricarboxylic acid (TCA) cycle and lipid metabolism. The inner mitochondrial membrane houses the protein complexes comprising the electron transport chain (ETC) (Hatefi 1985).

The roles for mitochondria in mammalian cells are ever expanding, and include important functions in innate immunity against microbial infection. While the role of mitochondria in oxidative phosphorylation is relatively well understood, their role in calcium homeostasis, lipid metabolism, apoptosis, aging, innate and adaptive immunity are still being elucidated (Duchen 2004a, b) (Fig. 10.1). More recently, there is also an appreciation for the role of mitochondria as key organelles in the metabolic syndrome characterized by diet-induced obesity and inflammation that manifests in conditions such as cardiovascular disease, diabetes, stroke, and cancer: all of which owe their severity to varying degrees of mitochondrial dysfunction. Thus, understanding mitochondrial physiology in order to restore optimal mitochondrial function has been the goal of several therapeutic strategies aimed at curing

S. Thomas • M. Gale Jr. (✉)
Department of Immunology, Center for Innate Immunity and Immune Disease,
University of Washington School of Medicine, Seattle, WA, USA
e-mail: mgale@uw.edu

D.M. Hockenbery (ed.), *Mitochondria and Cell Death*, Cell Death in Biology and Diseases, DOI 10.1007/978-1-4939-3612-0_10

X.-M. Yin and Z. Dong (Series eds.), *Cell Death in Biology and Diseases*

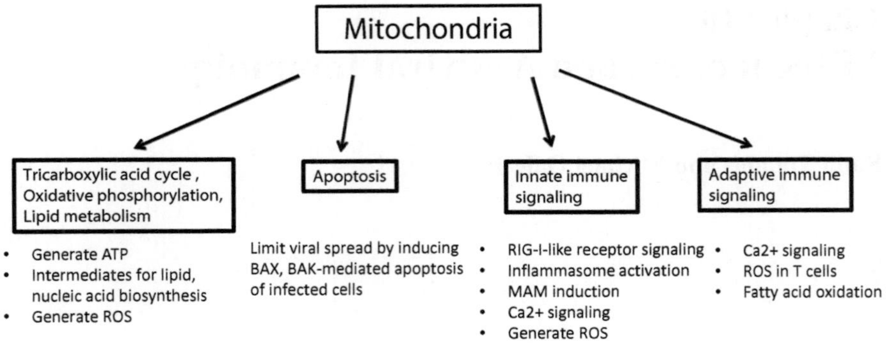

Fig. 10.1 Role of mitochondria in mammalian cells. Tricarboxylic acid cycle and Oxidative phosphorylation: Mitochondria are vital organelles that generate ATP and biosynthetic molecules necessary for cellular maintenance and proliferation. Apoptosis: During viral infection, BAX/BAK-mediated apoptosis initiated by mitochondria, limits viral spread by eliminating infected cells that would otherwise serve as viral reservoirs. Innate Immune signaling: During the innate immune response to viral infection, mitochondria serve as a platform for assembly for proteins involved in RLR signaling and inflammasome activation. Formation of MAMs allow for more efficient activation of processes such as Ca^{2+} signaling and induction of innate immune effector genes. Adaptive Immune signaling: Adaptive immune response is dependent on mitochondria for biosynthetic molecules, Ca^{2+} flux, and ROS generation within activated lymphocytes

metabolic diseases (Sorriento et al. 2014). In this chapter we present a contemporary overview of the mitochondria and mitochondria metabolic actions in infection and immunity, with focus on innate antiviral immunity.

10.2 Mitochondrial Metabolism

Mitochondria are best known for the generation of ATP and the TCA cycle. These processes also generate metabolic products that drive cellular responses involved in innate immunity and immune signaling, including reactive oxygen species. This process starts with glucose that is transported into the cytoplasm and is broken down to pyruvate by the enzymes of the glycolytic pathway. Pyruvate can then be converted into acetyl CoA in the mitochondrial matrix. Acetyl CoA then enters the TCA cycle that generates reduced nicotinamide adenine dinucleotide (NADH) and reduced flavin adenine dinucleotide ($FADH_2$) that are used to generate ATP by the mitochondria through OXPHOS (Duchen 2004b; Kadenbach 2012). During OXPHOS, electrons are transferred from NADH and $FADH_2$ through components of the electron transport chain (ETC) located in the inner mitochondrial membrane. NADH and $FADH_2$ transfer electrons to complex I and II, respectively, and subsequently through complexes III and IV, where the electrons mediate the combination

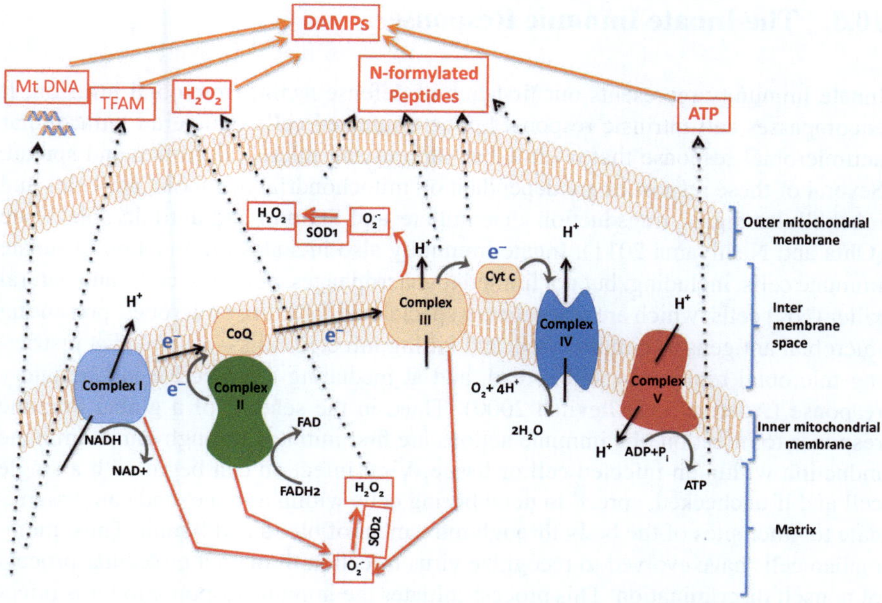

Fig. 10.2 Oxidative phosphorylation and sources of danger associated molecular patterns (DAMPs): The five complexes of the electron transport chain are located on the inner mitochondrial membrane. The process of transfer of electrons from NADH (to complex I) and from $FADH_2$ to complex II, and subsequently III, IV and to the final electron acceptor, O_2, is coupled to the pumping of H^+ ions from the matrix to the intermembrane space. The resulting electrochemical gradient that is generated is used by complex V, also called ATP synthase, to generate ATP from ADP and inorganic phosphate. Damage to mitochondria could release DAMPs that can trigger inflammatory responses. DAMPs are shown in *red*: mitochondrial DNA, mitochondrial transcription factor (TFAM), *N*-formylated peptides, mitochondrial ATP and H_2O_2 (generated from superoxides (O_2^-) by the action of the superoxide dismutase enzymes SOD1 and SOD2

of H^+ and oxygen to form water. The process of electron transport through the ETC is coupled to the pumping of H^+ ions from the mitochondrial matrix to the intermembrane space between the outer and inner mitochondrial membranes, generating an electrochemical gradient known as the membrane potential. H^+ ions are pumped back to the matrix through complex V or ATP synthase, where this process is coupled to the combination of ADP and Pi to form ATP (Hatefi 1985; Saraste 1999) (Fig. 10.2). The electrons that leak from complexes I, II, and III during this process combine with oxygen to form superoxide radicals. Superoxide is converted into H_2O_2 by superoxide dismutase. H_2O_2 can freely cross membranes and is detoxified by the glutathione peroxidase. H_2O_2, superoxides, and other intermediates that form transiently are collectively referred to as reactive oxygen species (ROS) (Kohchi et al. 2009). ROS products, along with microbial products are potent inducers of the innate immune response, and are discussed later in this chapter.

10.3 The Innate Immune Response

Innate immunity represents our first line of defense against microbial invasion. It encompasses cell-intrinsic responses, where infected cells initiate an intracellular antimicrobial response that is aimed at suppressing microbial growth and spread. Several of these responses are dependent on mitochondrial metabolic activities and intracellular signal transduction that initiate and support the antimicrobial state (Ohta and Nishiyama 2011). Innate immunity also includes the function of innate immune cells, including, but not limited to macrophages, dendritic cells, and natural killer (NK) cells, which are respectively specialized for killing microbes, presenting microbial antigens to lymphocytes, and killing infected cells—all aimed at restricting microbial replication and spread, and at mediating a protective inflammatory response (Aderem and Ulevitch 2000). Thus, in the scheme of a global immune response to infection, the immune actions are first initiated through innate immune induction within an infected cell or tissue. Viral infection can begin with a single cell and if unchecked, spread to neighboring cells within a tissue, and can disseminate to other sites of the body through movement of blood and lymph. Thus, mammalian cells have evolved to recognize virus infection through a molecular process of nonself discrimination. This process initiates the immune response to virus infection (Takeuchi and Akira 2009).

The process of initiating the innate immune response to virus infection starts when the host cell recognizes conserved molecular patterns within pathogens termed pathogen-associated molecular patterns (PAMPs) (Janeway 1989). PAMPs include nucleic acid, proteins, and lipid/protein complexes that are of viral origin or viral association (Kawai and Akira 2010). Innate immune recognition of these PAMPs is mediated by cellular proteins called pattern recognition receptors (PRRs) (Janeway 1989), that bind to cognate PAMPs, leading to an antiviral response, which when effective can confer viral clearance and result in a self-limiting infection. This scenario likely happens to us on a daily, if not hourly, basis to protect us from viral infection and disease. Problematically, under certain conditions several PRRs might also recognize and react with endogenous (self) molecules that might resemble PAMPs, leading to an inflammatory response, termed sterile inflammation, that occurs in the absence of microbial infection (Rock et al. 2010). Given the wide variety of pathogens that a host may encounter, including viral pathogens, an equally diverse innate response is required to recognize and mount a defense against potential microbial threats. The broad categories of PRRs include, but are not limited to, Toll-like receptors (TLRs), the nucleotide-binding oligomerization domain (NOD)-like receptors (NLRs), the C type lectin receptors (CLRs), and (RIG-I) (pronounced RIG-eye)-like receptors (RLRs) (Ireton and Gale 2011; Freed and Gale 2014; Brennan and Bowie 2010). PRRs are expressed on the cell surface, cytoplasm and in endosomes to facilitate the recognition of both extracellular and intracellular pathogen-derived PAMPs. The TLRs expressed on the plasma membrane are TLR1, TLR2, TLR4, TLR5, and TLR6, and have been shown to bind microbial components, including those of viral origin. The TLRs predominantly located in endocytic compartments are TLR3, TLR7, TLR8, and TLR9, and these PRRs bind bacterial or viral nucleic acids (Kawai and Akira 2010; West et al. 2011).

The NLRs are intracellular sensors of PAMPs and danger-associated molecular patterns (DAMPs). NLRs are composed of 20 members that recognize microbial PAMPs or molecules released following cellular stress such as DAMPs. The NLR family members include proteins such as NOD-1, NOD-2, pyrin domain containing 1 (NLRP1), NLRP3, and ice protease activating factor 4 (IPAF, also called NLRC4). NLRP3, the best understood NLR, upon activation, forms a macromolecular complex that mediates the maturation and secretion of proinflammatory IL-1β and IL-18 via the activation of caspase-1. The other NLRs except for NOD-1 and NOD-2, can also stimulate the secretion of IL-1β and IL-18 upon activation (Sutterwala et al. 2014; Yu and Finlay 2008). The NLRP3 inflammasome has been shown to assemble and engage MAVS in response to RNA virus infection, leading to the secretion of active IL-1β (Subramanian et al. 2013). Viral infection-mediated inflammasome activation has been reported in response to both DNA (Barlan et al. 2011; Kanneganti et al. 2006; Nour et al. 2011; Muruve et al. 2008) and RNA viruses (Kanneganti et al. 2006; Franchi and Nunez 2008; Rajan et al. 2011; Negash et al. 2013; Ramos et al. 2012; Kaushik et al. 2012). However the mechanisms by which viral PAMPs stimulate inflammasomes are unclear. While it's unclear whether direct interactions between viral PAMPs and inflammasomes occur, one report (Mitoma et al. 2013) describes an indirect interaction whereby DHX33, a RNA helicase, binds viral RNA and activates NLRP3 (Fig. 10.3).

The CLRs are receptors that recognize sugar/carbohydrate moieties on proteins. CLRs may recognize the carbohydrate portion as PAMPs or they may bind self-proteins and affect tissue homeostasis and repair. DC-specific ICAM3-grabbing non-integrin (DC-SIGN) receptor is an important member of this family (Robinson et al. 2006).

RNA viruses are recognized largely by the RLRs: The RLR family includes RIG-I itself, melanoma differentiation-associated gene-5 (MDA5) and Laboratory of Genetics and Physiology 2 (LGP2) (Ireton and Gale 2011). While RIG-I and MDA5 sense viral RNA and initiate downstream signaling (Errett et al. 2013; Jiang et al. 2011), LGP2 is believed to have a regulatory function in RLR signaling (Loo and Gale 2011). During the response to West Nile Virus (WNV) infection, LGP2 did not greatly affect innate immune signaling but was crucially important for CD8 T cell responses (Suthar et al. 2012). Recently, LGP2 was shown to help in MDA5–RNA interactions, thereby enhancing MDA5-mediated antiviral signaling (Loo and Gale 2011; Bruns et al. 2014). RIG-I and MDA5 have two N- terminal caspase recruitment domains (CARDs) that are required for signaling, an internal DExD/H box RNA helicase domain and a C-terminal domain involved in viral RNA detection and autoregulation. LGP2 is a non CARD-containing member of the RLR family. RIG-I can recognize viral RNA based on several PAMP motifs including 5′ triphosphate, double stranded (ds) RNA structure, poly-uridine signatures, and 3′ overhang (Saito et al. 2008). MDA5, on the other hand, although its PAMP is not well defined, is thought to recognize complex and long double-stranded RNA (dsRNA) (Pichlmair et al. 2009). Following RNA binding, the RLR translocate from

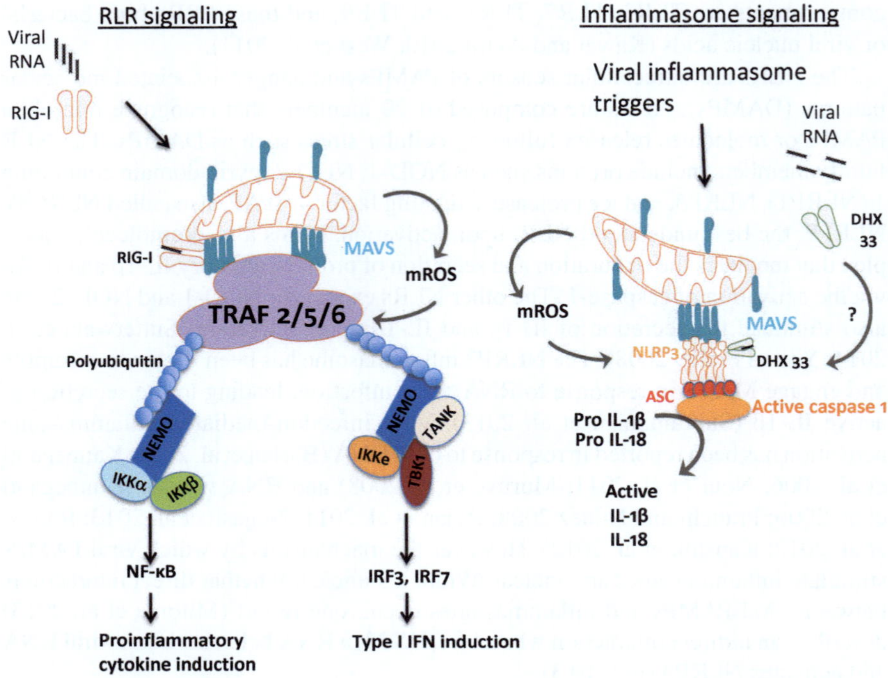

Fig. 10.3 RLR and inflammasome signaling processes during viral RNA infection. RLR signaling: During RNA virus infection the sensing of viral RNA by RIG-I/MDA5 leads to their activation and recruitment to MAVS on intracellular membranes including mitochondria where MAVS aggregates to form large-prion-like complexes, and MAM where the RLR/MAVS complexes assemble for signaling. Activated MAVS recruits the E3 ligases TRAF 2/5/6. The E3 ligases polyubiquitinate TRAF2 and other proteins and recruit NEMO. NEMO, in turn, recruits the kinases, IKKα and β, which can lead to NF-κB activation and proinflammatory cytokine induction. NEMO can also recruit IKKε and TBK1 leading to the phosphorylation and subsequent activation followed by nuclear translocation of IRF3 and IRF7 and induction of type I IFN. Inflammasome activation: The NLRP3 inflammasome is sensitive to viral infection. One possible mechanism of NLRP3 inflammasome activation is via the binding of viral RNA by the RNA helicase, DHX33, leading to the activation of caspase 1 and production of mature IL-1β. In addition, mitochondrial ROS produced as a consequence of viral infection activates both the RLR and NLRP3 inflammasome pathways

the cytosol to mitochondria-associated ER membranes (MAM) to interact with the MAM-anchored and mitochondrial membrane-anchored mitochondrial antiviral signaling (MAVS) protein, leading to innate immune activation. For RLRs, TNF receptor-associated factor (TRAF) family proteins and other adaptors facilitate this signaling by engaging downstream protein kinases that activate the transcription factors interferon regulatory factor-3 (IRF3) (Daffis et al. 2007) and IRF7 (Daffis et al. 2008; Erickson and Gale 2008; Chowdhury et al. 2014) (Fig. 10.3). In myeloid DC and macrophages responding to WNV infection, there is some redundancy with IRF5 also being able to induce IFNβ in a MAVS-dependent manner in the absence of IRF3 and IRF7 (Lazear et al. 2013; Daffis et al. 2009).

DNA virus infection can be recognized by another set of PRRs including TLR9, which binds CpG motifs in DNA, and cGMP-AMP (cGAMP) synthase) (cGAS), which binds to dsDNA and catalyzes production of a 2′-5′ dinucleotide that binds to signal transducer of interferon genes (STING) to drive IRF3 or IRF7 activation (Ablasser et al. 2013). In all cases, when activated, IRF3/7 homodimerize and translocate to the nucleus. Along with activated NF-κB, these factors then bind to specific response elements, leading to the expression of type I IFN, IRF3 target genes and NF-κB-responsive genes. Subsequently, secreted IFN acts both in an autocrine and paracrine manner to induce the expression of IFN-stimulated genes (ISGs). Type I IFN signaling is largely mediated by the JAK-STAT pathway, leading to interferon-stimulated gene factor 3 (ISGF3) expression that binds to the IFN-stimulated response element (ISRE) within the promoter of ISGs. ISGs include inflammatory cytokines and chemokines as well as immune-modulatory genes that impart control of virus replication, spread and regulation of the adaptive immune response. This establishes the "antiviral state" both in virus-infected and bystander cells (Sadler and Williams 2008; Schoggins et al. 2011). Thus, successful innate immune response will limit viral replication while serving to program the adaptive immune response toward antiviral actions.

10.4 Mitochondria and Innate Immunity

Mitochondria and the mitochondria-associated membrane (MAM) play major roles in innate immune signaling against RNA viruses because each functions specifically in RLR signaling. Active RIG-I and MDA5 signal innate immunity through direct association with the adaptor MAVS that is critical for antiviral responses against RNA viruses (Kumar et al. 2006; Sun et al. 2006). MAVS is a 540 amino acid protein with a predicted molecular weight of 56 kDa. MAVS contains an N-terminal CARD domain, a proline-rich region (PRR) and a C-terminal transmembrane domain. MAVS is anchored on the outer membrane of the mitochondria, peroxisomes and MAM. MAM is a specialized extension of the endoplasmic reticulum membrane that makes contacts with the mitochondrial outer membrane (Fig. 10.4) and is discussed later in this chapter (Horner et al. 2011). RIG-I binding to MAVS is dependent upon K63-linked ubiquitination of RIG-I by the E3 ubiquitin ligase tripartite Motif 25 (TRIM 25). Binding of K63 polyubiquitin chains to RIG-I or MDA5 eventually leads to activation of IRF3 (Jiang et al. 2012). Proteins such as human oligoadenylate synthetase L (OASL) that mimic polyubiquitin also enhance RIG-I antiviral activity (Zhu et al. 2014). RIG-I interacts with TRIM25 and the mitochondrial targeting chaperone protein, 14-3-3ε, to form a translocon that delivers activated RIG-I to MAVS on membranes (Liu et al. 2012). The proline-rich region of MAVS is involved in interacting with TNFR-associated factor (TRAF) family members, though it appears that additional contact points for MAVS–TRAF interactions may exist on MAVS (Seth et al. 2005; Xu et al. 2005; Saha et al. 2006), possibly in the CARD region (Tang and Wang 2009; Tang and Wang 2010)

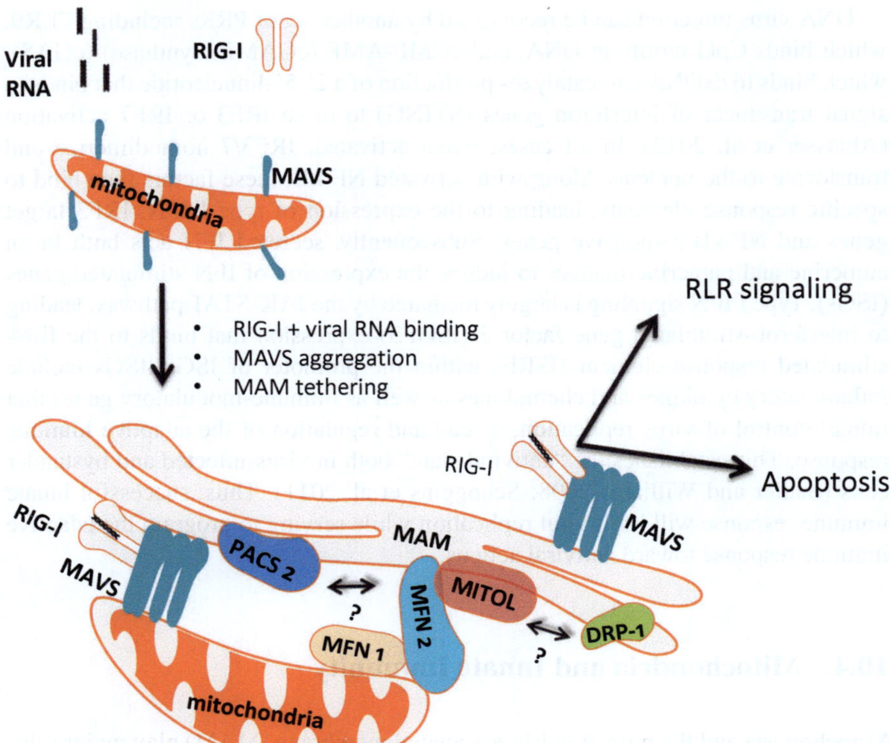

Fig. 10.4 Proteins associated with mitochondria-associated ER membranes (MAM) tethering. Upon activation by viral infection, mitochondria become activated and are found in close apposition with ER membranes in structures called MAM. MAM allow members of signaling pathways such as RLR signaling and apoptosis induction to come together as a complex to allow for more efficient signal transduction, such as during Ca^{2+} signaling, by forming an "innate immune synapse." Several proteins have been implicated in the formation of MAM in nonviral systems. How these proteins may interact with MAVS and with one another, especially in the context of viral infections is being defined. MFN2 has been directly implicated in tethering ER to mitochondria. The other proteins that are implicated as MAM constituents may have indirect roles; they might help stabilize MAM or catalyze its formation. MFN1 is located on the outer mitochondrial membrane and interacts with MAVS. Mitochondrial ubiquitin ligase (MITOL) ubiquitinates MFN2 and enhances MAM formation indirectly. Dynamin related Protein 1 (DRP1) is a cytosolic GTPase that has been shown to colocalize with MAM, while phosphofurin acidic cluster sorting protein 2 (PACS2) regulates MAM formation and stability (Adapted from Vance JE (2014) Biochim Biophys Acta 1841: 595–609)

(Fig. 10.2). Interaction with RIG-I or MDA5 occurs through the N terminal CARD domain, leading to MAVS oligomerization. These rod-shaped prion-like aggregates appear to form self-perpetuating filaments that amplify signaling, albeit they have only been identified in vitro in cell-free systems and are not yet demonstrated to exist in intact mammalian cells (Hou et al. 2011). MAVS is known to interact with upwards of 30 proteins (Kumar et al. 2006; Belgnaoui et al. 2011). Subsets of these proteins assemble into signaling complexes or signalosomes. The formation of these "MAVS signalosomes" is assisted by proteins such as the outer mitochondrial membrane protein, tripartite motif 14 (TRIM14) that after interacting with MAVS, undergoes polyubiquitination at Lys-365. Upon activation, it recruits NF-κB enhancer modulator (NEMO) to MAVS and allows for the activation of NF-κB and IRFs (Zhou et al. 2014). The ubiquitination and activation of NEMO is promoted by TRAF2, 5, and 6, which are also recruited by MAVS (Liu et al. 2013) (Fig. 10.3). Ubiquitously expressed TV-1 (UXT-V1) is another protein that aids in the formation of the MAVS signalosome by binding TRAF3 and thus facilitates the interaction between TRAF 3 and MAVS (Huang et al. 2012).

Differential usage of TRAF family members as adaptors in interactions with MAVS may regulate IRF3/IRF7/NF-κB and proinflammatory cytokine/ type I IFN (IFN-I) production. While MAVS interaction with TRAF6 and TRAF5 (Tang and Wang 2010) along with TNF receptor associated death domain (TRADD) via receptor interacting protein1 (RIP1) and FAS-associated death domain (FADD) (Michallet et al. 2008) can lead to NF-κB activation, MAVS interaction with TRAF 2/3 (Saha et al. 2006; Liu et al. 2013), TRADD, and TRAF family-associated NF-κB activator (TANK) promotes IKKε and/or TANK binding kinase1 (TBK1)-mediated IRF3 phosphorylation and (IFN-I) production. The nature of the interactions between TRAF and molecules of the RLR pathway and how they are regulated are unclear. A recent report has identified an interaction between the mitochondria-nucleus shuttling protein, FK506-binding protein 51 (FKBP51), TRAF6, and TRAF3 as being important for signaling the expression of (IFN-I) induced by cytosolic dsRNA. This suggests a potential point of regulation, via FKBP51, imposed on the RLR pathway during viral infection (Akiyama et al. 2014).

STING and Translocase of mitochondria-70 (TOM70) are shown to be mitochondrial cofactors for innate immune signaling by MAVS, although this activity is likely to be pathogen-specific (Ishikawa and Barber 2008). TOM 70, along with TOM 20, forms a protein channel on the outer mitochondrial membrane through which nuclear encoded mitochondrial proteins are imported into the mitochondria. TOM 70 interacts directly with MAVS, and this interaction is enhanced in particular during Sendai virus infection (Liu et al. 2010) and is a crucial positive regulator of RLR signaling (Liu et al. 2010). TOM 70 also interacts with HSP90 (Bhangoo et al. 2007), and this tethering allows for close association between HSP90, MAVS, TBK1, and IRF3, presumably allowing for efficient signaling. STING was originally identified as an endoplasmic reticulum (ER)-resident protein that mediates IRF3 activation and stimulator of interferon following recognition of viral nucleic acid (Ishikawa and Barber 2008; Zhong et al. 2008; Sun et al. 2009). Decreased STING expression increased susceptibility to vesicular stomatitis virus (VSV, a RNA virus) as well as herpes simplex virus (HSV, a DNA virus) infection. Subsequent work has more clearly defined a role for STING in intracellular DNA sensing and its role in RNA virus protection may be indirect (Brunette et al. 2012; Wu et al. 2013; Schoggins et al. 2014). STING-mediated antiviral responses are dependent on association with the ER translocon protein known as translocon associated protein (TRAP), SEC61 translocon and TBK1, which leads to phosphorylation of IRF3 and IRF7. Thus, close ER-mitochondrial interactions such as those imposed by MAM, are crucial in antiviral signaling.

As noted above, mitochondria produce ROS as a result of their metabolic processes. ROS can impact several signaling pathways to regulate innate immune defenses during virus infection. The cellular stress induced by viral infection is associated with ROS production as observed in human adenovirus (HAdv) (McGuire et al. 2011), hepatitis C virus (HCV) (Nishina et al. 2008), Human Immunodeficiency Virus (HIV) (Kruman et al. 1998), and Epstein Barr virus (EBV) (Lassoued et al. 2008) infections. Mitochondrial ROS (mROS) increases RLR signaling and IFNβ production, which consequently decreases viral replication. While the mechanistic details of how mROS might regulate the RLR pathway during viral infection are unclear, several proteins that regulate this process have been identified. The NOD-like protein, NLRX1 and the globular domain of complement 1q receptor (gc1qR) interact with components of oxidative phosphorylation and increase ROS production (Tattoli et al. 2008; Saha et al. 2013) whereas the cytochrome c oxidase complex subunit COX5B, coordinates with the autophagy pathway through autophagy-related gene 5 (ATG5), to suppress MAVS signaling and decrease ROS production (Zhao et al. 2012).

A crucial requirement for MAVS to function as an adaptor in the RLR signaling pathway is the existence of optimum mitochondrial membrane potential (Koshiba et al. 2011). Fibroblasts lacking both mitofusin 1 (MFN1) and mitofusin 2 (MFN2) and which have low membrane potential, or cells in which the mitochondrial membrane potential is pharmacologically dissipated using protonophores such as carbonyl cyanide m-chlorophenyl hydrazone (CCCP), support TLR3 signaling but not RLR signaling. Interestingly, RLR signaling via MAVS is not itself dependent on de novo ATP production, although many of the RLR signaling proteins bind or hydrolyze ATP during their signaling actions. This relationship suggests that the requirement for mitochondrial membrane potential for MAVS activity reflects needed alterations in mitochondria membrane and MAM interactions that likely form a functional membrane-bound signalosome platform (Koshiba et al. 2011).

10.5 Mitochondria in Sterile Inflammation

Release of mitochondrial components into the extracellular space induces inflammatory reactions (Fig. 10.2). Trauma-induced cell-damage releases DAMPs such as *N*-formyl peptides, mitochondrial DNA, mitochondrial transcriptional factor (TFAM), mitochondrial ATP, mitochondrial ROS or high mobility group box 1 (HMGB-1) that lead to pathophysiological consequences (Wenceslau et al. 2014; Kono and Rock 2008). HMGB-1, a ubiquitous nuclear protein, functions as a DAMP that can induce apoptosis in Jurkat T cells by lowering the levels of MFN2 that may be required to maintain optimum Ca^{2+} levels in T cells (Wu et al. 2014; Castanier et al. 2010). In addition, HMGB-1 may play a more direct role in regulating MAVS function by acting as `universal sentinels' of nucleic acid detection (Yanai et al. 2009). Mitochondrial DAMPs activate the NLRP3 inflammasome and amplify the inflammatory response by releasing IL-1β and promoting neutrophil infiltration following trauma (Ma et al. 2014). MAVS is a critical adaptor in addition to apoptosis-associated speck-like protein containing a caspase recruitment domain (ASC) in mediating the production of IL-1β from the NLRP3 inflammasome (Subramanian et al. 2013). Mitochondrial ROS may be a major trigger of NLRP3 inflammasome activation. Therefore, inhibition of the NLRP3 inflammasome and the use of ROS inhibitors such as MitoTEMPO can be protective following traumatic events such as intracerebral hemorrhage (Ma et al. 2014). Such inflammatory responses have been observed in immune cells such as gamma-delta cells that upon incubation with mitochondrial DAMPS secrete IL-1β, IL-10 and IL-6 (Schwacha et al. 2014). TLR9 engagement of DNA from damaged cells by immune cells results in an inflammatory response mediated by MyD88 and TRAF6 that eventually results in cell death. However, in non-regenerative terminally differentiated non-immune cells such as cardiomyocytes and neurons, an alternative pathway has been reported where TLR9 binds the sarco/endoplasmic reticulum Ca^{2+}-ATPase 2 (SERCA2) protein, reduces its activity, and leads to reduced mitochondrial Ca^{2}. This process then decreases mitochondrial ATP levels and protects the cells from cell death (Shintani et al. 2014). A pathway to prevent inappropriate activation following recognition of endogenous RNA by the RLR pathway has been reported. The superkiller viralicidic activity 2-like (SKIV2L) RNA exosome potently suppress the activation of the RLR pathway and prevents development of autoimmunity. The importance of this mechanism of handling immunostimulatory RNA is indicated by the finding that deficiencies in SKIV2L have been associated with certain immune disorders (Eckard et al. 2014).

10.6 Mitochondrial and MAM Dynamics in Innate Immunity

Mitochondria are dynamic organelles whose function is regulated by alterations in size, number and interactions with other organelles such as ER and peroxisomes by mechanisms such as fusion, fission and autophagy (Campello and Scorrano 2010). Mitochondrial fusion is regulated by mitofusin1 (MFN1), mitofusin2 (MFN2) as well as by optic atrophy A1 (OPA1) (Chan 2006), whereas mitochondrial fission is regulated by dynamin related protein 1 (DRP1) and fission 1 (FIS1) (Chan 2006). In addition to mediating mitochondrial fusion, mitofusin2 (MFN2) is crucial for mitochondria–MAM interactions (Horner et al. 2011; Horner and Gale 2013). These mitochondrial processes are conserved across organisms and cell types and are critical for cellular function. For example, both mammalian DRP1 and the homologous yeast dynamin related GTPase 1 (DNM1), regulate mitochondrial fission and a DRP1 mutation in mice is implicated in severe developmental defects (Sesaki et al. 2014). Knockdown of MFN1 and OPA1 (thereby suppressing fusion) results in decreased IRF3 and NF-κB activation, whereas knockdown of DRP1 and FIS1 potentiates IRF3 and NF-κB signaling (Castanier et al. 2010). Thus, fused mitochondria, in general, support enhanced RLR signaling, while fragmented mitochondria do not. Mitochondrial fusion events and formation of MAM support RLR signaling as described earlier. The role for MFN2 in RLR signaling, however, has been unclear. MFN2 is enriched at points of contact between ER and mitochondria in the MAM (Fig. 10.4) and, as such, increases in MAM are associated with enhanced RLR signaling. This supportive role for MFN2 in RLR signaling is potentially at odds with the finding that RLR signaling is inhibited when MFN2 is overexpressed (Yasukawa et al. 2009). It was shown in this study that MFN2 overexpression inhibited RIG-I and MDA5-mediated signaling. It is possible to reconcile these opposing roles if they occur at different times following activation, with the MAM-supportive activity of MFN2 occurring early during the response and the inhibitory role occurring at a point when the RLR pathway needs to be suppressed. Alternatively, MFN2 could be interacting with different, as yet undiscovered, accessory proteins in the above two situations. However, neither of the above two scenarios has been demonstrated and both await experimental evidence.

A crucial part of the dynamic nature of the interactions between mitochondria and the ER is through MAM. As noted above, these membranes are closely associated with ER and mitochondria. MAM was initially identified through electron microscopy studies and later confirmed by cell fractionation studies to contain marker proteins that were highly enriched for this special membrane but not present in high levels in either ER or mitochondria (Cui et al. 1993). Several MAM or MAM-associated proteins have been identified but only a few have been shown to be involved in tethering MAM to mitochondria (Fig. 10.4) (Vance 2014): MFN2 has been directly implicated in tethering ER to mitochondria (de Brito and Scorrano 2008), while MFN1 has been shown to be an outer mitochondrial membrane protein (de Brito and Scorrano 2008) that interacts with MAVS (Castanier et al. 2010). Mitochondrial ubiquitin ligase (MITOL) ubiquitinates MFN2 and enhances MAM formation indirectly (Sugiura et al. 2013). DRP1 is a cytosolic GTPase that has been shown to colocalize with MAM (Friedman et al. 2011) and phosphofurin acidic cluster protein 2 (PACS2) regulates MAM formation and stability (Youker et al. 2009) (Fig. 10.4). Despite these advances, identifying unique markers for MAM has been difficult, as the nature of the MAM seems to vary with cell type and activation status (Vance 2014). It is therefore conceivable that there may be cell type specific enrichment of these proteins or a combination of these and other proteins in the MAM during different viral infections.

MAMs are conserved in organisms ranging from yeast to mammals and have been identified as important in several functions such as importing the lipid, phosphatidylserine (PS) into mitochondria, mitochondrial morphology, calcium homeostasis and promoting cell survival and autophagy (Vance 2014). Mitochondria are important organelles that regulate calcium homeostasis. Recent studies have revealed that the MAM is a crucial part of Ca^{2+} regulation, as transport of Ca^{2+} from the ER to the mitochondria appears to occur at points of close apposition between the two organelles and conditions that disrupt this contact not only disrupt Ca^{2+} homeostasis, but also ATP production and cell survival (Rowland and Voeltz 2012).

Viral proteins, including the exon 1 protein from cytomegalovirus (CMV) and the HCV core protein have been identified as being associated with MAM (Williamson and Colberg-Poley 2009; Williamson et al. 2012). The pathological significance of localization of viral proteins in MAMs is not clear. Whether these viral proteins are passively transported to the MAM from the ER or whether this represents an active strategy on the part of the virus to subvert host defense needs to be elucidated. The compartmentalization of host proteins in the mitochondria versus the MAM has functional relevance as seen in the case of MAVS cleavage by the viral protease, NS3/4A, during HCV infection (Horner et al. 2011; Horner et al. 2012; Lin et al. 2006). MAVS is localized on mitochondria and peroxisomes as well as in the MAM in HCV-infected cells. Peroxisomal MAVS mounts a rapid response mediated by the transcription factor IRF1 (Ding and Robek 2014). This early response is mediated by the JAK-STAT pathway and resembles a type III IFN response (Ding and Robek 2014; Odendall et al. 2014), while mitochondrial MAVS mounts an IFN-I response with delayed kinetics (Odendall et al. 2014; Dixit et al. 2010). The localization of MAVS in the MAM potentially allows for MAVS to be sterically removed from mitochondria-associated negative regulators, such as NLRX1, that block CARD–CARD interactions between MAVS and recruited RIG-I, thus facilitating potent antiviral signaling. However, during HCV infection, MAM-associated MAVS is specifically subject to cleavage by the HCV NS3/4A protease, whereas the mitochondria-associated MAVS is not. Since RIG-I is recruited to the MAM after infection and HCV targets and cleaves MAM-associated MAVS, the virus thereby effectively blocks IFNβ production to support persistent viral replication and chronic infection. Overall these studies implicate the MAM as a crucial signaling platform from which MAVS and RLRs function to drive innate antiviral immunity.

Apart from being involved as an adaptor in RLR signaling, mediating IFNβ production, activation of MAVS and MAM signaling can have deleterious effects. This outcome was shown in the study by Mukherjee et al. (2013), where they demonstrated that pediatric viral encephalitis attributed to La Crosse Virus (LACV), a bunyavirus, occurs via upregulation of the adaptor protein, sterile alpha and TIR-containing motif 1 (SARM1), a molecule directly involved in neuronal damage, that is induced upon MAVS activation by LACV infection. SARM1-mediated cell death was associated with induced oxidative stress responses and mitochondrial damage. These observations provide an innate immune signaling mechanism for virus-induced neuronal death, and indicate that MAM and mitochondrial signaling of MAVS requires strict control to avoid deleterious effects. The work also reveals potential targets of MAM/mitochondrial signaling that might be considered for development of therapeutics to manage the deleterious effects of such encephalitic viral infections

10.7 Negative Regulation of Mitochondrial Function and Antiviral Signaling

Viruses trigger multiple PRR pathways upon encountering the host. These pathways act in concert to eliminate the virus. Signaling cross talk between the RLR pathway, NLR pathways and caspase pathways have been identified as mediating both positive and negative regulation of RLR signaling (Ramos and Gale 2011). RLR signaling pathways may be antagonized at the level of PAMP recognition itself where, for example, the ebola virus VP35 protein sequesters dsRNA and prevents viral recognition by RIG-I (Cardenas et al. 2006). The influenza A virus NS1 protein binds RIG-I itself and prevents it from binding its downstream partners (Mibayashi et al. 2007). Negative regulation of mitochondria-mediated innate immunity is exerted at several levels, the most basic of which is the process of mitophagy, whereby the autophagy-related proteins, autophagy-related gene 5 (ATG 5) and/or ATG 12 remove damaged mitochondria and consequently suppress RLR signaling (Tal et al. 2009). Viruses such as measles virus exploit the mitophagy pathway to increase mitophagy and thereby decrease RLR signaling and enhance their replication (Xia et al. 2014). Severe fever with thrombocytopenia syndrome virus (SFTSV), an emerging bunyavirus, targets downstream molecules of the RLR pathway by inducing the formation of inclusion bodies that encapsulate IKKε and TBK1, thereby preventing IRF3 activation and IFNβ production (Ning et al. 2014). IKKε is a novel target for the nucleoprotein (NP) of arena viruses such as Lymphocytic Choriomeningitis Virus (LCMV). LCMV NP can bind the kinase domain of IKKε, and block its autocatalytic activity as well as sequester it away from MAVS and therefore inhibit its ability to phosphorylate IRF3, thereby down-regulating activation of the RLR pathway. However, LCMV NP does not bind the closely related kinase, TBK-1, indicating that the virus specifically targets IKKε (Pythoud et al. 2012). Mitochondrial E3 ubiquitin protein ligase 1 (MUL1) is another negative regulator of RIG-I signaling that interacts with MAVS on mitochondria and catalyzes RIG-I post-translational modifications that inhibit RIG-I signaling. Depletion of MUL1 potentiated RIG-I-mediated NF-κB and IFNβ reporter activity and enhanced responses to poly IC and Sendai virus. Thus, MUL1 is a novel regulator of the RIG-I-like receptor-dependent antiviral response (Jenkins et al. 2013).

MAVS is a major target for the host to regulate antiviral signaling. As noted earlier, the NOD-like protein NLRX1 that has been reported to be localized in the mitochondrial matrix as well as the outer membrane prevents CARD–CARD interactions between MAVS and RLR, thus inhibiting the RLR pathway (Arnoult et al. 2009). Globular head domain of complement component 1q (gC1qR) also directly binds and inhibits MAVS (Xu et al. 2009). Poly (rC)-binding protein 2 (PCBP2), an endogenous protein, similarly recruits atrophin interacting protein 4 (AIP4), catalyzing K48-linked polyubiquitination and degradation of MAVS (You et al. 2009). Similarly, another endogenous protein, PSMA7, the α4 subunit of the 20S proteasome, also induces the proteasomal degradation of MAVS (Jia et al. 2009). Though the precise mechanisms by which some of the above endogenous proteins operate to suppress MAVS signaling are unclear (Koshiba 2013), the variety and the redundancy in the mode of action of some of these molecules suggests that the RLR pathway is tightly regulated to not only avoid aberrant activation but also to allow for quick down-regulation once the viral threat has been eliminated.

Specifically targeting MAVS to suppress innate antiviral immunity is a strategy employed by several viruses. The hepatitis A virus protease 3ABC, GB virus-B and the HCV protease NS3/4A all cleave and inactivate MAVS, though they cleave MAVS at different sites, likely implicating each site in specific roles of innate immune signaling (Horner and Gale 2013; Li et al. 2005; Chen et al. 2007). Ubiquitination induction and consequent proteasomal degradation is a strategy adopted by Hepatitis B virus X protein that interacts with MAVS and promotes its degradation. While respiratory syncytial virus exploits mitochondrial MAVS to suppress innate immunity (Goswami et al. 2013), UBX-domain containing family member, UBXN1, inhibits MAVS oligomerization in several RNA virus infections including vesicular stomatitis virus, Sendai virus, and dengue virus, thereby inhibiting RLR-mediated antiviral signaling (Wang et al. 2013a). The enterovirus 71 protease 2Apro targets MAVS to inhibit antiviral type I interferon responses (Wang et al. 2013b), while influenza A virus Enhancer of zeste homolog 2 is a negative regulator (Chen et al. 2013) that binds to MAVS and interferes with the interaction between MAVS and RIG-I. Polo-like kinase is another negative regulator of MAVS that disrupts the interaction between MAVS and TRAF3 and thereby negatively regulates IFNβ induction (Vitour et al. 2009). HIV infection results in damage to mitochondrial proteins involved in energy metabolism. Levels of a mitochondrial chaperone, prohibitin, are suppressed in HIV patients. Paradoxically, highly active antiretroviral therapy (HAART) further lowers prohibitin levels in HIV patients (Ciccosanti et al. 2010; Rossi et al. 2014). This understanding of the role of prohibitin in mitochondrial health and the deleterious effects of HAART on prohibitin could lead to the design of better antiretrovirals with fewer nonspecific effects. The numerous viral evasive strategies targeting MAVS that are outlined above point to the pivotal role that MAVS plays in antiviral immunity.

10.8 Apoptosis and Mitochondrial Antiviral Immunity

Mitochondria orchestrate the intrinsic pathway of apoptosis (Wang and Youle 2009). While the primary triggers that drive mitochondria-mediated apoptosis are still not well understood, the earliest detectable change is a increase in outer membrane permeability that triggers the release of proapoptotic molecules such as cytochrome c (Liu et al. 1996), second mitochondria-derived activator of caspase/Direct inhibitor of apoptosis-binding protein with low PI (Smac/DIABLO) (Du et al. 2000; Verhagen et al. 2000), Omi/Htr serine peptidase 2 (HtrA2) (Faccio et al. 2000; Hegde et al. 2002; Martins et al. 2002; Suzuki et al. 2001), endonuclease G (EndoG) (Li et al. 2001), and apoptosis-inducing factor (AIF) (Susin et al. 1996). This permeabilization is produced by activation of the proapoptotic molecules, BAX and/or BAK ultimately leading to activation of caspase 9, leading to activation of caspase 3, the effector caspase that acts on apoptotic substrates leading to apoptosis. Activity of the BCL2 family of antiapoptotic molecules such as BCL2 and BCL-X_L inhibits the activity of BAX and or BAK and thus inhibits apoptosis (Wang and Youle 2009).

Viruses have evolved several strategies to evade mitochondria-induced apoptosis to keep cells viable for longer following infection and thereby continue to propagate virus, or to induce premature apoptosis in innate sentinel cells and thus evade detection. In cytomegalovirus-infected cells, viral mitochondria-localized inhibitor of apoptosis (vMIA) can bind mitochondrial BAX, but not BAK, and thus prevent apoptosis (Arnoult et al. 2004). In influenza A virus-infected macrophages, mitochondrial NLRX1 prevents PB1-F2-induced premature apoptosis by binding to the protein (Jaworska et al. 2014). In a recent report, MAVS was shown to mediate apoptosis following Semliki forest virus infection in a BAX/BAK-independent, but caspase 8/caspase 3 dependent manner. This form of apoptosis did not require IFNβ, IRF3/IRF7, TRAF2 or FADD (El Maadidi et al. 2014). MAVS has also been shown to induce apoptosis in a RLR-independent manner following Sendai virus, HCV and severe acute respiratory syndrome corona virus (SARS CoV) infections (Lei et al. 2009; Huang et al. 2014). IRF3 can also mediate apoptosis by binding to BAX and translocating to mitochondria. Sendai virus infection of IRF3-/- mice leads to persistent infection (Chattopadhyay et al. 2013), suggesting that IRF3-mediated apoptotic removal of infected cells represents an important means of containing viral infection. HCV subverts mitochondria-induced apoptosis by stimulating parkin-mediated removal of damaged mitochondria by mitophagy (Kim et al. 2014), and increasing the rate of mitochondrial fission via Ser 616 phosphorylation of DRP1. Experimental silencing of DRP1 or parkin caused a significant increase in apoptotic signaling, evidenced by increased cytochrome c release from mitochondria, caspase 3 activity, and cleavage of poly(ADP-ribose) polymerase. These results suggest that HCV-induced mitochondrial fission and mitophagy serve to attenuate apoptosis and may contribute to persistent HCV infection (Kim et al. 2014).

10.9 Mitochondria and Adaptive immunity

The role for mitochondrial metabolism in supporting cell growth and proliferation has been well studied in tumor cells. Among immune cells, the relationship between metabolism and function has been better studied in macrophages (O'Neill and Hardie 2013; Shapiro et al. 2011) and T cells (Fox et al. 2005; van der Windt and Pearce 2012; Gerriets and Rathmell 2012). In B cells, the role for mitochondria in maintaining self-tolerance by inducing apoptosis following inappropriate activation has been established (Bouchon et al. 2000; Deming and Rathmell 2006). In dendritic cells, TLR-induced maturation and immunogenicity are dictated by cellular metabolism such that DC cultured with or without glucose, for example, program T cells toward drastically different fates (Everts and Pearce 2014; Everts et al. 2014).

Upon recognizing antigen presented by antigen-presenting cells, T cells proliferate rapidly over about a 5–10 day period and exert their effector functions. Following this, they decline in numbers, and a small fraction of the originally proliferated cells, called memory cells, are maintained for months to years. Cellular metabolic requirements placed on mitochondria change at different phases of the above process. Mammalian target of rapamycin (mTOR) is an evolutionarily conserved serine/threonine kinase that integrates the diverse signals that T cells receive during this process to turn on transcription programs that support the varying metabolic demands (Waickman and Powell 2012).

Several studies have examined the requirement for various aspects of mitochondrial function in T cells during adaptive responses. At the proliferative effector phase response to LCMV infection, the mitochondrial membrane potential was maximal, pointing to a role for mitochondria during the exponential expansion of T cells (Zhu et al. 2014). Quiescent naïve and memory cells, while maintaining low levels of glycolysis, rely predominantly on oxidative phosphorylation (OXPHOS) for their energy and biosynthetic needs. As they undergo antigen-driven proliferation, effector cells make a necessary shift to predominantly use glycolysis while maintaining a low level of OXPHOS. This low level of OXPHOS-mediated ATP production is vital for effector function, as specifically inhibiting mitochondrial ATP production using oligomycin blocks T cell activation and proliferation (Chang et al. 2013). The role for mitochondria-derived ROS in T cell function was studied in a mouse model with T cells deficient for ubiquinol-cytochrome c reductase (UQCRFS1−/−). These T cells had a defect in their ability to generate ROS from complex III of mitochondria (Sena et al. 2013) and resulted in their poor proliferation and cytotoxic potential owing to a deficiency in activation of nuclear factor of activated T cells (NFAT) and subsequent IL-2 production. Interestingly, T cells exposed to elevated levels of ROS showed a deficient ability to respond to influenza A virus and impaired development (Case et al. 2011), suggesting that an optimal threshold of ROS is required for T cells to function normally. Another study that examined the role of mitochondrial metabolism in memory T cells found that IL-15 stimulated mitochondrial biogenesis and expression of carnitine palmitoyl transferase 1a (CPT1a), an enzyme that controlled a rate-limiting step in mitochondrial fatty acid metabolism. This gave the memory cells, but not naïve or effector cells, a higher mitochondrial spare respiratory capacity (SRC) and enhanced ability to transport fatty acids into the mitochondria for oxidation (van der Windt et al. 2012). This enabled memory cells, uniquely, to mount a rapid response upon reexposure to antigen (van der Windt et al. 2012). Our study of the role of MAVS in CD8 T cells during viral infection uncovered a role for MAVS in regulating cell proliferation, calcium homeostasis, and cytokine production, which indicates that MAVS has a crucial role in supporting effector functions in T cells (unpublished data). Thus in T cells, mitochondria have important roles in OXPHOS, lipid metabolism, ROS generation, and Ca^{2+} flux (Pearce et al. 2013) during the adaptive immune response.

10.10 Conclusion

Mitochondria stand at the crossroads of innate and adaptive immunity. They not only provide energy and biosynthetic molecules to support the rapid cellular proliferation that is characteristic of the adaptive response, but also provide a platform on which large complexes of antiviral molecules can be assembled to mount effective innate immune responses. Because of their importance to immunity, mitochondria are targeted for regulation of their associated proteins by several viruses. Their roles in antiviral immunity are complex with each cellular subset requiring a different aspect of mitochondrial metabolic function at distinct differentiation states and in responding to different pathogens. The involvement of mitochondria in response to infection as well as in noninfectious metabolic processes make them attractive targets for therapeutic interventions aimed at restoring or enhancing their function. The field of metabolic immunity that focuses on understanding how metabolism affects immune cell fate is relatively new and is expected to progress to reveal insights that will direct new approaches and strategies for enhancing the immune response to microbial pathogens and vaccination.

Acknowledgments S. T. and M. G. are supported by NIH grants AI00625, AI93019, and AI074973. We apologize to colleagues whose research could not be cited here owing to space constraints. We thank Courtney Wilkins, Amina Negash, John Errett, and Priya Handa for critical review of the manuscript.

References

Ablasser A, Goldeck M, Cavlar T, Deimling T, Witte G, Rohl I, Hopfner KP, Ludwig J, Hornung V (2013) cGAS produces a 2′-5′-linked cyclic dinucleotide second messenger that activates STING. Nature 498(7454):380–384

Aderem A, Ulevitch RJ (2000) Toll-like receptors in the induction of the innate immune response. Nature 406(6797):782–787

Akiyama T, Shiraishi T, Qin J, Konno H, Akiyama N, Shinzawa M, Miyauchi M, Takizawa N, Yanai H, Ohashi H et al (2014) Mitochondria-nucleus shuttling FK506-binding protein 51 interacts with TRAF proteins and facilitates the RIG-I-like receptor-mediated expression of type I IFN. PLoS One 9(5):e95992

Arnoult D, Bartle LM, Skaletskaya A, Poncet D, Zamzami N, Park PU, Sharpe J, Youle RJ, Goldmacher VS (2004) Cytomegalovirus cell death suppressor vMIA blocks Bax- but not Bak-mediated apoptosis by binding and sequestering Bax at mitochondria. Proc Natl Acad Sci U S A 101(21):7988–7993

Arnoult D, Soares F, Tattoli I, Castanier C, Philpott DJ, Girardin SE (2009) An N-terminal addressing sequence targets NLRX1 to the mitochondrial matrix. J Cell Sci 122(Pt 17):3161–3168

Barlan AU, Griffin TM, McGuire KA, Wiethoff CM (2011) Adenovirus membrane penetration activates the NLRP3 inflammasome. J Virol 85(1):146–155

Belgnaoui SM, Paz S, Hiscott J (2011) Orchestrating the interferon antiviral response through the mitochondrial antiviral signaling (MAVS) adapter. Curr Opin Immunol 23(5):564–572

Bhangoo MK, Tzankov S, Fan AC, Dejgaard K, Thomas DY, Young JC (2007) Multiple 40-kDa heat-shock protein chaperones function in Tom70-dependent mitochondrial import. Mol Biol Cell 18(9):3414–3428

Bonawitz ND, Clayton DA, Shadel GS (2006) Initiation and beyond: multiple functions of the human mitochondrial transcription machinery. Mol Cell 24(6):813–825

Bouchon A, Krammer PH, Walczak H (2000) Critical role for mitochondria in B cell receptor-mediated apoptosis. Eur J Immunol 30(1):69–77

Brennan K, Bowie AG (2010) Activation of host pattern recognition receptors by viruses. Curr Opin Microbiol 13(4):503–507

Brunette RL, Young JM, Whitley DG, Brodsky IE, Malik HS, Stetson DB (2012) Extensive evolutionary and functional diversity among mammalian AIM2-like receptors. J Exp Med 209(11):1969–1983

Bruns AM, Leser GP, Lamb RA, Horvath CM (2014) The innate immune sensor LGP2 activates antiviral signaling by regulating MDA5-RNA interaction and filament assembly. Mol Cell 55(5):771–781

Campello S, Scorrano L (2010) Mitochondrial shape changes: orchestrating cell pathophysiology. EMBO Rep 11(9):678–684

Cardenas WB, Loo YM, Gale M Jr, Hartman AL, Kimberlin CR, Martinez-Sobrido L, Saphire EO, Basler CF (2006) Ebola virus VP35 protein binds double-stranded RNA and inhibits alpha/beta interferon production induced by RIG-I signaling. J Virol 80(11):5168–5178

Case AJ, McGill JL, Tygrett LT, Shirasawa T, Spitz DR, Waldschmidt TJ, Legge KL, Domann FE (2011) Elevated mitochondrial superoxide disrupts normal T cell development, impairing adaptive immune responses to an influenza challenge. Free Radic Biol Med 50(3):448–458

Castanier C, Garcin D, Vazquez A, Arnoult D (2010) Mitochondrial dynamics regulate the RIG-I-like receptor antiviral pathway. EMBO Rep 11(2):133–138

Chan DC (2006) Mitochondrial fusion and fission in mammals. Annu Rev Cell Dev Biol 22:79–99

Chang CH, Curtis JD, Maggi LB Jr, Faubert B, Villarino AV, O'Sullivan D, Huang SC, van der Windt GJ, Blagih J, Qiu J et al (2013) Posttranscriptional control of T cell effector function by aerobic glycolysis. Cell 153(6):1239–1251

Chattopadhyay S, Fensterl V, Zhang Y, Veleeparambil M, Yamashita M, Sen GC (2013) Role of interferon regulatory factor 3-mediated apoptosis in the establishment and maintenance of persistent infection by Sendai virus. J Virol 87(1):16–24

Chen Z, Benureau Y, Rijnbrand R, Yi J, Wang T, Warter L, Lanford RE, Weinman SA, Lemon SM, Martin A et al (2007) GB virus B disrupts RIG-I signaling by NS3/4A-mediated cleavage of the adaptor protein MAVS. J Virol 81(2):964–976

Chen S, Sheng C, Liu D, Yao C, Gao S, Song L, Jiang W, Li J, Huang W (2013) Enhancer of zeste homolog 2 is a negative regulator of mitochondria-mediated innate immune responses. J Immunol 191(5):2614–2623

Chowdhury JB, Kim H, Ray R, Ray RB (2014) Hepatitis C virus NS5A protein modulates IRF-7-mediated interferon-alpha signaling. J Interferon Cytokine Res 34(1):16–21

Ciccosanti F, Corazzari M, Soldani F, Matarrese P, Pagliarini V, Iadevaia V, Tinari A, Zaccarelli M, Perfettini JL, Malorni W et al (2010) Proteomic analysis identifies prohibitin down-regulation as a crucial event in the mitochondrial damage observed in HIV-infected patients. Antivir Ther 15(3):377–390

Cui Z, Vance JE, Chen MH, Voelker DR, Vance DE (1993) Cloning and expression of a novel phosphatidylethanolamine N-methyltransferase. A specific biochemical and cytological marker for a unique membrane fraction in rat liver. J Biol Chem 268(22):16655–16663

Daffis S, Samuel MA, Keller BC, Gale M Jr, Diamond MS (2007) Cell-specific IRF-3 responses protect against West Nile virus infection by interferon-dependent and -independent mechanisms. PLoS Pathog 3(7):e106

Daffis S, Samuel MA, Suthar MS, Keller BC, Gale M Jr, Diamond MS (2008) Interferon regulatory factor IRF-7 induces the antiviral alpha interferon response and protects against lethal West Nile virus infection. J Virol 82(17):8465–8475

Daffis S, Suthar MS, Szretter KJ, Gale M Jr, Diamond MS (2009) Induction of IFN-beta and the innate antiviral response in myeloid cells occurs through an IPS-1-dependent signal that does not require IRF-3 and IRF-7. PLoS Pathog 5(10):e1000607

de Brito OM, Scorrano L (2008) Mitofusin 2 tethers endoplasmic reticulum to mitochondria. Nature 456(7222):605–610
Deming PB, Rathmell JC (2006) Mitochondria, cell death, and B cell tolerance. Curr Directions Autoimmun 9:95–119
Ding S, Robek MD (2014) Peroxisomal MAVS activates IRF1-mediated IFN-lambda production. Nat Immunol 15(8):700–701
Dixit E, Boulant S, Zhang Y, Lee AS, Odendall C, Shum B, Hacohen N, Chen ZJ, Whelan SP, Fransen M et al (2010) Peroxisomes are signaling platforms for antiviral innate immunity. Cell 141(4):668–681
Du C, Fang M, Li Y, Li L, Wang X (2000) Smac, a mitochondrial protein that promotes cytochrome c-dependent caspase activation by eliminating IAP inhibition. Cell 102(1):33–42
Duchen MR (2004a) Roles of mitochondria in health and disease. Diabetes 53(Suppl 1):S96–S102
Duchen MR (2004b) Mitochondria in health and disease: perspectives on a new mitochondrial biology. Mol Asp Med 25(4):365–451
Eckard SC, Rice GI, Fabre A, Badens C, Gray EE, Hartley JL, Crow YJ, Stetson DB (2014) The SKIV2L RNA exosome limits activation of the RIG-I-like receptors. Nat Immunol 15(9):839–845
El Maadidi S, Faletti L, Berg B, Wenzl C, Wieland K, Chen ZJ, Maurer U, Borner C (2014) A novel mitochondrial MAVS/Caspase-8 platform links RNA virus-induced innate antiviral signaling to Bax/Bak-independent apoptosis. J Immunol 192(3):1171–1183
Erickson AK, Gale M Jr (2008) Regulation of interferon production and innate antiviral immunity through translational control of IRF-7. Cell Res 18(4):433–435
Errett JS, Suthar MS, McMillan A, Diamond MS, Gale M Jr (2013) The essential, nonredundant roles of RIG-I and MDA5 in detecting and controlling West Nile virus infection. J Virol 87(21):11416–11425
Everts B, Pearce EJ (2014) Metabolic control of dendritic cell activation and function: recent advances and clinical implications. Front Immunol 5:203
Everts B, Amiel E, Huang SC, Smith AM, Chang CH, Lam WY, Redmann V, Freitas TC, Blagih J, van der Windt GJ et al (2014) TLR-driven early glycolytic reprogramming via the kinases TBK1-IKKvarepsilon supports the anabolic demands of dendritic cell activation. Nat Immunol 15(4):323–332
Faccio L, Fusco C, Chen A, Martinotti S, Bonventre JV, Zervos AS (2000) Characterization of a novel human serine protease that has extensive homology to bacterial heat shock endoprotease HtrA and is regulated by kidney ischemia. J Biol Chem 275(4):2581–2588
Fox CJ, Hammerman PS, Thompson CB (2005) Fuel feeds function: energy metabolism and the T-cell response. Nat Rev Immunol 5(11):844–852
Franchi L, Nunez G (2008) The Nlrp3 inflammasome is critical for aluminium hydroxide-mediated IL-1beta secretion but dispensable for adjuvant activity. Eur J Immunol 38(8):2085–2089
Freed EO, Gale M Jr (2014) Antiviral innate immunity: editorial overview. J Mol Biol 426(6):1129–1132
Friedman JR, Lackner LL, West M, DiBenedetto JR, Nunnari J, Voeltz GK (2011) ER tubules mark sites of mitochondrial division. Science 334(6054):358–362
Gerriets VA, Rathmell JC (2012) Metabolic pathways in T cell fate and function. Trends Immunol 33(4):168–173
Goswami R, Majumdar T, Dhar J, Chattopadhyay S, Bandyopadhyay SK, Verbovetskaya V, Sen GC, Barik S (2013) Viral degradasome hijacks mitochondria to suppress innate immunity. Cell Res 23(8):1025–1042
Hatefi Y (1985) The mitochondrial electron transport and oxidative phosphorylation system. Annu Rev Biochem 54:1015–1069
Hegde R, Srinivasula SM, Zhang Z, Wassell R, Mukattash R, Cilenti L, DuBois G, Lazebnik Y, Zervos AS, Fernandes-Alnemri T et al (2002) Identification of Omi/HtrA2 as a mitochondrial apoptotic serine protease that disrupts inhibitor of apoptosis protein-caspase interaction. J Biol Chem 277(1):432–438

Horner SM, Gale M Jr (2013) Regulation of hepatic innate immunity by hepatitis C virus. Nat Med 19(7):879–888

Horner SM, Liu HM, Park HS, Briley J, Gale M Jr (2011) Mitochondrial-associated endoplasmic reticulum membranes (MAM) form innate immune synapses and are targeted by hepatitis C virus. Proc Natl Acad Sci U S A 108(35):14590–14595

Horner SM, Park HS, Gale M Jr (2012) Control of innate immune signaling and membrane targeting by the Hepatitis C virus NS3/4A protease are governed by the NS3 helix alpha0. J Virol 86(6):3112–3120

Hou F, Sun L, Zheng H, Skaug B, Jiang QX, Chen ZJ (2011) MAVS forms functional prion-like aggregates to activate and propagate antiviral innate immune response. Cell 146(3):448–461

Huang Y, Liu H, Ge R, Zhou Y, Lou X, Wang C (2012) UXT-V1 facilitates the formation of MAVS antiviral signalosome on mitochondria. J Immunol 188(1):358–366

Huang Y, Liu H, Li S, Tang Y, Wei B, Yu H, Wang C (2014) MAVS-MKK7-JNK2 defines a novel apoptotic signaling pathway during viral infection. PLoS Pathog 10(3):e1004020

Ireton RC, Gale M Jr (2011) RIG-I like receptors in antiviral immunity and therapeutic applications. Viruses 3(6):906–919

Ishikawa H, Barber GN (2008) STING is an endoplasmic reticulum adaptor that facilitates innate immune signalling. Nature 455(7213):674–678

Janeway CA Jr (1989) Approaching the asymptote? Evolution and revolution in immunology. Cold Spring Harb Symp Quant Biol 54(Pt 1):1–13

Jaworska J, Coulombe F, Downey J, Tzelepis F, Shalaby K, Tattoli I, Berube J, Rousseau S, Martin JG, Girardin SE et al (2014) NLRX1 prevents mitochondrial induced apoptosis and enhances macrophage antiviral immunity by interacting with influenza virus PB1-F2 protein. Proc Natl Acad Sci U S A 111(20):E2110–E2119

Jenkins K, Khoo JJ, Sadler A, Piganis R, Wang D, Borg NA, Hjerrild K, Gould J, Thomas BJ, Nagley P et al (2013) Mitochondrially localised MUL1 is a novel modulator of antiviral signaling. Immunol Cell Biol 91(4):321–330

Jia Y, Song T, Wei C, Ni C, Zheng Z, Xu Q, Ma H, Li L, Zhang Y, He X et al (2009) Negative regulation of MAVS-mediated innate immune response by PSMA7. J Immunol 183(7):4241–4248

Jiang F, Ramanathan A, Miller MT, Tang GQ, Gale M Jr, Patel SS, Marcotrigiano J (2011) Structural basis of RNA recognition and activation by innate immune receptor RIG-I. Nature 479(7373):423–427

Jiang X, Kinch LN, Brautigam CA, Chen X, Du F, Grishin NV, Chen ZJ (2012) Ubiquitin-induced oligomerization of the RNA sensors RIG-I and MDA5 activates antiviral innate immune response. Immunity 36(6):959–973

Kadenbach B (2012) Introduction to mitochondrial oxidative phosphorylation. Adv Exp Med Biol 748:1–11

Kanneganti TD, Body-Malapel M, Amer A, Park JH, Whitfield J, Franchi L, Taraporewala ZF, Miller D, Patton JT, Inohara N et al (2006) Critical role for Cryopyrin/Nalp3 in activation of caspase-1 in response to viral infection and double-stranded RNA. J Biol Chem 281(48):36560–36568

Kaushik DK, Gupta M, Kumawat KL, Basu A (2012) NLRP3 inflammasome: key mediator of neuroinflammation in murine Japanese encephalitis. PLoS One 7(2):e32270

Kawai T, Akira S (2010) The role of pattern-recognition receptors in innate immunity: update on Toll-like receptors. Nat Immunol 11(5):373–384

Kim SJ, Syed GH, Khan M, Chiu WW, Sohail MA, Gish RG, Siddiqui A (2014) Hepatitis C virus triggers mitochondrial fission and attenuates apoptosis to promote viral persistence. Proc Natl Acad Sci U S A 111(17):6413–6418

Kohchi C, Inagawa H, Nishizawa T, Soma G (2009) ROS and innate immunity. Anticancer Res 29(3):817–821

Kono H, Rock KL (2008) How dying cells alert the immune system to danger. Nat Rev Immunol 8(4):279–289

Koshiba T (2013) Mitochondrial-mediated antiviral immunity. Biochim Biophys Acta 1833(1): 225–232

Koshiba T, Yasukawa K, Yanagi Y, Kawabata S (2011) Mitochondrial membrane potential is required for MAVS-mediated antiviral signaling. Sci Signal 4(158):ra7

Kruman II, Nath A, Mattson MP (1998) HIV-1 protein Tat induces apoptosis of hippocampal neurons by a mechanism involving caspase activation, calcium overload, and oxidative stress. Exp Neurol 154(2):276–288

Kumar H, Kawai T, Kato H, Sato S, Takahashi K, Coban C, Yamamoto M, Uematsu S, Ishii KJ, Takeuchi O et al (2006) Essential role of IPS-1 in innate immune responses against RNA viruses. J Exp Med 203(7):1795–1803

Lassoued S, Ben Ameur R, Ayadi W, Gargouri B, Ben Mansour R, Attia H (2008) Epstein-Barr virus induces an oxidative stress during the early stages of infection in B lymphocytes, epithelial, and lymphoblastoid cell lines. Mol Cell Biochem 313(1–2):179–186

Lazear HM, Lancaster A, Wilkins C, Suthar MS, Huang A, Vick SC, Clepper L, Thackray L, Brassil MM, Virgin HW et al (2013) IRF-3, IRF-5, and IRF-7 coordinately regulate the type I IFN response in myeloid dendritic cells downstream of MAVS signaling. PLoS Pathog 9(1):e1003118

Lei Y, Moore CB, Liesman RM, O'Connor BP, Bergstralh DT, Chen ZJ, Pickles RJ, Ting JP (2009) MAVS-mediated apoptosis and its inhibition by viral proteins. PLoS One 4(5):e5466

Li LY, Luo X, Wang X (2001) Endonuclease G is an apoptotic DNase when released from mitochondria. Nature 412(6842):95–99

Li XD, Sun L, Seth RB, Pineda G, Chen ZJ (2005) Hepatitis C virus protease NS3/4A cleaves mitochondrial antiviral signaling protein off the mitochondria to evade innate immunity. Proc Natl Acad Sci U S A 102(49):17717–17722

Lin R, Lacoste J, Nakhaei P, Sun Q, Yang L, Paz S, Wilkinson P, Julkunen I, Vitour D, Meurs E et al (2006) Dissociation of a MAVS/IPS-1/VISA/Cardif-IKKepsilon molecular complex from the mitochondrial outer membrane by hepatitis C virus NS3-4A proteolytic cleavage. J Virol 80(12):6072–6083

Liu X, Kim CN, Yang J, Jemmerson R, Wang X (1996) Induction of apoptotic program in cell-free extracts: requirement for dATP and cytochrome c. Cell 86(1):147–157

Liu XY, Wei B, Shi HX, Shan YF, Wang C (2010) Tom70 mediates activation of interferon regulatory factor 3 on mitochondria. Cell Res 20(9):994–1011

Liu HM, Loo YM, Horner SM, Zornetzer GA, Katze MG, Gale M Jr (2012) The mitochondrial targeting chaperone 14-3-3epsilon regulates a RIG-I translocon that mediates membrane association and innate antiviral immunity. Cell Host Microbe 11(5):528–537

Liu S, Chen J, Cai X, Wu J, Chen X, Wu YT, Sun L, Chen ZJ (2013) MAVS recruits multiple ubiquitin E3 ligases to activate antiviral signaling cascades. eLife 2:e00785

Loo YM, Gale M Jr (2011) Immune signaling by RIG-I-like receptors. Immunity 34(5):680–692

Ma Q, Chen S, Hu Q, Feng H, Zhang JH, Tang J (2014) NLRP3 inflammasome contributes to inflammation after intracerebral hemorrhage. Ann Neurol 75(2):209–219

Martins LM, Iaccarino I, Tenev T, Gschmeissner S, Totty NF, Lemoine NR, Savopoulos J, Gray CW, Creasy CL, Dingwall C et al (2002) The serine protease Omi/HtrA2 regulates apoptosis by binding XIAP through a reaper-like motif. J Biol Chem 277(1):439–444

McGuire KA, Barlan AU, Griffin TM, Wiethoff CM (2011) Adenovirus type 5 rupture of lysosomes leads to cathepsin B-dependent mitochondrial stress and production of reactive oxygen species. J Virol 85(20):10806–10813

Mibayashi M, Martinez-Sobrido L, Loo YM, Cardenas WB, Gale M Jr, Garcia-Sastre A (2007) Inhibition of retinoic acid-inducible gene I-mediated induction of beta interferon by the NS1 protein of influenza A virus. J Virol 81(2):514–524

Michallet MC, Meylan E, Ermolaeva MA, Vazquez J, Rebsamen M, Curran J, Poeck H, Bscheider M, Hartmann G, Konig M et al (2008) TRADD protein is an essential component of the RIG-like helicase antiviral pathway. Immunity 28(5):651–661

Mitoma H, Hanabuchi S, Kim T, Bao M, Zhang Z, Sugimoto N, Liu YJ (2013) The DHX33 RNA helicase senses cytosolic RNA and activates the NLRP3 inflammasome. Immunity 39(1): 123–135

Mukherjee P, Woods TA, Moore RA, Peterson KE (2013) Activation of the innate signaling molecule MAVS by bunyavirus infection upregulates the adaptor protein SARM1, leading to neuronal death. Immunity 38(4):705–716

Muruve DA, Petrilli V, Zaiss AK, White LR, Clark SA, Ross PJ, Parks RJ, Tschopp J (2008) The inflammasome recognizes cytosolic microbial and host DNA and triggers an innate immune response. Nature 452(7183):103–107

Negash AA, Ramos HJ, Crochet N, Lau DT, Doehle B, Papic N, Delker DA, Jo J, Bertoletti A, Hagedorn CH et al (2013) IL-1beta production through the NLRP3 inflammasome by hepatic macrophages links hepatitis C virus infection with liver inflammation and disease. PLoS Pathog 9(4):e1003330

Ning YJ, Wang M, Deng M, Shen S, Liu W, Cao WC, Deng F, Wang YY, Hu Z, Wang H (2014) Viral suppression of innate immunity via spatial isolation of TBK1/IKK{epsilon} from mitochondrial antiviral platform. J Mol Cell Biol 6(4):324–337

Nishina S, Hino K, Korenaga M, Vecchi C, Pietrangelo A, Mizukami Y, Furutani T, Sakai A, Okuda M, Hidaka I et al (2008) Hepatitis C virus-induced reactive oxygen species raise hepatic iron level in mice by reducing hepcidin transcription. Gastroenterology 134(1):226–238

Nour AM, Reichelt M, Ku CC, Ho MY, Heineman TC, Arvin AM (2011) Varicella-zoster virus infection triggers formation of an interleukin-1beta (IL-1beta)-processing inflammasome complex. J Biol Chem 286(20):17921–17933

Odendall C, Dixit E, Stavru F, Bierne H, Franz KM, Durbin AF, Boulant S, Gehrke L, Cossart P, Kagan JC (2014) Diverse intracellular pathogens activate type III interferon expression from peroxisomes. Nat Immunol 15(8):717–726

Ohta A, Nishiyama Y (2011) Mitochondria and viruses. Mitochondrion 11(1):1–12

O'Neill LA, Hardie DG (2013) Metabolism of inflammation limited by AMPK and pseudo-starvation. Nature 493(7432):346–355

Pearce EL, Poffenberger MC, Chang CH, Jones RG (2013) Fueling immunity: insights into metabolism and lymphocyte function. Science 342(6155):1242454

Pichlmair A, Schulz O, Tan CP, Rehwinkel J, Kato H, Takeuchi O, Akira S, Way M, Schiavo G, Reis e Sousa C (2009) Activation of MDA5 requires higher-order RNA structures generated during virus infection. J Virol 83(20):10761–10769

Pythoud C, Rodrigo WW, Pasqual G, Rothenberger S, Martinez-Sobrido L, de la Torre JC, Kunz S (2012) Arenavirus nucleoprotein targets interferon regulatory factor-activating kinase IKKepsilon. J Virol 86(15):7728–7738

Rajan JV, Rodriguez D, Miao EA, Aderem A (2011) The NLRP3 inflammasome detects encephalomyocarditis virus and vesicular stomatitis virus infection. J Virol 85(9):4167–4172

Ramos HJ, Gale M Jr (2011) RIG-I like receptors and their signaling crosstalk in the regulation of antiviral immunity. Curr Opin Virol 1(3):167–176

Ramos HJ, Lanteri MC, Blahnik G, Negash A, Suthar MS, Brassil MM, Sodhi K, Treuting PM, Busch MP, Norris PJ et al (2012) IL-1beta signaling promotes CNS-intrinsic immune control of West Nile virus infection. PLoS Pathog 8(11):e1003039

Robinson MJ, Sancho D, Slack EC, LeibundGut-Landmann S, Reis e Sousa C (2006) Myeloid C-type lectins in innate immunity. Nat Immunol 7(12):1258–1265

Rock KL, Latz E, Ontiveros F, Kono H (2010) The sterile inflammatory response. Annu Rev Immunol 28:321–342

Rossi L, Bonuccelli L, Iacopetti P, Evangelista M, Ghezzani C, Tana L, Salvetti A (2014) Prohibitin 2 regulates cell proliferation and mitochondrial cristae morphogenesis in planarian stem cells. Stem Cell Rev 10(6):871–887

Rowland AA, Voeltz GK (2012) Endoplasmic reticulum-mitochondria contacts: function of the junction. Nat Rev Mol Cell Biol 13(10):607–625

Sadler AJ, Williams BR (2008) Interferon-inducible antiviral effectors. Nat Rev Immunol 8(7):559–568

Sagan L (1967) On the origin of mitosing cells. J Theor Biol 14(3):255–274

Saha SK, Pietras EM, He JQ, Kang JR, Liu SY, Oganesyan G, Shahangian A, Zarnegar B, Shiba TL, Wang Y et al (2006) Regulation of antiviral responses by a direct and specific interaction between TRAF3 and Cardif. EMBO J 25(14):3257–3263
Saha P, Chowdhury AR, Dutta S, Chatterjee S, Ghosh I, Datta K (2013) Autophagic vacuolation induced by excess ROS generation in HABP1/p32/gC1qR overexpressing fibroblasts and its reversal by polymeric hyaluronan. PLoS One 8(10):e78131
Saito T, Owen DM, Jiang F, Marcotrigiano J, Gale M Jr (2008) Innate immunity induced by composition-dependent RIG-I recognition of hepatitis C virus RNA. Nature 454(7203): 523–527
Saraste M (1999) Oxidative phosphorylation at the fin de siecle. Science 283(5407):1488–1493
Schoggins JW, Wilson SJ, Panis M, Murphy MY, Jones CT, Bieniasz P, Rice CM (2011) A diverse range of gene products are effectors of the type I interferon antiviral response. Nature 472(7344):481–485
Schoggins JW, MacDuff DA, Imanaka N, Gainey MD, Shrestha B, Eitson JL, Mar KB, Richardson RB, Ratushny AV, Litvak V et al (2014) Pan-viral specificity of IFN-induced genes reveals new roles for cGAS in innate immunity. Nature 505(7485):691–695
Schwacha MG, Rani M, Zhang Q, Nunez-Cantu O, Cap AP (2014) Mitochondrial damage-associated molecular patterns activate gammadelta T-cells. Innate immun 20(3):261–268
Sena LA, Li S, Jairaman A, Prakriya M, Ezponda T, Hildeman DA, Wang CR, Schumacker PT, Licht JD, Perlman H et al (2013) Mitochondria are required for antigen-specific T cell activation through reactive oxygen species signaling. Immunity 38(2):225–236
Sesaki H, Adachi Y, Kageyama Y, Itoh K, Iijima M (2014) In vivo functions of Drp1: Lessons learned from yeast genetics and mouse knockouts. Biochim Biophys Acta 1842(8): 1179–1185
Seth RB, Sun L, Ea CK, Chen ZJ (2005) Identification and characterization of MAVS, a mitochondrial antiviral signaling protein that activates NF-kappaB and IRF 3. Cell 122(5):669–682
Shapiro H, Lutaty A, Ariel A (2011) Macrophages, meta-inflammation, and immuno-metabolism. Sci World J 11:2509–2529
Shintani Y, Drexler HC, Kioka H, Terracciano CM, Coppen SR, Imamura H, Akao M, Nakai J, Wheeler AP, Higo S et al (2014) Toll-like receptor 9 protects non-immune cells from stress by modulating mitochondrial ATP synthesis through the inhibition of SERCA2. EMBO Rep 15(4):438–445
Sorriento D, Pascale AV, Finelli R, Carillo AL, Annunziata R, Trimarco B, Iaccarino G (2014) Targeting mitochondria as therapeutic strategy for metabolic disorders. Sci World J 2014:604–685
Subramanian N, Natarajan K, Clatworthy MR, Wang Z, Germain RN (2013) The adaptor MAVS promotes NLRP3 mitochondrial localization and inflammasome activation. Cell 153(2): 348–361
Sugiura A, Nagashima S, Tokuyama T, Amo T, Matsuki Y, Ishido S, Kudo Y, McBride HM, Fukuda T, Matsushita N et al (2013) MITOL regulates endoplasmic reticulum-mitochondria contacts via Mitofusin2. Mol Cell 51(1):20–34
Sun Q, Sun L, Liu HH, Chen X, Seth RB, Forman J, Chen ZJ (2006) The specific and essential role of MAVS in antiviral innate immune responses. Immunity 24(5):633–642
Sun W, Li Y, Chen L, Chen H, You F, Zhou X, Zhou Y, Zhai Z, Chen D, Jiang Z (2009) ERIS, an endoplasmic reticulum IFN stimulator, activates innate immune signaling through dimerization. Proc Natl Acad Sci U S A 106(21):8653–8658
Susin SA, Zamzami N, Castedo M, Hirsch T, Marchetti P, Macho A, Daugas E, Geuskens M, Kroemer G (1996) Bcl-2 inhibits the mitochondrial release of an apoptogenic protease. J Exp Med 184(4):1331–1341
Suthar MS, Ramos HJ, Brassil MM, Netland J, Chappell CP, Blahnik G, McMillan A, Diamond MS, Clark EA, Bevan MJ et al (2012) The RIG-I-like receptor LGP2 controls CD8(+) T cell survival and fitness. Immunity 37(2):235–248
Sutterwala FS, Haasken S, Cassel SL (2014) Mechanism of NLRP3 inflammasome activation. Ann N Y Acad Sci 1319(1):82–95

Suzuki Y, Imai Y, Nakayama H, Takahashi K, Takio K, Takahashi R (2001) A serine protease, HtrA2, is released from the mitochondria and interacts with XIAP, inducing cell death. Mol Cell 8(3):613–621
Taanman JW (1999) The mitochondrial genome: structure, transcription, translation and replication. Biochim Biophys Acta 1410(2):103–123
Takeuchi O, Akira S (2009) Innate immunity to virus infection. Immunol Rev 227(1):75–86
Tal MC, Sasai M, Lee HK, Yordy B, Shadel GS, Iwasaki A (2009) Absence of autophagy results in reactive oxygen species-dependent amplification of RLR signaling. Proc Natl Acad Sci U S A 106(8):2770–2775
Tang ED, Wang CY (2009) MAVS self-association mediates antiviral innate immune signaling. J Virol 83(8):3420–3428
Tang ED, Wang CY (2010) TRAF5 is a downstream target of MAVS in antiviral innate immune signaling. PLoS One 5(2):e9172
Tattoli I, Carneiro LA, Jehanno M, Magalhaes JG, Shu Y, Philpott DJ, Arnoult D, Girardin SE (2008) NLRX1 is a mitochondrial NOD-like receptor that amplifies NF-kappaB and JNK pathways by inducing reactive oxygen species production. EMBO Rep 9(3):293–300
van der Windt GJ, Pearce EL (2012) Metabolic switching and fuel choice during T-cell differentiation and memory development. Immunol Rev 249(1):27–42
van der Windt GJ, Everts B, Chang CH, Curtis JD, Freitas TC, Amiel E, Pearce EJ, Pearce EL (2012) Mitochondrial respiratory capacity is a critical regulator of CD8+ T cell memory development. Immunity 36(1):68–78
Vance JE (2014) MAM (mitochondria-associated membranes) in mammalian cells: lipids and beyond. Biochim Biophys Acta 1841(4):595–609
Verhagen AM, Ekert PG, Pakusch M, Silke J, Connolly LM, Reid GE, Moritz RL, Simpson RJ, Vaux DL (2000) Identification of DIABLO, a mammalian protein that promotes apoptosis by binding to and antagonizing IAP proteins. Cell 102(1):43–53
Vitour D, Dabo S, Ahmadi Pour M, Vilasco M, Vidalain PO, Jacob Y, Mezel-Lemoine M, Paz S, Arguello M, Lin R et al (2009) Polo-like kinase 1 (PLK1) regulates interferon (IFN) induction by MAVS. J Biol Chem 284(33):21797–21809
Waickman AT, Powell JD (2012) mTOR, metabolism, and the regulation of T-cell differentiation and function. Immunol Rev 249(1):43–58
Wang C, Youle RJ (2009) The role of mitochondria in apoptosis*. Ann Rev Genet 43:95–118
Wang P, Yang L, Cheng G, Yang G, Xu Z, You F, Sun Q, Lin R, Fikrig E, Sutton RE (2013a) UBXN1 interferes with Rig-I-like receptor-mediated antiviral immune response by targeting MAVS. Cell Rep 3(4):1057–1070
Wang B, Xi X, Lei X, Zhang X, Cui S, Wang J, Jin Q, Zhao Z (2013b) Enterovirus 71 protease 2Apro targets MAVS to inhibit anti-viral type I interferon responses. PLoS Pathog 9(3):e1003231
Wenceslau CF, McCarthy CG, Szasz T, Spitler K, Goulopoulou S, Webb RC (2014) Mitochondrial damage-associated molecular patterns and vascular function. Eur Heart J 35(18):1172–1177
West AP, Shadel GS, Ghosh S (2011) Mitochondria in innate immune responses. Nat Rev Immunol 11(6):389–402
Williamson CD, Colberg-Poley AM (2009) Access of viral proteins to mitochondria via mitochondria-associated membranes. Rev Med Virol 19(3):147–164
Williamson CD, DeBiasi RL, Colberg-Poley AM (2012) Viral product trafficking to mitochondria, mechanisms and roles in pathogenesis. Infect Disord Drug target 12(1):18–37
Wu J, Sun L, Chen X, Du F, Shi H, Chen C, Chen ZJ (2013) Cyclic GMP-AMP is an endogenous second messenger in innate immune signaling by cytosolic DNA. Science 339(6121):826–830
Wu ZS, Yao YM, Hong GL, Xu XP, Liu Y, Dong N, Zheng JY, Lu ZQ, Zhao GJ, Zhu XM et al (2014) Role of mitofusin-2 in high mobility group box-1 protein-mediated apoptosis of T cells in vitro. Cell Physiol Biochem 33(3):769–783

Xia M, Gonzalez P, Li C, Meng G, Jiang A, Wang H, Gao Q, Debatin KM, Beltinger C, Wei J (2014) Mitophagy enhances oncolytic measles virus replication by mitigating DDX58/RIG-I-like receptor signaling. J Virol 88(9):5152–5164

Xu LG, Wang YY, Han KJ, Li LY, Zhai Z, Shu HB (2005) VISA is an adapter protein required for virus-triggered IFN-beta signaling. Mol Cell 19(6):727–740

Xu L, Xiao N, Liu F, Ren H, Gu J (2009) Inhibition of RIG-I and MDA5-dependent antiviral response by gC1qR at mitochondria. Proc Natl Acad Sci U S A 106(5):1530–1535

Yanai H, Ban T, Wang Z, Choi MK, Kawamura T, Negishi H, Nakasato M, Lu Y, Hangai S, Koshiba R et al (2009) HMGB proteins function as universal sentinels for nucleic-acid-mediated innate immune responses. Nature 462(7269):99–103

Yasukawa K, Oshiumi H, Takeda M, Ishihara N, Yanagi Y, Seya T, Kawabata S, Koshiba T (2009) Mitofusin 2 inhibits mitochondrial antiviral signaling. Sci Signal 2(84):ra47

You F, Sun H, Zhou X, Sun W, Liang S, Zhai Z, Jiang Z (2009) PCBP2 mediates degradation of the adaptor MAVS via the HECT ubiquitin ligase AIP4. Nat Immunol 10(12):1300–1308

Youker RT, Shinde U, Day R, Thomas G (2009) At the crossroads of homoeostasis and disease: roles of the PACS proteins in membrane traffic and apoptosis. Biochem J 421(1):1–15

Yu HB, Finlay BB (2008) The caspase-1 inflammasome: a pilot of innate immune responses. Cell Host Microbe 4(3):198–208

Zhao Y, Sun X, Nie X, Sun L, Tang TS, Chen D, Sun Q (2012) COX5B regulates MAVS-mediated antiviral signaling through interaction with ATG5 and repressing ROS production. PLoS Pathog 8(12):e1003086

Zhong B, Yang Y, Li S, Wang YY, Li Y, Diao F, Lei C, He X, Zhang L, Tien P et al (2008) The adaptor protein MITA links virus-sensing receptors to IRF3 transcription factor activation. Immunity 29(4):538–550

Zhou Z, Jia X, Xue Q, Dou Z, Ma Y, Zhao Z, Jiang Z, He B, Jin Q, Wang J (2014) TRIM14 is a mitochondrial adaptor that facilitates retinoic acid-inducible gene-I-like receptor-mediated innate immune response. Proc Natl Acad Sci U S A 111(2):E245–E254

Zhu J, Zhang Y, Ghosh A, Cuevas RA, Forero A, Dhar J, Ibsen MS, Schmid-Burgk JL, Schmidt T, Ganapathiraju MK et al (2014) Antiviral activity of human OASL protein is mediated by enhancing signaling of the RIG-I RNA sensor. Immunity 40(6):936–948

Chapter 11
Mitochondrial Permeabilization: From Lethality to Vitality

Andrew Oberst, Gabriel Ichim, and Stephen W.G. Tait

11.1 Introduction

11.1.1 Engagement of MOMP

Apoptotic cell death is a highly regulated pathway of cell suicide that is fundamentally required for development, tissue homeostasis, immunity, and tumor suppression (Green 2011). Tens of billions of cells are deleted from our bodies by apoptosis each day, and nearly all of these die following engagement of the mitochondrial pathway of apoptosis. This cell death pathway has been described in great detail, and other chapters of this book discuss its intricacies; here, we present only a general overview.

The pathways of apoptosis are generally broadly divided into "intrinsic" and "extrinsic" pathways; these names refer to the origin of the apoptosis-initiating signals. Intrinsic apoptosis is initiated by signals from within the dying cell, such as DNA damage, aberrant proliferation or cell cycle dysregulation, ER stress, or lack of sufficient trophic support. These signals converge on members of the BCL2 protein family. This group of proteins comprises both pro- and anti-apoptotic members, and their complex interactions determine whether mitochondrial permeabilization will occur. Upon activation, the BCL2 proteins BAX and BAK form pores in the

A. Oberst (✉)
Department of Immunology, University of Washington, Seattle, WA, USA
e-mail: oberst@uw.edu

G. Ichim • S.W.G. Tait (✉)
Cancer Research-UK Beatson Institute, Institute of Cancer Sciences,
University of Glasgow, Glasgow, UK
e-mail: stephen.tait@glasgow.ac.uk

D.M. Hockenbery (ed.), *Mitochondria and Cell Death*, Cell Death in Biology and Diseases, DOI 10.1007/978-1-4939-3612-0_11

X.-M. Yin and Z. Dong (Series eds.), *Cell Death in Biology and Diseases*

mitochondrial outer membrane; these proteins are activated by BH3-only proteins such as BIM or BID, and this activation is opposed by the anti-apoptotic family members BCL2, BCLXL, and MCL1. These activation and inhibition interactions are influenced by other family members, and have been exhaustively described elsewhere (Chipuk et al. 2010; Czabotar et al. 2014). For our purposes, it is sufficient to understand that BAX and BAK-mediated pore formation in the outer mitochondrial membrane is the key event that triggers mitochondrial outer membrane permeabilization (or MOMP).

Extrinsic apoptosis is triggered by signaling events that originate outside the dying cell, mediated by engagement of a subset of the TNF receptor superfamily termed the "death receptors." These include TNFR1, the TRAIL receptors DR4 and DR5, and Fas (also called CD95) (Dickens et al. 2012). Engagement of these receptors by their cognate ligands can lead to the activation of caspase 8. While in some conditions caspase 8 can directly engage a downstream caspase cascade (Jost et al. 2009), in most cells the apoptotic signal must be amplified via the mitochondrial pathway. To accomplish this, caspase 8 cleaves the activating BCL2 protein BID; this cleavage event triggers BID-mediated activation of BAX and/or BAK, and thereby engages MOMP.

11.1.2 Consequences of MOMP

Once activated, BAX and/or BAK insert into the mitochondrial outer membrane, oligomerize, and allow proteins sequestered within the intermembrane space to escape. Notably, this event generally occurs in a rapid, coordinated and complete manner, meaning that all mitochondria in a cell become permeabilized within a few minutes; however, as discussed below this is not always the case (Goldstein et al. 2000). Once mitochondria are permeabilized, intermembrane space proteins gain access to the cytosol. These include cytochrome *c*, an iron-containing, essential component of the respiratory machinery, as well as the proteins SMAC (also called DIABLO) and OMI (also called HTRA2) (Tait and Green 2010). Importantly, the inner mitochondrial membrane, which is responsible for maintaining the ion gradient that drives oxidative phosphorylation, remains intact following MOMP (Cookson and Silverstone 1976).

Once cytochrome *c*, SMAC and OMI gain access to the cytosol, they coordinate the activation of the caspase proteolytic cascade. This occurs by binding of the adapter APAF1 to cytochrome *c*, which in turn recruits caspase 9 to form a heptameric complex called the "apoptosome." Recruitment of caspase 9 to the apoptosome triggers its activation through dimerization. Once activated, caspase 9 cleaves and activates the downstream executioner caspases 3 and 7. These potent enzymes cleave hundreds of targets within the cell, leading to the ordered destruction, condensation, and phagocytic elimination of the dying cell (McIlwain et al. 2013). SMAC and OMI contribute to this process by binding to and inactivating the IAPs, a family of E3 ubiquitin ligases that keep caspases in check in healthy cells. MOMP thereby simultaneously activates the executioners (caspases) and inactivates their inhibitors (IAPs), leading to rapid and coordinated destruction of the cell.

11.1.3 Why MOMP at All?

The description above is cursory, largely because MOMP is such a well-described phenomenon. However, our familiarity with MOMP as a part of the apoptotic pathway obscures what a fundamentally weird process it is: That cells posses proteins whose sole function is to permeabilize and destroy an essential organelle—the mitochondria—is itself quite striking. That the key effect of this event is the release of an otherwise unexceptional member of the metabolic machinery, cytochrome *c*, and that this of all things is the precipitating event in the initiation of the apoptotic cascade is truly remarkable. Why would the signaling be wired this way?

Hints to the answer to this question can be gleaned by examining the proteins that orchestrate MOMP, and their homologues. Strikingly, the cytochrome *c* binding adapter APAF1 bears significant homology to members of the NOD-like receptor (NLR) family of innate immune sensors. These proteins bind to pathogen-associated molecular patterns (PAMPs) such as bacterial flagellin or toxin molecules, and in so doing, alert cells to the presence of these invaders, initiating immune responses (Kanneganti 2010; Man and Kanneganti 2015). Many of the NLRs signal by forming circular protein complexes termed inflammasomes, which recruit the inflammatory protease caspase 1. This in turn can cause cell death via a non-apoptotic program termed "pyroptosis," and can also promote the processing and maturation of the cytokines IL-1β and IL-18.

The similarities between PAMP-mediated activation of NLRs and cytochrome *c* mediated activation of APAF1 are notable: the structures of the apoptosome and the inflammasome are similar, both recruit caspases via interactions mediated by CARD domains, and both can trigger cell death. But why would cytochrome *c* be standing in for bacterial PAMPs in these two parallel pathways?

Of course, mitochondria entered eukaryotic cells as prokaryotic symbionts about 1.5 billion years ago, bringing with them a unique set of metabolic tools that allowed host cells to more efficiently extract energy from carbohydrate bonds. One of the key tricks that bacteria brought to the cell was the ability to orchestrate complex redox chemistry through proteins that coordinated inorganic molecules; the ancestral cytochrome *c* was such a protein. Thus, prior to establishment of mitochondria as integral symbionts, cytochrome *c* would have been an excellent PAMP, a molecule that would signal the presence of a prokaryotic invader. This observation leads to the idea that the mitochondrial pathway of apoptosis may represent a regulated form of an ancient pathway of pathogen-induced cell death. Further bolstering this idea is the observation that BAX and BAK, the BCL2 family members responsible for permeabilizing the outer mitochondrial membrane, bear structural homology to bacterial pore-forming toxins called colicins (Lazebnik 2001). But how do we make sense of these observations?

11.1.4 A Long Time Ago, in an Ocean not So Very Far Away…

Here is a scenario that provides a plausible—albeit speculative—answer to the questions above. A few billion years ago, single-celled organisms lived and grew in large clonally similar populations. Being single-celled, they had no need to develop pathways of cell death for developmental or homeostatic purposes. However, altruistic suicide in response to invasion of these cells by other organisms evolved as a means to deprive the invaders of replicative niches. This provided a selective advantage to clonal groups that developed these pathways, since death of a few individuals allowed survival of the population as a whole. In order to induce cell death in response to invasion, these organisms developed sensors that would recognize key reactive molecules in the invaders, such as the iron-coordinating molecules these invaders encoded. These single-celled organisms sequestered the invaders inside vesicles within their cytoplasm, and the invaders responded by producing pore-forming toxins in an effort to escape these membranes. It was war!

Flashing forward a billion years, cooler heads have prevailed; our single-celled protagonists have discovered that their invaders encode metabolic enzymes that allow them to use nutrients more efficiently, and they've harnessed this new efficiency by sequestering the invaders inside vacuoles and providing them with the building blocks they need to produce energy. These are proto-mitochondria. Genetic exchanges between the host and (former) invader take place, and the sensors that triggered cell death in response to cytochrome *c*-like molecules lie dormant. However, as time goes by these single-celled organisms begin to band together and to specialize, eventually forming multicellular communities. As these become more sophisticated, the death of some cells is required for the proto-organism to survive. To achieve this, evolution fashions an elegant solution: the pathways that used to trigger cell death in response to invading bacteria are dusted off and reconfigured to allow controlled engagement of these same pathways by the molecules contained within the now-symbiotic bacteria, i.e., the mitochondria. The bacterial toxins that had formerly been used to puncture the vacuoles enclosing the invaders are reconfigured to perform this task in a controlled way, becoming the BCL2 family. And the innate immune sensor that recognized the PAMP cytochrome *c* was reactivated, becoming APAF1; thus, the cell was able to control the emergence of a former PAMP into its cytosol, and thereby reformat a pathway of innate immune sensing and defensive suicide into a regulated process for organismal health. A summary of parallels between MOMP and bacterial sensing are presented in Fig. 11.1a–c.

As an aside, it is interesting to note that the mitochondrial pathway is not present in all animals, being absent in nematodes and apparently having been lost in *Drosophila melanogaster* (Bender et al. 2012; Oberst et al. 2008). These organisms likely represent different mechanisms by which the evolutionary process reconfig-

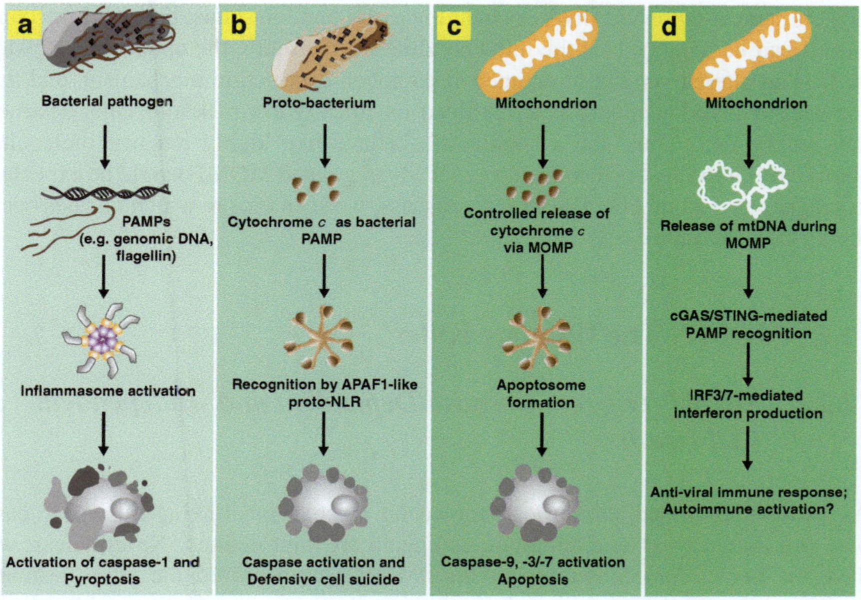

Fig. 11.1 Parallels between bacterial innate immunity and MOMP. (**a**) Modern pathways of inflammasome activation. PAMPs from invading bacterial pathogens are sensed by members of the NLR family, forming inflammasomes. These activate caspase 1 to trigger pyroptotic cell death. (**b**) A hypothetical ancient pathway for "defensive cell suicide" in response to detection of a cytochrome *c* like molecule encoded by proto-bacteria. (**c**) Modern pathways of MOMP-driven apoptosis, as a reconfiguration of the pathway depicted in (**b**) following establishment of symbiosis. This allows programmed cell death in response to controlled release of cytochrome *c*. (**d**) When caspase activation is prevented, mitochondrial DNA released during MOMP can activate the cGAS/STING pathway, leading to IFN production and establishment of an anti-viral immune state

ured its innate immune pathways for regulated cell death. For example, in the nematode *Caenorhabditis elegans*, the BCL2 homologue (CED9) binds to and sequesters the APAF1 homologue (CED4) on the mitochondria; once displaced, CED4 can activate the caspase homologue CED3 (Chinnaiyan et al. 1997; Spector et al. 1997). This configuration may represent a different PAMP-sensing pathway, one centered around the BCL2-like bacterial pore-forming toxins. It is fascinating to note, however, that despite the absence of MOMP, the *C. elegans* cell death pathways remain arrayed around the mitochondria.

11.1.5 Mitochondria: The Enemy Within

This long and speculative interlude should illustrate a key idea: that mitochondria can be thought of as foreign matter within our cells. Our innate immune system remains on high alert, and is capable of sensing invaders through the recognition of triggers such as cytosolic DNA or changing redox and ion status; both of these conditions can be triggered by the mitochondria. Furthermore, the apoptotic pathway itself is an altered version of an ancient pathogen-destruction mechanism, and the enzymes involved in apoptosis are evolved specifically to wreak molecular destruction on the cell. Thus, the mitochondria behave like highly reactive molecular bombs within the host cytosol, and as such, dysregulated MOMP would be expected to trigger detrimental effects. Here, we discuss two scenarios in which this has been shown to be the case.

11.2 MOMP: The Ultimate Killer

11.2.1 MOMP Triggers Caspase-Dependent and -Independent Cell Death

As we have discussed, the intricate molecular mechanisms linking MOMP to caspase protease activity and apoptotic cell death are well defined. Nevertheless, we have also known for a long time that once MOMP has occurred, the cell is destined to die irrespective of caspase activity (Tait et al. 2014). The ability of MOMP to trigger caspase-dependent apoptosis or, in its absence, caspase-independent cell death (CICD) directly relates to the extent of mitochondrial permeabilization. Following a cell death trigger, MOMP is widespread and rapid leading to robust caspase activity (Fig. 11.2a). Because of its extensive, and likely irreversible nature, MOMP ensures cell death even in the absence of caspase activity, by triggering massive mitochondrial dysfunction and metabolic catastrophe (Lartigue et al. 2009). In vivo studies have shown that CICD often represents an effective cell death modality; for example, mice deficient in MOMP-dependent caspase activity (*Apaf1* or *Casp9* knockout) display grossly normal development, contrasting to the much severe developmental defects observed in MOMP-deficient mice (BAX/BAK knockout) (Cecconi et al. 1998; Kuida et al. 1998; Lindsten et al. 2000). These findings argue that CICD can effectively substitute for apoptosis under caspase-deficient conditions, however, whether CICD is engaged under pathophysiological conditions remains an open question. Addressing this, comparative morphological analysis suggests that up to 10 % of cells dying during development may engage CICD, although definitive assessment necessitates combined single-cell analysis of MOMP and caspase activity in vivo (Chautan et al. 1999).

11.2.2 In Death, Timing Is Everything

If MOMP usually kills cells irrespective of caspase activity, then why have caspases and dedicated activation machinery? One likely reason is that caspase activity dramatically enhances the speed of cell death. Following MOMP, caspases typically

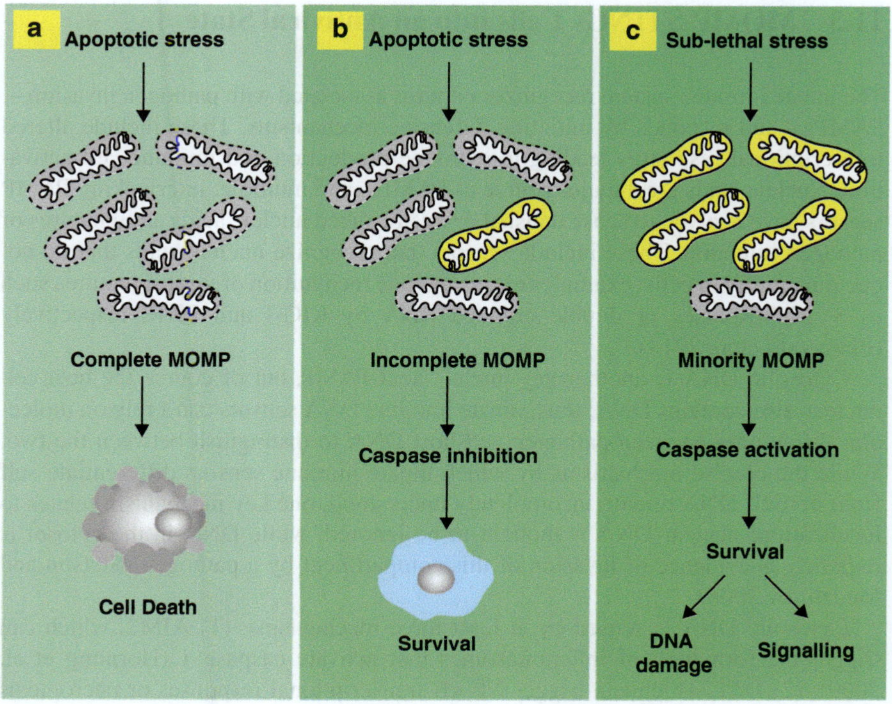

Fig. 11.2 Three degrees of MOMP drive lethal and nonlethal functions. (**a**) During apoptosis, MOMP often occurs in all mitochondria triggering massive caspase activation and rapid cell death. (**b**) Following an apoptotic stress, MOMP can sometimes be incomplete, allowing mitochondria to remain intact. In specific settings, cells can survive dependent on these intact mitochondria, provided caspase activity is lacking. (**c**) Following a sublethal stress, MOMP can be engaged in a minority of mitochondria. Although minority MOMP induces caspase activity, this fails to kill the cell. Instead, caspase activity can cause DNA damage and contribute to other non-apoptotic caspase signaling functions

kill the cell within a matter of minutes, whereas, in the absence of caspase activity, cells can remain viable for hours or even days. In some contexts, the kinetics of cell death are crucial; as discussed, while many aspects of embryonic development in *Apaf1* or *Casp9*-deficient mice are grossly normal, mice lacking these proteins often display massive forebrain outgrowth. This was recently shown to be due to the prolonged survival of morphogen (FGF-8) producing cells, that otherwise would have died rapidly via apoptosis (Nonomura et al. 2013). This extended window of FGF-8 production wreaks havoc on the developing brain, contributing to forebrain outgrowth. Besides enhancing the kinetics of cell death, as we will now discuss, a second reason as to why caspases are activated following MOMP is distinct from their role in cellular execution.

11.3 MOMP STINGs Cells into an Antiviral State

The innate immune system recognizes patterns associated with pathogen invasion—PAMPs—and responds by initiating defensive mechanisms. These include altered transcriptional responses leading to cytokine production and immune cell activation. Nucleic acids are a major source of PAMPs, and multiple, interlocking PAMP sensors have evolved to sense unusual or mislocalized nucleic acids as signatures of pathogen infection. These include sensors that recognize nucleic acids that do not occur naturally in cells; examples of this include recognition of viral signatures such as 5′ triphosphates or double-stranded RNA by RIG-I and PKR, respectively (Fitzgerald et al. 2014).

Cytosolic DNA is another key nucleic acid PAMP, but of course the host cell nucleus also contains DNA; thus, innate immune DNA sensors can't rely on molecular differences between pathogen and host DNA to distinguish between the two. While the precise mechanisms by which innate immune sensors differentiate self from nonself DNA remain incompletely understood, one key mechanism relates to localization: nuclear DNA is thought to be ignored, while DNA in the cytosol is regarded as evidence of invasion of this compartment by a pathogen (Stetson and Medzhitov 2006).

Cytosolic DNA is sensed by at least three mechanisms: (1) AIM2, which can trigger the formation of inflammasomes that activate caspase 1 (Hornung et al. 2009); (2) DAI, which can trigger NF-κB transcriptional responses or necroptosis via interaction with the RIP kinases (Rebsamen et al. 2009); or (3) the cGAS/STING pathway, which triggers interferon responses via activation of IRF3/7 signaling (Cai et al. 2014; Zhang et al. 2014; Li et al. 2013). Notably, while the AIM2 and STING pathways were initially described as mediators of antiviral responses, recent work has indicated that they also play key roles in responding to bacterial infection (Meunier et al. 2015; Karki et al. 2015; Belhocine and Monack 2012; Watson et al. 2012). In this latter context, it is thought that release of DNA from damaged bacteria leads to their activation. Indeed, recent studies have shown that cells possess specific mechanisms to ensure that bacterial PAMPs gain access to the cytosol, presumably to promote their sensing by these and other mechanisms. Many species of bacteria attempt to "hide" in vacuolar compartments, and it appears that cells have countered this strategy by evolving mechanisms to disrupt bacteria-containing vacuoles. One such strategy involves a family of cellular GTPases called the guanylate binding proteins (GBPs), which disrupt bacterial vacuoles to induce release and recognition of bacterial DNA and other PAMPs (Meunier et al. 2015). Another involves recognition of secreted bacterial DNA by the STING pathway, which in turn unleashes the autophagy machinery to promote bacterial clearance (Watson et al. 2012).

The purposeful release of bacteria and their contents from vacuoles by cellular factors may sound familiar; as we've already discussed, mitochondria can be thought of as ancient bacteria enclosed within cellular membranes. MOMP is, in essence, the disruption of the vacuole (the mitochondrial outer membrane) surrounding the proto-bacterium (the mitochondrial inner membrane.) And of course, like bacterial invaders, mitochondria have their own genomes.

These observations set the stage for recent reports that MOMP can lead to release of mitochondrial genomic DNA (mtDNA), which is sensed by the STING pathway to trigger production of type-I interferon (Rongvaux et al. 2014; White et al. 2014). This rather interesting observation can be put into context with our earlier discussion of cytochrome *c* as a vestigial PAMP: mitochondria do not contain other bacterium-specific PAMPs, such as LPS or peptidoglycans, possibly because these were lost during the establishment of symbiosis. But because DNA is an irreducible and irreplaceable molecule rather than a bacterium-specific "pattern," and because mitochondria continue to contain their own vestigial genomes, mtDNA remains able to aberrantly trigger the cell's DNA sensing machinery (Fig. 11.1d).

These recent studies have demonstrated that when downstream caspase 9-mediated apoptosis is prevented (either by knockout of *Casp9*, double-knockout of *Casp3* and *Casp7*, or chemical inhibition of all caspases), MOMP induction triggers interferon responses that depend on the cGAS/STING pathway (Rongvaux et al. 2014; White et al. 2014). This observation explains the unexpected finding that mice in which caspase-9 is deleted in the endothelial lineage are resistant to viral infection; they have an elevated basal interferon signature mediated by sensing of cytosolic DNA released from the mitochondria.

These reports provide clear evidence for mtDNA as a PAMP, but they also raise questions. One obvious issue is with the release of mtDNA itself; mtDNA resides in the mitochondrial matrix, inside the impermeable inner membrane. MOMP itself does not lead to permeabilization of this membrane, so how DNA escapes—and if this process is regulated in some way—remains unclear. A broader question is: When might this happen under physiological conditions? Normally MOMP leads to rapid cell death, and these interferon responses only manifest when this is prevented by ablation or inhibition of the caspases. So, is sensing of mitochondrial DNA a biologically relevant driver of the interferon response? We can speculate that minority MOMP (discussed further below) may be able to release enough DNA to promote IFN responses, without sufficient cytochrome *c* release to drive full-blown apoptosis.

Another interesting recent observation that is likely to prove relevant is the finding that knockout of a mtDNA packaging protein, TFAM, promotes escape of mtDNA into the cytosol and also triggers a STING-dependent antiviral response (West et al. 2015). This finding indicates that mtDNA stress, which is a feature of several human diseases, can lead to mislocalization of mtDNA and thus to elevated IFN signaling. It may therefore be the case that limited MOMP could similarly act to a mitochondrial stressor.

While these effects have been explored in the context of potentiated antiviral signaling, they may also have detrimental effects. Several human mutations have been described that lead to aberrant sensing of cytosolic DNA derived from genomic retro-elements (Rice et al. 2009; Crow et al. 2006a, b; Volkman and Stetson 2014; Stetson et al. 2008). This in turn leads to STING-dependent IFN responses that cause lupus-like autoimmune diseases collectively called Aicardi–Goutieres syndrome. It is therefore possible that mitochondrial DNA could also contribute to such deleterious IFN responses, either due to incomplete MOMP or other forms of mitochondrial stress. It is also unclear whether DNA released from the mitochondria may trigger other cytosolic DNA sensing pathways, such as AIM2-mediated inflammasome activation and concurrent IL-1β maturation, which could likewise contribute to autoimmune and inflammatory responses.

11.4 Cell Survival Following MOMP

11.4.1 Survival: A Fate Worse Than Death?

Seminal live-cell imaging studies initially revealed MOMP as a binary all or nothing event (Goldstein et al. 2000). However, a more nuanced view of MOMP has since emerged. Specifically, during apoptosis some mitochondria have been found to evade MOMP, a process called incomplete MOMP or iMOMP (Tait et al. 2010) (Fig. 11.2b). The ability of mitochondria to avoid permeabilization becomes important under caspase-deficient settings. While widespread MOMP normally triggers cell death regardless of caspase activity, the presence of intact mitochondria can serve to repopulate the cell, restoring mitochondrial function and allowing cell survival. Survival under these conditions may be important in various pathophysiological situations, for example by enabling tumor cells to evade therapy-induced cell death or by allowing the survival of post-mitotic cells such as neurons that are required lifelong. More recently, it has emerged that sublethal levels of apoptotic stress can engage MOMP in a minority of mitochondria—so-called minority MOMP (Fig. 11.2c). Importantly, minority MOMP still engages caspase activity but at levels which are insufficient to kill the cell (Ichim et al. 2015). Rather than killing, sublethal caspase activity causes DNA damage and genomic instability that can actually promote oncogenic transformation and tumorigenesis (Ichim et al. 2015; Liu et al. 2015). Why might this be important? To consider this we must first take a step back. Apoptosis is widely considered a potent anticancer mechanism—tumor suppressors prevent cancer by inducing apoptosis and anticancer therapies often kill cancer cells by apoptosis (Czabotar et al. 2014). These recent findings demonstrate that the same core apoptotic machinery that prevents cancer may also have the potential to promote cancer, if cell death is not executed. The ability of minority MOMP-induced DNA damage and genomic instability may, in part, provide a mechanistic basis for recurrent reports demonstrating an oncogenic role for apoptotic signaling (Labi and Erlacher 2015). Developing this idea further, an unfortunate consequence of some anticancer therapies is that they increase the likelihood of developing leukemia in later life. The chromosomal translocations that underlie these leukemias have been associated with sublethal caspase activity, which potentially may be initiated via minority MOMP (Hars et al. 2006). From these findings, a desirable therapeutic goal will be to enhance minority MOMP-induced tumor cell killing while preventing unwanted DNA damage in healthy tissue.

11.4.2 *MOMP in Nonlethal Signaling Functions*

By triggering limited caspase activity, minority MOMP may also play various physiological signaling functions. Various non-apoptotic functions for caspase protease activity have been described in processes that range from cell cycle to migration (Connolly et al. 2014). Perhaps their non-apoptotic role is best supported in neurons (Hyman and Yuan 2012). Modification of neuronal synaptic plasticity is important for both the development of the central nervous system as well as information storage. Non-apoptotic caspase activity modulates synaptic plasticity in different ways, for example by causing retraction of neuronal dendrites that form synapses (so-called pruning) (Cusack et al. 2013; Wang et al. 2014; Nikolaev et al. 2009). Additionally, non-apoptotic caspase activity has been shown to regulate the endocytosis of receptors that control synaptic plasticity (Li et al. 2010). In both settings, MOMP-dependent caspase activity has been implicated. The ability of neurons to activate caspases at non-apoptotic levels is likely due to two reasons. First, due to their large size, the spatial separation between permeabilized, caspase-activating mitochondria at the synapse and the rest of the cell likely restrains propagation of a pro-apoptotic signal. Secondly, neurons are remarkably resistant to cell death following MOMP, such that they must develop a competency to die (Deshmukh et al. 2000; Martinou et al. 1999).

11.5 Concluding Remarks

MOMP was discovered as an integral part of the apoptotic program, and a defining characteristic of MOMP was its rapid, all-or-nothing character. However, further analysis has demonstrated that MOMP is not always the end of a cell's life, and that MOMP-related phenomena can trigger unforeseen and detrimental cellular processes independent of cell death. These findings make more sense when viewed in the context of the evolution of the mitochondria themselves, and of the mitochondrial pathway of cell death as a "special case" of an ancient pathogen-sensing response. Future studies will no doubt shed additional light on the regulation of these fascinating processes.

References

Belhocine K, Monack DM (2012) Francisella infection triggers activation of the AIM2 inflammasome in murine dendritic cells. Cell Microbiol 14:71–80. doi:10.1111/j.1462-5822.2011.01700.x

Bender CE et al (2012) Mitochondrial pathway of apoptosis is ancestral in metazoans. Proc Natl Acad Sci U S A 109:4904–4909. doi:10.1073/pnas.1120680109

Cai X, Chiu YH, Chen ZJ (2014) The cGAS-cGAMP-STING pathway of cytosolic DNA sensing and signaling. Mol Cell 54:289–296. doi:10.1016/j.molcel.2014.03.040

Cecconi F, Alvarez-Bolado G, Meyer BI, Roth KA, Gruss P (1998) Apaf1 (CED-4 homolog) regulates programmed cell death in mammalian development. Cell 94:727–737

Chautan M, Chazal G, Cecconi F, Gruss P, Golstein P (1999) Interdigital cell death can occur through a necrotic and caspase-independent pathway. Curr Biol 9:967–970

Chinnaiyan AM, O'Rourke K, Lane BR, Dixit VM (1997) Interaction of CED-4 with CED-3 and CED-9: a molecular framework for cell death. Science 275:1122–1126

Chipuk JE, Moldoveanu T, Llambi F, Parsons MJ, Green DR (2010) The BCL-2 family reunion. Mol Cell 37:299–310. doi:10.1016/j.molcel.2010.01.025

Connolly PF, Jager R, Fearnhead HO (2014) New roles for old enzymes: killer caspases as the engine of cell behavior changes. Front Physiol 5:149. doi:10.3389/fphys.2014.00149

Cookson J, Silverstone T (1976) 5-Hydroxytryptamine and dopamine pathways in mania: a pilot study of fenfluramine and pimozide. Br J Clin Pharmacol 3:942–943

Crow YJ et al (2006a) Mutations in genes encoding ribonuclease H2 subunits cause Aicardi-Goutieres syndrome and mimic congenital viral brain infection. Nat Genet 38:910–916. doi:10.1038/ng1842

Crow YJ et al (2006b) Mutations in the gene encoding the 3'-5' DNA exonuclease TREX1 cause Aicardi-Goutieres syndrome at the AGS1 locus. Nat Genet 38:917–920. doi:10.1038/ng1845

Cusack CL, Swahari V, Hampton Henley W, Michael Ramsey J, Deshmukh M (2013) Distinct pathways mediate axon degeneration during apoptosis and axon-specific pruning. Nat Commun 4:1876. doi:10.1038/ncomms2910

Czabotar PE, Lessene G, Strasser A, Adams JM (2014) Control of apoptosis by the BCL-2 protein family: implications for physiology and therapy. Nat Rev Mol Cell Biol 15:49–63. doi:10.1038/nrm3722

Deshmukh M, Kuida K, Johnson EM Jr (2000) Caspase inhibition extends the commitment to neuronal death beyond cytochrome c release to the point of mitochondrial depolarization. J Cell Biol 150:131–143

Dickens LS, Powley IR, Hughes MA, MacFarlane M (2012) The 'complexities' of life and death: death receptor signalling platforms. Exp Cell Res 318:1269–1277. doi:10.1016/j.yexcr.2012.04.005

Fitzgerald ME, Rawling DC, Vela A, Pyle AM (2014) An evolving arsenal: viral RNA detection by RIG-I-like receptors. Curr Opin Microbiol 20:76–81. doi:10.1016/j.mib.2014.05.004

Goldstein JC, Waterhouse NJ, Juin P, Evan GI, Green DR (2000) The coordinate release of cytochrome c during apoptosis is rapid, complete and kinetically invariant. Nat Cell Biol 2:156–162. doi:10.1038/35004029

Green DR (2011) Means to an end: apoptosis and other cell death mechanisms. Cold Spring Harbor, New York, NY

Hars ES, Lyu YL, Lin CP, Liu LF (2006) Role of apoptotic nuclease caspase-activated DNase in etoposide-induced treatment-related acute myelogenous leukemia. Cancer Res 66:8975–8979. doi:10.1158/0008-5472.CAN-06-1724

Hornung V et al (2009) AIM2 recognizes cytosolic dsDNA and forms a caspase-1-activating inflammasome with ASC. Nature 458:514–518. doi:10.1038/nature07725

Hyman BT, Yuan J (2012) Apoptotic and non-apoptotic roles of caspases in neuronal physiology and pathophysiology. Nat Rev Neurosci 13:395–406. doi:10.1038/nrn3228

Ichim G et al (2015) Limited mitochondrial permeabilization causes DNA damage and genomic instability in the absence of cell death. Mol Cell 57:860–872. doi:10.1016/j.molcel.2015.01.018

Jost PJ et al (2009) XIAP discriminates between type I and type II FAS-induced apoptosis. Nature 460:1035–1039. doi:10.1038/nature08229, nature08229 [pii]

Kanneganti TD (2010) Central roles of NLRs and inflammasomes in viral infection. Nat Rev Immunol 10:688–698. doi:10.1038/nri2851

Karki R et al (2015) Concerted activation of the AIM2 and NLRP3 inflammasomes orchestrates host protection against Aspergillus infection. Cell Host Microbe 17:357–368. doi:10.1016/j.chom.2015.01.006

Kuida K et al (1998) Reduced apoptosis and cytochrome c-mediated caspase activation in mice lacking caspase 9. Cell 94:325–337
Labi V, Erlacher M (2015) How cell death shapes cancer. Cell Death Dis 6:e1675. doi:10.1038/cddis.2015.20
Lartigue L et al (2009) Caspase-independent mitochondrial cell death results from loss of respiration, not cytotoxic protein release. Mol Biol Cell 20:4871–4884. doi:10.1091/mbc.E09-07-0649
Lazebnik Y (2001) Why do regulators of apoptosis look like bacterial toxins? Curr Biol 11:R767–R768
Li Z et al (2010) Caspase-3 activation via mitochondria is required for long-term depression and AMPA receptor internalization. Cell 141:859–871. doi:10.1016/j.cell.2010.03.053
Li XD et al (2013) Pivotal roles of cGAS-cGAMP signaling in antiviral defense and immune adjuvant effects. Science 341:1390–1394. doi:10.1126/science.1244040
Lindsten T et al (2000) The combined functions of proapoptotic Bcl-2 family members bak and bax are essential for normal development of multiple tissues. Mol Cell 6:1389–1399
Liu X et al (2015) Caspase-3 promotes genetic instability and carcinogenesis. Mol Cell 58:284–296. doi:10.1016/j.molcel.2015.03.003
Man SM, Kanneganti TD (2015) Regulation of inflammasome activation. Immunol Rev 265:6–21. doi:10.1111/imr.12296
Martinou I et al (1999) The release of cytochrome c from mitochondria during apoptosis of NGF-deprived sympathetic neurons is a reversible event. J Cell Biol 144:883–889
McIlwain DR, Berger T, Mak TW (2013) Caspase functions in cell death and disease. Cold Spring Harb Perspect Biol 5:a008656. doi:10.1101/cshperspect.a008656
Meunier E et al (2015) Guanylate-binding proteins promote activation of the AIM2 inflammasome during infection with Francisella novicida. Nat Immunol 16:476–484. doi:10.1038/ni.3119
Nikolaev A, McLaughlin T, O'Leary DD, Tessier-Lavigne M (2009) APP binds DR6 to trigger axon pruning and neuron death via distinct caspases. Nature 457:981–989. doi:10.1038/nature07767
Nonomura K et al (2013) Local apoptosis modulates early mammalian brain development through the elimination of morphogen-producing cells. Dev Cell 27:621–634. doi:10.1016/j.devcel.2013.11.015
Oberst A, Bender C, Green DR (2008) Living with death: the evolution of the mitochondrial pathway of apoptosis in animals. Cell Death Differ 15:1139–1146. doi:10.1038/cdd.2008.65
Rebsamen M et al (2009) DAI/ZBP1 recruits RIP1 and RIP3 through RIP homotypic interaction motifs to activate NF-kappaB. EMBO Rep 10:916–922. doi:10.1038/embor.2009.109
Rice GI et al (2009) Mutations involved in Aicardi-Goutieres syndrome implicate SAMHD1 as regulator of the innate immune response. Nat Genet 41:829–832. doi:10.1038/ng.373
Rongvaux A et al (2014) Apoptotic caspases prevent the induction of type I interferons by mitochondrial DNA. Cell 159:1563–1577. doi:10.1016/j.cell.2014.11.037
Spector MS, Desnoyers S, Hoeppner DJ, Hengartner MO (1997) Interaction between the *C. elegans* cell-death regulators CED-9 and CED-4. Nature 385:653–656. doi:10.1038/385653a0
Stetson DB, Medzhitov R (2006) Recognition of cytosolic DNA activates an IRF3-dependent innate immune response. Immunity 24:93–103. doi:10.1016/j.immuni.2005.12.003
Stetson DB, Ko JS, Heidmann T, Medzhitov R (2008) Trex1 prevents cell-intrinsic initiation of autoimmunity. Cell 134:587–598. doi:10.1016/j.cell.2008.06.032
Tait SW, Green DR (2010) Mitochondria and cell death: outer membrane permeabilization and beyond. Nat Rev Mol Cell Biol 11:621–632. doi:10.1038/nrm2952
Tait SW et al (2010) Resistance to caspase-independent cell death requires persistence of intact mitochondria. Dev Cell 18:802–813. doi:10.1016/j.devcel.2010.03.014
Tait SW, Ichim G, Green DR (2014) Die another way--non-apoptotic mechanisms of cell death. J Cell Sci 127:2135–2144. doi:10.1242/jcs.093575
Volkman HE, Stetson DB (2014) The enemy within: endogenous retroelements and autoimmune disease. Nat Immunol 15:415–422. doi:10.1038/ni.2872

Wang JY et al (2014) Caspase-3 cleavage of dishevelled induces elimination of postsynaptic structures. Dev Cell 28:670–684. doi:10.1016/j.devcel.2014.02.009

Watson RO, Manzanillo PS, Cox JS (2012) Extracellular M. tuberculosis DNA targets bacteria for autophagy by activating the host DNA-sensing pathway. Cell 150:803–815. doi:10.1016/j.cell.2012.06.040

West AP et al (2015) Mitochondrial DNA stress primes the antiviral innate immune response. Nature 520:553–557. doi:10.1038/nature14156

White MJ et al (2014) Apoptotic caspases suppress mtDNA-induced STING-mediated type I IFN production. Cell 159:1549–1562. doi:10.1016/j.cell.2014.11.036

Zhang X et al (2014) The cytosolic DNA sensor cGAS forms an oligomeric complex with DNA and undergoes switch-like conformational changes in the activation loop. Cell Rep 6:421–430. doi:10.1016/j.celrep.2014.01.003

Index

A
Acidosis, 25–26
Adaptive immunity, 202–203
Aging, 121–122
A kinase anchoring protein 1 (AKAP1), 91
Allen, B.G., 111–122
Al-Nassar, I., 32
Amyotrophic lateral sclerosis (ALS), 116
Antioxidants
  catalase, 116
  glutaredoxins, 117–118
  glutathione peroxidases, 117
  GSH, 116–117
  NADPH, 119
  peroxiredoxin isoforms, 117–118
  SOD, 116
  thioredoxin, 118
Antiviral immunity
  adaptive immunity, 202–203
  apoptosis, 201–202
  innate immunity
    adaptor, 199
    autophagy, 197
    CLRs, 191
    cytomegalovirus, 199
    DNA virus infection, 193
    ER, 198
    fibroblasts, 196
    fission, 197
    fusion, 197
    HCV infection, 199
    MAM dynamics, 197–199
    MAVS, 194, 195
    NLRs, 191
    PAMPs, 190
    RNA viruses, 191, 192
    ROS, 196
    STING and translocase, 195, 196
    toll-like receptors, 190, 191
    TRAF interactions, 193
    viral infection, 190
  metabolism, 188–189
  negative regulation, 200–201
  oxidative phosphorylation, 187–188
  role of, 187
  sterile inflammation, 196–197
  tricarboxylic acid, 187
Apoptosis
  antiviral immunity, 201–202
  C. elegans, 85
  definition, 69, 85
  GRB, 50–51
  IMS, 48
  mitochondrial remodeling
    (*see* Mitochondrial remodeling)
Apoptosis-inducing factor (AIF), 56–57
Arginyl transferase 1 (ATE1), 177
ATP depletion, 23–25

B
Bak. *See* Bcl-2 antagonist killer 1 (Bak)
Bax. *See* Bcl-2-associated X protein (Bax)
B-cell lymphoma-2 (Bcl-2) protein
  Bak
    activation of, 75
    BH3 domain, 76
    IMS proteins, 70
    oligomerization, 76
    OMM, 69

D.M. Hockenbery (ed.), *Mitochondria and Cell Death*, Cell Death in Biology and Diseases, DOI 10.1007/978-1-4939-3612-0

X.-M. Yin and Z. Dong (Series eds.), *Cell Death in Biology and Diseases*

B-cell lymphoma-2 (Bcl-2) protein (*cont.*)
Bax
activation of, 75
BH3 domain, 76
C-terminal membrane anchor, 77
IMS proteins, 70
oligomerization, 76
OMM, 69
pore-forming activity, 75–77
BH1-4, 70
BH3, regulation, 71–73
Bok, 70
“embedded together” model, 71
globular protein fold, 70
hydrophobic groove, 70
mitochondrial dynamics, 77–78
MOMP, 70
pro-apoptotic, 70
retrotranslocation, 73–75
B cell receptor-associated protein 31 (BAP31), 94
Bcl-2 antagonist killer 1 (Bak)
activation, 75
BH3 domain, 76
IMS proteins, 70
oligomerization, 76
OMM, 69
Bcl-2-associated X protein (Bax)
activation, 75
BH3 domain, 76
C-terminal membrane anchor, 77
IMS proteins, 70
oligomerization, 76
OMM, 69
pore-forming activity, 76
Bcl-2 homology (BH) domains 1-4 (BH1-4, 70
Bcl-2 homology motif (BH3), 71–73
Bernardi, P., 1–11
BH3-interacting domain death agonist (Bid), 72
Bielas, J.H., 131–143
Bok, 70
Boyman, L., 17–35

C
Ca^{2+} release units (CRUs), 22
Cardiac ischemia
energetic and ionic imbalances
acidosis, 25–26
ATP depletion, 23–25
$[Ca^{2+}]_i$ dysregulation, 27–29
$[Na^{+}]_i$ dysregulation, 26–27
functions, 17
IR injury (*see* Ischemia-reperfusion (IR) injury)
ischemic preconditioning
beneficial effects of, 18
mechanism, 33–34
mPTP, 33–34
Caspase-independent cell death (CICD), 58, 218
Ceramide and its transport protein (CERT), 103
c-Jun N terminal kinase (JNK), 92
Correll, R.N., 19
Cristae and crista junction remodeling
cytochrome c, 96
OPA1
disruption, 97
isoforms, 96, 97
knock-down, 97
localization, 96
prohibitin, 100
proteolytic cleavage, 100
siRNA committed cells, 96
structural remodeling, 98–100
Crompton, M., 32
Cyclophilin D, 8
Cytochrome c, 52–53

D
Death-inducing signalling complex (DISC), 50
Dynamin related protein 1 (DRP1), 87, 90, 96, 102

E
Edlich, F., 69–78
Eigel, B.N., 27
Electron transport chains (ETCs), 114
Ellisman, M.H., 85–104
Endoplasmic reticulum (ER), 93–96
Endoplasmic reticulum associated degradation (ERAD), 150
ER-localized inverted formin 2 (INF2), 96

F
Fernandez-Mosquera, L., 169–182
Fieni, F., 19
Flanagan, L., 45–58
Fontaine, E., 1–11

G
Gale, M.Jr., 187–204
Gasparre, G., 140
Germain, D., 149–164

Glucose-6-phosphate dehydrogenase (G6PD), 119
Glutaredoxins, 117–118
Glutathione (GSH), 116–117, 154
Glutathione peroxidases, 117
Granzyme B (GRB), 50–51

H

Hadley, R.W., 27
Haworth, R.A., 2, 4
Halestrap, A.P., 33
Hockenbery, D.M., 85
Homoplasmy, 132
Hoppins, S., 102
Hunter, D.R., 2, 4

I

Ichim, G., 213–223
IFN-stimulated genes (ISGs), 193
Innate immunity
  adaptor, 199
  autophagy, 197
  CLRs, 191
  cytomegalovirus, 199
  DNA virus infection, 193
  ER, 198
  fibroblasts, 196
  fission, 197
  fusion, 197
  HCV infection, 199
  MAM dynamics, 197–199
  MAVS, 194, 195
  NLRs, 191
  PAMPs, 190
  RNA viruses, 191, 192
  ROS, 196
  STING and translocase, 195, 196
  toll-like receptors, 190, 191
  TRAF interactions, 193
  viral infection, 190
Intermembrane space (IMS)
  AIF, 56–57
  ATP-binding cassette, 47
  CICD, 58
  cristae morphology, 46
  cytochrome c, 52–53
  endonuclease G, 57
  extrinsic pathway, 50
  functions, 47–48
  GRB, 50–51
  intrinsic pathway, 48–49
  lipid composition, 46
  mechanisms, 48
  mitochondrial carrier family, 47
  OMI/HTRA2, 55
  vs. OMM, 46
  SMAC, 53–54
  structural changes, 51–52
  UPR
    antioxidant machinery, 154
    electron transport chain, 153
    estrogen receptors, 159
    hypoxia, 159
    inner membrane, 153
    OMI/HTRA2, 156
    oxidation, 153
    protein quality control, 153–154
  VDAC, 47
Inter-organelle signaling, 180–181
Ischemia-reperfusion (IR) injury
  Ca^{2+}movement, 21–23
  cytosolic Ca^{2+}signals, 18–21
  mitochondria, role of, 29–32
  mitochondrial Ca^{2+}signals, 18–21
  mPTP
    Ca^{2+}, 32–33
    molecular identity, 32
  myocardial infarction, 17
  preconditioning, 17
Ischemic preconditioning (IP)
  beneficial effects, 18
  beneficial outcome, 17
  mechanism, 34
  mPTP, 33–34
Ishikawa, K., 141

K

Kerr, J.F., 85
Kokoszka, J.E., 32
Kuwana, T., 103

L

Lederer, W.J., 17–35
Lucantoni, F., 45–58
Lymphocytic Choriomeningitis Virus (LCMV), 200

M

Martinou, J.-C., 69–78
Membrane anchor (MA), 70
Metabolic oxidation–reduction (redox) reactions
  cellular functions, 111

Metabolic oxidation–reduction (redox) reactions (*cont.*)
nonequilibrium steady-state, 111
oxidative damage, 112
oxidative metabolism and gene expression, 113
oxidative stress, 112–113
ROS
aging, 121–122
antioxidants (*see* Antioxidants)
cancer, 119–120
free radicals, 114
intracellular signaling, 114
sources of, 114–115
superoxide and nitric oxide, 114
targets of, 116
Mitchell, P., 1
Mitochondria-associated membrane (MAM), 193, 194
Mitochondrial antiviral signaling (MAVS), 200
Mitochondrial Ca^{2+} uniporter (MCU), 22
Mitochondrial DNA (mtDNA) mutations
in cancer
aberrant cellular metabolism, 133
biomarkers, 142–143
displacement loop, 134–135
ETC subunits, 135–136
insertions and deletions, 136–137
methodology, 133–134
point mutations, 136
stringencies, 134
tRNA, 136
detection and definitions, 131–133
electron transport chain, 131
exposure, 139–140
genomic drift, 138–139
metabolism and human health, 133
selective pressure, 137–138
tumorigenic effects
cancer establishment and adaptation, 141
metastatic disease, 141
prevalence, 140
resistance to apoptosis, 142
Mitochondrial E3 ubiquitin protein ligase 1 (MUL1), 200
Mitochondrial fission factor (MFF), 87
Mitochondrial inner-membrane topology, 88–91
Mitochondrial Na^+/Ca^{2+} exchanger (NCLX), 22
Mitochondrial outer membrane permeabilization (MOMP), 46, 70
altruistic suicide, 216
apoptotic cell death, 213
BCL2 proteins, 213
Caenorhabditis elegans, 217
caspase-independent cell death, 218
cell death, 218, 219
consequences, 214–215
cytochrome c, 215
cytosolic DNA, 217
Drosophila melanogaster, 216
IMS (*see* Intermembrane space (IMS))
intrinsic and extrinsic pathways, 213
minority, 222
NLRs signal, 215
nonlethal signaling functions, 223
seminal live-cell imaging, 222
STINGs cells, 220–222
Mitochondrial permeability transition pore (mPTP)
Ca^{2+}, 23, 32–33
deleterious openings, in heart, 20
IP, 33–34
molecular identity, 32
properties of, 20
Mitochondrial remodeling
BAX/BAK-dependent pathway, 86
cardiolipin, 100–101, 103
cristae and crista junction remodeling (*see* Cristae and crista junction remodeling)
ER constriction, 93–96
inner-membrane topology, 88–91
mitochondrial fission, 86–88
mitochondrial outer membrane, 101–103
signaling pathways convergent, 91–93
Mitochondrial signaling
ATP and ADP, 175
Ca^{2+} signaling, 175, 176
cytoplasmic Ca^{2+} concentration, 174
definition, 171
energy factory, 169
identity, 173
inter-organelle signaling, 180–181
interpretation outside, 173
metabolites, 177
mtDNA, 178
NAD^+ and NADH, 175
NRF1, 169
origin, 172
peptides, 178
PINK1, 174
reactive oxygen species, 176
retrograde communication, 169
RNS, 176
superoxide, 174
termination, 173–174
threshold and tissue-specificity, 179–180

tissue-specificity, 170
vesicles, 178
Molkentin, J.D., 19
Mukherjee, P., 199
Murry, C.E., 17
Myoclonic Epilepsy with Ragged Red Fibers (MERRF) syndrome, 136

N
Na^+/Ca^{2+} exchanger (NCLX), 22, 23, 32
Na^+/H^+ exchanger (NHE1), 27
NHE1, 27, 30
Notch intercellular domain (NICD), 92
Nunnari, J., 102

O
Oberst, A., 213–223
Optic atrophy A1 (OPA1)
disruption, 97
isoforms, 96, 97
knock-down, 97
localization, 96
prohibitin, 100
proteolytic cleavage, 100
siRNA committed cells, 96
structural remodeling, 98–100
Outer mitochondrial membrane (OMM), 46, 47, 69, 73, 75
Oxidative phosphorylation (OXPHOS), 203

P
Pattern recognition receptors (PRRs), 190–191
Perkins, G.A., 85–104
Permeability transition pore (PTP)
Ca^{2+} and Pi, 1
cell death, 5–7
chemiosmotic theory, 1
inner membrane
active swelling, 3
glycohydrolase and PARP, 4
mitochondrial matrix volume, 2, 3
passive swelling, 3
reactive oxygen species, 4
respiratory inhibition, 4
swelling experiments, 2
TCA cycle, 3
nonlethal functions
ATP production, 10–11
calcium homeostasis, 7–9
transient uptake-release, 9–10
reversible and irreversible consequences, 5
transient and permanent, 5
transport systems, 4
Prehn, J.H.M., 45–58
Prostate cancer, 141
Protein kinase A (PKA), 91, 92, 94

R
Raimundo, N., 169–182
Reactive nitrogen species (RNS), 114, 176
Reactive oxygen species (ROS), 18, 31
aging, 121–122
antioxidants
catalase, 116
glutaredoxins, 117–118
glutathione peroxidases, 117
GSH, 116–117
NADPH, 119
peroxiredoxin isoforms, 117–118
SOD, 116
thioredoxin, 118
cancer, 119–120
free radicals, 114
intracellular signaling, 114
sources of, 114–115
superoxide and nitric oxide, 114
targets of, 116
Reperfusion, 29
Retrograde signaling cascade (RTG), 156

S
Salas, A., 134
Second mitochondria-derived activator of caspases (SMAC), 53–54
Severe fever with thrombocytopenia syndrome virus (SFTSV), 200
Shidara, Y., 141
Single-celled organisms single-celled organisms, 216
Spitz, D.R., 111–122
Superoxide dismutase (SOD, 116

T
Tait, S.W.G., 213–223
Thioredoxin, 118
Thomas, S., 187–204

U
Unfolded protein response (UPR)
aging and development, 161–162
AKT and AMPK, 160

Unfolded protein response (UPR) (*cont.*)
antioxidant machinery, 152
breast cancer cells, 163
CHOP, 157
endonucleaseG, 157
endoplasmic reticulum
ATF6, 149
ERAD, 150
identification, 149
IRE, 149
PERK, 149
proteotoxic stress, 163
vs. UPR^{mt}, 163
FOXO3A, 157
IMS
antioxidant machinery, 154
electron transport chain, 153
estrogen receptors, 159
hypoxia, 159
inner membrane, 153
OMI/HTRA2, 156
oxidation, 153
protein quality control, 153–154
limiting accumulation, 155, 156
mammalian cells, 159
mechanisms, 150
oxidative stress, 161
ROS levels, 158
SIRT3, 158
stress responses, 156–157
validation, 162
mitophagy, 155
protein quality control, 151–152
ROS, 150, 151

V
Valente, W.J., 131–143
Vaux, D.L., 85
Voltage-dependent anion channel (VDAC), 47

W
Warburg, O., 150
Wilkinson, J.C., 56
Williams, G.S.B., 17–35

X
X-linked inhibitor of apoptosis protein (XIAP), 54

Y
Yambire, K.F., 169–182

MIX
Papier aus verantwortungsvollen Quellen
Paper from responsible sources
FSC® C105338

If you have any concerns about our products,
you can contact us on
ProductSafety@springernature.com

In case Publisher is established outside the EU,
the EU authorized representative is:
Springer Nature Customer Service Center GmbH
Europaplatz 3, 69115 Heidelberg, Germany

Printed by Libri Plureos GmbH
in Hamburg, Germany